T0338450

INTRODUCTION TO NUMERICAL GEODYNAMIC MODELLING

Until now, numerical modelling of geodynamic processes has been the domain of highly trained mathematicians with long experience of numerical and computational techniques. Now, for the first time, students and new researchers in the Earth Sciences can learn the basic theory and applications from a single, accessible reference text.

Assuming only minimal prerequisite mathematical training (simple linear algebra and derivatives) the author provides a solid grounding in the basic mathematical theory and techniques, including continuum mechanics and partial differential equations, before introducing key numerical and modelling methods. Eight well-documented and state-of-the-art visco-elasto-plastic, 2D models are then presented, which allow robust modelling of key dynamic processes such as subduction, lithospheric extension, collision, slab break-off, intrusion emplacement, mantle convection and planetary core formation.

Incorporating 47 practical exercises and 67 MATLAB examples (for which codes are available online at www.cambridge.org/gerya) this textbook provides a user-friendly introduction for graduate courses or self-study, and encourages readers to experiment with geodynamic models first hand.

TARAS GERYA was awarded a Ph.D. in 1990 from the Moscow State University and went on to become a Senior Researcher and Head of the Laboratory of Metamorphism at the Institute of Experimental Mineralogy, Russian Academy of Sciences, Moscow. He was awarded a Habilitation in petrology before moving to the Ruhr University of Bochum, Germany in 2000 as an Alexander von Humboldt Foundation Research Fellow. In 2004 he took up a position as Senior Research Scientist in the Department of Earth Sciences at ETH-Zurich, Switzerland, while continuing to be an Adjunct Professor in the Geology Department of Moscow State University. He was awarded a Habilitation in numerical geodynamic modelling by ETH-Zurich in 2008 and the Golden Owl Prize 2008 from ETH students for teaching of continuum mechanics and numerical geodynamic modelling. Dr Gerya is the author of over 50 papers on geodynamic modelling in leading peer-review journals.

INTRODUCTION TO NUMERICAL GEODYNAMIC MODELLING

TARAS V. GERYA

*Department of Earth Sciences, Swiss Federal Institute of
Technology (ETH-Zurich)*

CAMBRIDGE
UNIVERSITY PRESS

University Printing House, Cambridge CB2 8BS, United Kingdom

Published in the United States of America by Cambridge University Press, New York

Cambridge University Press is part of the University of Cambridge.

It furthers the University's mission by disseminating knowledge in the pursuit of education, learning and research at the highest international levels of excellence.

www.cambridge.org
Information on this title: www.cambridge.org/9780521887540

First published 2010
Reprinted with corrections 2011

A catalogue record for this publication is available from the British Library

Library of Congress Cataloguing in Publication data
Gerya, Taras.
Introduction to numerical geodynamic modelling / Taras Gerya.
p.   cm.
Includes bibliographical references and index.
ISBN 978-0-521-88754-0 (hardback)
1. Geophysics – Mathematical models.   2. Geodynamics – Mathematical models.   I. Title.
QE501.4.M38G47   2010
550.1'5118 – dc22   2009036645

ISBN 978-0-521-88754-0 Hardback

Additional resources for this publication at www.cambridge.org/9780521887540

Contents

Acknowledgements

In relation to this book I'd like to acknowledge many people and I'll try to do this in chronological order. I'm grateful to my wife Irina for her inspiration and support. I'm grateful to my Ph.D. supervisor and good friend of mine, Leonid Perchuk for suggesting that I start with numerical modelling in 1995 (a long time ago, indeed, but I feel like it was yesterday). I'm grateful to Alexander Simakin for explaining to me in a few words what numerical modelling is about, when I had just started to learn it and was really puzzled about what to do with all these PDEs written in textbooks (he told me that I simply have *to compose and solve altogether as many linear equations as I have unknowns* and this is really the main idea behind numerical modelling). I'm grateful to Roberto Weinberg and Harro Schmeling for their excellent paper about polydiapirs published in 1992 which introduced me to the marker-in-cell techniques when I had just started. I'm grateful to Alexey Polyakov for suggesting that I use upwind differences for solving the temperature equation when I was programming my first thermomechanical code. I'm grateful to Walter Maresch and Bernhard Stöckhert for cooperating with me on modelling of subduction processes which is a challenging topic and stimulated a lot of my code developments. I'm grateful to David Yuen – my continuous co-author in numerics – for our long-term join work and friendship (after we met in 2001 at AGU in San Francisco) and for many great suggestions concerning this book. I'm grateful to Paul Tackley for telling me about the fully staggered grid in 2002 (I was using the half-staggered one before that time) and for introducing me to multigrid in 2005 as well as for joint studies and good suggestions concerning a proposal for this book. I'm grateful to Jean-Pierre Burg for inviting me to ETH-Zurich and cooperating with me on challenging modelling projects (which again triggered many code developments) and for being a very careful and constructive first reader of this book. I'm grateful to Yuriy Podladchikov for many stimulating discussions, continuous healthy criticism and challenging suggestions (for example, adding elasticity and plasticity to my codes that 'spoiled' six months of my life). I'm grateful to Boris Kaus for arguing and discussing with me about numerics which we both like so much (although he is more inclined toward finite elements while I like finite differences) and for great detailed comments and suggestions on the initial version of this book. I'm grateful to James Connolly for fruitful work on coupling of thermodynamics and phase petrology with thermomechanical experiments (what I call petrological-thermomechanical numerical modelling). I'm grateful to David May for very creative checking of the first book version and many good hints about its content. I'm grateful to my son Bogdan for the computer and graphic assistance, to my parents Lyudmila and Viktor and my entire family for the moral support. I'm grateful to all my students and co-authors for bright ideas and great work done together. Finally, I'm grateful for the generous support of my numerical modelling projects by Alexander von Humboldt foundation fellowships and ETH (TH -12/04–1, TH -12/05–3, TH -08 07–3) and SNF (200021–113672/1, Topo-4D, 4D-Adamello) research grants.

Introduction

Theory: What is this book? What this book is not. Get started. Seven golden rules to learn the topic. Short history of geodynamics and numerical geodynamic modelling. Few words about programming and visualisation. Nine programming rules.
Exercises: Starting with MATLAB. Visualisation exercise.

What is this book?

This book is a practical, hands-on introduction to numerical geodynamic modelling for inexperienced people, i.e. for young students and newcomers from other fields. It does not require much background in mathematics or physics and is therefore written with a maximum amount of simple technical details. If you are inexperienced – this book is for you!

What this book is not

This book is not a treatise or a compendium of knowledge for experienced researchers. It does not contain large overviews of existing numerical techniques and only simple approaches are explained. If you are experienced in numerical methods, read Chapter 17 first and then decide if you wish to read about the technical details presented in previous chapters.

Get started

Already decided?! Then let us get started! In recent decades numerical modelling has become an essential approach in geosciences in general and in geodynamics in particular. This is a very natural process ('instinctive evolution') since direct human observation scales are extremely limited in both time and space (depth) and rapid progress in computer technology offers every day new and exceptional possibilities to explore sophisticated mathematical models and this is true in every discipline, and even industrial applications. Numerical modelling in geosciences is widely used for both testing and generating hypotheses and strongly pushing geology from an observational, intuitive to a deductive, predictive natural science. Geo-modelling and geo-visualisation play a strong role in relating different branches of geosciences. Therefore, it has become necessary to have some knowledge about numerical techniques before planning and conducting state-of-the-art interdisciplinary research in any branch of geosciences. In this respect, geodynamics is traditionally 'infected' by numerical modelling and promotes the progress of numerical methods in geosciences.

Before starting with numerical modelling we should consider one of the very popular 'myths' among geologists, who often declare (or think) something like:

Numerical modelling is very complicated; it is too difficult for people with a traditional geological background and should be performed by mathematicians.

I used to think like that before I started. I always remember my feeling when I heard for the first time the expression, 'Navier–Stokes equation'. 'Ok, forget it! This is hopeless,' I thought at that time, and that was wrong. Therefore, let me formulate the seven 'golden rules' elaborated during my learning experience.

Golden Rule 1. *Numerical modelling is simple and is based on simple mathematics.*

All you need to know is:

- linear algebra,
- derivatives.

Most of the 'complicated' mathematical knowledge is learned in school before we even start to study at university! I often say to my students that all is needed is:

strong MOTIVATION,
usual MATHS,
clear EXPLANATIONS,
regular EXERCISES.
Motivation is most important, indeed . . .

Golden Rule 2. *When numerical modelling looks complicated see Rule 1.*

Golden Rule 3. *Numerical modelling consists of solving partial differential equations (PDEs).*

There are only a few equations to learn (e.g. Lynch, 2005). They are generally not complicated, but it is essential to learn and understand them gradually and properly.

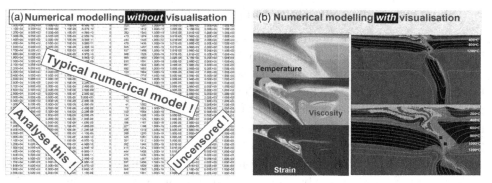

Fig. Introduction.1 Rule 6: Visualisation is important!

For example, to model the broad spectrum of geodynamic processes discussed in this book, it is necessary to know three principal conservation PDEs only:

- the equation of continuity (conservation of mass),
- the equation of motion (conservation of momentum – *Navier–Stokes equation!*)
- the temperature equation (conservation of heat).
  So, only *three* equations have to be understood and not tens or hundreds of them!

Golden Rule 4. *Read books on numerical methods several times.*

There are many excellent books on numerical methods. Most of these books are, however, written for physicists and engineers and need effort to be 'digested' by people with a traditional geological background.

Golden Rule 5. *Repeat transformations of equations involved in numerical modelling.*

These transformations are generally standard and trivial, but repeating them develops a familiarity with the PDEs (maybe you will even start to like them . . .), and allows you to understand the structure of the different PDEs. This book, by the way, is full of such trivial detailed transformations – follow them too.

Golden Rule 6. *Visualisation is important!*

Without proper visualisation of results, almost nothing can be done with numerical modelling (Fig. Introduction.1). Modellers often spend more time on visualisation than on computing and programming.

Golden Rule 7. *Ask!*

This is the most efficient way of learning. In numerical geodynamic modelling also many small hints and details exist. They are extremely important, but rarely discussed in publications (in contrast to this book).

Short history of geodynamics and numerical geodynamic modelling

The numerical modelling approaches discussed in this book are adopted for solving *thermomechanical* geodynamic problems. Geodynamics – dynamics of the Earth – is a core geological subject that was very actively progressing during the last century, especially since the establishment of plate tectonics in the 1960s. This was a really great time for geology that 'drifted' strongly from a descriptive (qualitative) field, to a predictive (quantitative) physical science. The overall history of the development of geodynamics was not, indeed, very 'dynamic' but rather slow and complicated. A brilliant introduction to this field (which I strongly recommend you to read) is written by Donald L. Turcotte and Gerald Schubert (2002). According to this introduction and other sources, the following steps were historic in understanding the Earth as a dynamic system:

1620: *Francis Bacon pointed out the similarity in shape between the west coast of Africa and the east coast of South America.*

This was about 400 years ago (!) and several centuries were needed to start interpreting this similarity.

1665: *Athanasius Kircher, in his two-volume* 'Mundus subterraneus', *probably the first printed work on geophysics and vulcanology, held that much of the phenomena on earth were due to the fact that there is 'fire' under the terra firma.*

This was, indeed, very unusual teaching for those days (about 350 years ago!) and very much in line with the thermal origin of the mantle convection.

Early part of eighteenth century: *Gottfried Wilhelm Leibniz proposed that the Earth has a molten core and anticipated the igneous nature of the mantle.*

The understanding of the Earth as a hot layered planetary body. One should really have a vision to guess this around 300 years ago!

Latter part of the nineteenth century: *Establishing the fluid-like behaviour of the Earth's mantle based on gravity studies: mountain ranges have low-density roots.*

This crucial finding was 'coupled' to Earth dynamics only one hundred years later and was not explored in the continental drift hypothesis.

1895–1915: *The unforeseen discovery of radioactivity.*

That 'killed' the concept of progressive dissipation of the heat of the Earth, and then the correlative contraction as the mechanism for orogenic stresses. It also changed the estimation of the age of the Earth and stratas by an order of magnitude . . . All this forced further serious rethinking of geological concepts about dynamic processes shaping the Earth.

1910: *Frank B. Taylor, Continental Drift hypothesis.*

The real beginning of 'drifting' toward plate tectonics, still a long way to go.

1912–1946: *Alfred Wegener, further developed the Continental Drift hypothesis, and showed a correspondence of the geological provinces, relict mountain ranges and fossil types. Driving forces – tidal/rotation of the Earth. Single protocontinent – Pangea.*

The principal questions are considered to be, 'why do continents move?' and 'what are the driving forces?' and not yet, 'how do continents move?' and 'what is the movement mechanism?'

1916: *Gustaaf Adolf Frederik Molengraaff proposed mid-ocean ridges to be formed by seafloor spreading as the result of the movement of continents in order to account for the opening of the Atlantic Ocean as well as the East Africa Rift.*

The mid-ocean ridges were 're-discovered' for plate tectonics 40 years later . . .

1924: *Harold Jeffreys showed the insufficiency of Wegener's forces to move continents.*

Computing forces for testing a geodynamic hypothesis is one of the core principles of modern geodynamics as well! Another point to learn – opposition to the Continental Drift hypothesis using physical arguments was always strong and probably considerably delayed the theory of plate tectonics.

1931: *Arthur Holmes suggested that thermal convection in the Earth's mantle can drive continental drift.*

This crucial idea answered a question about driving forces, but not about the movement mechanisms. It was known from seismic studies, that the Earth's mantle is in a solid state and elastic deformation does not allow thousands of kilometres of motion of the continents.

1935: *N. A. Haskell established the fluid-like behaviour of the mantle (viscosity 10^{20} Pa s) based on the analysis of beach terraces in Scandinavia and the existence of post-glaciation rebound.*

Actually, this was also established earlier from inferring crustal roots. The question about the physical mechanisms of solid-state mantle deformation remains open.

1937: *Alexander du Toit suggested the existence of two protocontinents – Laurasia and Gondwanaland, separated by the Tethys ocean.*

This is a really dramatic story: geologists were continuously developing and supporting the Continental Drift hypothesis, but the general idea of large lateral displacements of continents was continuously rejected by geophysicists.

1950s: *Improved understanding of the worldwide network of mid-ocean ridges during the extensive exploration of the seafloor.*

Evidence is critically growing in line with Molengraaff's ideas . . .

1950s: *Finding mechanisms of solid-state creep of crystalline materials applicable, for example, for the flow of ice in glaciers.*

The answer to the second crucial question was finally found in material science!

Breakthrough! The Great 1960s have started!

1960s: *Palaeomagnetic studies, the finding of regular patterns of magnetic anomalies on the sea floor.*

1962: *Harry Hess suggested that the seafloor was created at the axis of the ridge.*

In fact, this was a refinement of the Molengraaff's hypothesis.

1965: *B. Gordon proposed the quantitative link between solid-state creep and mantle viscosity.*

1968: *Jason Morgan formulated the basic hypothesis of Plate Tectonics (mosaic of rigid plates in relative motion with respect to one another as a natural consequence of mantle convection).*

1968: *Isacks and co-workers attributed earthquakes, volcanoes and mountain building to plate boundaries.*

1967–1970: *Development and broad acceptance of Plate Tectonics.*

Before this time, continental drift was always opposed by geophysicists based on the rigidity of the solid elastic mantle and the 'absence' of physical mechanisms allowing horizontal displacements of thousands of kilometres for continents.

The crucial point that was finally understood by the geological community is that both viscous (i.e., fluid-like) and elastic (i.e., solid-like) behaviour is a characteristic of the Earth depending on the timescale of deformation: the Earth's mantle, which is elastic on a human timescale is viscous on geological timescales ($>10\,000$ years) and can be strongly internally deformed due to solid state creep. There is an amazing substance demonstrating a similar 'dual' viscous–elastic behaviour. This is silicon putty or 'silly putty' which is frequently used as an analogue of rocks in experimental tectonics. It deforms like clay in the hands, but when dropped on the floor it jumps up like a rubber ball (see animation **Silly_putty.mpg**).

Plate tectonics has largely established both a conceptual and a physical basis of geodynamics. The next rapid development of numerical methods of continuum mechanics in this field is the logical consequence of both theoretical and technological progress. Numerical modelling is a necessary tool for geodynamics since tectonic processes are too slow and too deep in the Earth to be observed directly. The snapshot-like history of 2D/3D numerical geodynamic modelling (1D models appeared even earlier!) looks as follows (partly subjective literature-web-search-based view, more details on this issue can be found in several overviews on mantle convection modelling: Richter, 1978; Schubert, 1992; Bercovici, 2007):

1970: *First 2D numerical models of subduction (Minear and Toksoz, 1970).* Exactly the time when the 'Plate Tectonics Era' had just started! The first subduction model was purely thermal, with a prescribed velocity field corresponding to the downgoing slab inclined at $45°$.

1971: *First 2D mantle thermal convection models (Torrance and Turcotte, 1971).* This paper discussed possible implications of mantle convection with temperature-dependent viscosity for continental drift. Thermomechanical models based on the stream function formulation for the mechanical part were explored. A rectangular model domain, with a temperature-dependent viscosity and resolution up to 22×16 nodal points was used.

1972, 1978: *First 2D numerical (finite-element) models of salt domes dynamics (Berner et al., 1972; Woidt, 1978).* Before that, geodynamic modelling studies of crustal diapirism used analytical and analogue modelling approaches. Paper by Woidt (1978) pointed out inconsistency of the numerical approach used by Berner *et al.* (1972).

1977–1980: *First 2D mantle thermal-chemical convection models (Keondzhyan and Monin,* 1977, 1980*).* A binary stratified medium was used to study the effects of compositional layering on mantle convection.

1978: *First numerical models of continental collision (Daignières* et al., 1978; *Bird,* 1978*).* Mechanical models exploring the finite-element approach.

1985–1986: *First 3D spherical mantle convection models (Baumgardner,* 1985, *Machetel* et al., 1986*).* The first 3D models were spherical and not Cartesian as one would expect. Also, for some reason, the first paper appeared in the *Journal of Statistical Physics,* which is not really a geophysical journal . . .

1988: *First 3D Cartesian mantle convection models (Cserepes* et al., 1988; *Houseman,* 1988*).*

Since the 1980s, numerical geodynamic modelling has been developing very rapidly in terms of both the number of various applications and numerical techniques explored. Geodynamic modelling now stands as one of the most dynamic and advanced fields of Earth Sciences.

A few words about programming and visualisation

In this book MATLAB is used for the exercises and for visualisation. This is a good language of choice for people starting with modelling as it allows both easy computing and visualisation. C and FORTRAN are often used for advanced studies that involve usage of supercomputers and computer clusters. In these studies, visualisation is mostly done as a post-processing step that allows independent use of specialised visualisation packages. In our short book, we are more interested to see results instantaneously, during computations. In addition, MATLAB greatly simplifies the solving of system of linear equations which is the core of numerical modelling.

In this book we will consider many example programs, since learning *to write programs (and not just using them)* is an essential part of numerical geodynamic modelling. There are nine important programming rules (which I call *Bug Rules*) which you should follow when writing your own programs.

Bug Rule 1: *Think before programming!* Think carefully about the algorithm of your new code and the most efficient way of making modifications to your old code – you will then develop the program faster and more efficiently and will not need too much code re-thinking and re-writing.

Bug Rule 2: *Comment!* Making comments in the code is essential to enable the code to be used, debugged and modified correctly. The ratio between comment lines and program lines in a good numerical code is larger than 1:1. Do not be lazy, explain every program line – this will save you a lot of time afterwards!

Bug Rule 3: *Programming makes bugs!* We always introduce *bugs* (i.e. programming errors) while writing a code. We typically introduce at least one bug when we modify one single line and we have to test the modified code until we find the bug!

Bug Rule 4: *Programming means debugging!* Be prepared that only 1% of the time will be spent on programming and 99% of your time will be for debugging.

Bug Rule 5: *Most difficult bugs are trivial ones!* There are three types of the most common bugs:

- errors in index (90% of your bugs!), e.g. $y = x(i, j) + z(12)$ instead of $y = x(j, i) - z(2)$
- errors in sign, e.g. $y = x + z$ instead of $y = x - z$ or $y = 1e - 19$ instead of $y = 1e + 19$
- errors in order of magnitude, e.g. $y = 0.0831$ instead of $y = 0.00831$.

Don't be surprised that finding these 'trivial' bugs will sometimes be very difficult (we simply don't see them) and will take a lot of time – this is normal.

Bug Rule 6: *If you see something strange– there is a bug!* Be suspicious, do not ignore even small strange things and discrepancies that you see when computing with your code – in 100% of cases you will find that a bug is the cause. Never try to convince yourself (although this is what we typically tend to do) that a single last digit discrepancy in results with the previous version of the code is due to computer accuracy – it is due to either old or new bugs!

Bug Rule 7: *Single bug can ruin 10 000-lines code!* We should really be motivated to carefully debug and test codes. Don't think that one single small error in the code can be ignored – it will spoil results of months of calculations.

Bug Rule 8: *Wrong model looks beautiful and realistic!* Often erroneous models do not look bad or strange and some of them are really beautiful. Therefore, be prepared that of the numerical modelling results you like, some are actually wrong...

Bug Rule 9: *Creating a good, correct and nicely working code is possible!* This is what should motivate us to follow the eight previous rules!

Units

In this book, the metre-kilogram-second (MKS) system is used in all basic equations as a standard, with only occasional specified deviations toward other conventional units widely used in geosciences (kbar, °C etc.).

How to use this book

Once again, this is a textbook which is primarily aimed at people inexperienced with numerical methods. Therefore, it is organised in a way that, after my learning and teaching experience, provides the easiest path for learning the basics of continuum mechanics and numerical geodynamic modelling. Follow it from one chapter to the next and do all the exercises. Do all the programming by yourself and study code examples ONLY when you are stuck or unsure what to do (all 67 quoted MATLAB codes are provided with this book). The complexity of the programming exercises gradually increases from one chapter to the next, introducing more and more complex aspects of continuum mechanics and numerical modelling. Just trust this way and *don't give up!*

Programming exercises and homework

Exercise Introduction.1
Open MATLAB and use it for the first time. Study the following (use MATLAB Help to read about various functions and operations):

(a) Defining variables, vectors and matrixes
(b) Using mathematical functions ($+$, $-$, *, $.^*$, $/$, $./$, $^\wedge$, $.^\wedge$, *exp*, *log10*, etc.)
(c) Opening/closing text files and loading/writing data from/to them (*fopen, fclose, fscanf, fprintf*)
(d) Plotting of data in 2D and 3D (*figure, plot, pcolor, surf, xlabel, ylabel, shading, light, lighting, axis, colorbar*)
(e) programming loops (*for, while, end*) and conditions (*if, else, end, switch, case,* &&, ||, ==, ~=, >, <, >=, <= etc.)

Exercise Introduction.2
Write your first MATLAB code for visualising *sin, cos* functions in 2D (*plot, pcolor, contour*) and 3D (*surf, light, lighting*). An example is in **Visualisation_is_important.m**.

1

The continuity equation

> **Theory:** Definition of a geological medium as a continuum. Field variables used for the representation of a continuum. Methods for definition of the field variables. Eulerian and Lagrangian points of view. Continuity equation in Eulerian and Lagrangian forms and their derivation. Advective transport term. Continuity equation for an incompressible fluid.
> **Exercises:** Computing the divergence of velocity field in 2D.

1.1 Continuum – what is it?

What we should understand from the very beginning is that geodynamics considers major rock units, such as the Earth's crust and mantle as *continuous geological media*. *Continuity* of any medium implies that, on a macroscopic scale, the material under consideration does not contain *mass-free voids or gaps* (there can indeed be pores or cavities but they are also filled with some continuous substances). Different physical properties of a continuum may vary at every geometrical point and we thus need a *continuous description*. In *continuum mechanics*, the physical properties of a continuum (*field properties*) are described by *field variables* such as pressure, temperature, density, velocity, etc. There are three major types of field variables:

scalars (e.g., pressure, temperature, density),
vectors (e.g., velocity, mass flux, heat flux),
tensors (e.g. stress, strain, strain rate).

Field variables can be represented in a *fully continuous* manner (analytical expressions, Fig. 1.1(a)) or in a *discrete-continuous* way (by arrays of values which characterise selected *nodal* geometrical points, Fig. 1.1(b–d)). In the latter case,

11

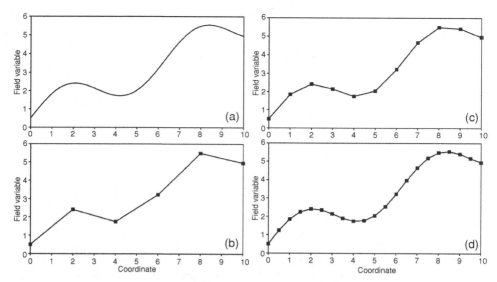

Fig. 1.1 Continuous (a) and discrete-continuous (b)–(d) representations of a field variable as a function of a coordinate. Note that in the case of the discrete-continuous representation with linear interpolation between nodal points, the representation accuracy notably increases with increasing number of nodal points (compare (a) with (b), (c) and (d)).

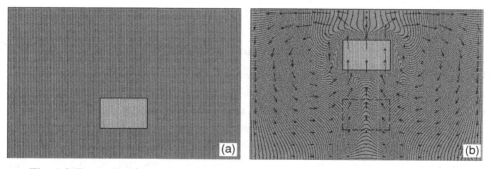

Fig. 1.2 Example of the deformation of a continuous medium (dark grey) due to the buoyant rise of a rigid block (light grey). (a) initial stage. (b) final stage with the corresponding velocity field indicated by arrows. Note that no voids are formed where the block was initially located (dashed contour). Individual black and white dots in (a) and (b) correspond to different Lagrangian points displaced by the flow. Diagrams are computed numerically with the code developed by Gerya and Yuen (2003a), and an animation is shown in file **Fig1_2.ppt** associated with this chapter.

various linear and non-linear *interpolation* rules are used to calculate values of field variables between the nodal points (Fig. 1.1(b)–(d)).

Continuity also implies that displacements of different portions of the medium are not fully independent. These displacements must proceed without creating macroscopic voids and gaps: if some rocks are displaced *from* a certain area (for example due to tectonic extrusion), other rocks must come *into* this area and substitute the displaced fragment (Fig. 1.2). In a way, this type of continuous behaviour is very similar to that of water or, generally, any fluid which can be described by *fluid mechanics* (e.g., Landau and Lifshitz, 1987; Kundu and Cohen, 2002). Since on long timescales rocks behave like slowly creeping fluids, geodynamic processes in the Earth's mantle, as for example mantle convection, are often also referred to as processes of geophysical fluid dynamics.

1.2 Continuity equation

Our qualitative, intuitive understanding of continuity can, indeed, be transformed into a quantitative mathematical formalism. This formalism is widely used in numerical geodynamic modelling in form of a *continuity equation* which describes the *conservation of mass* during the displacement of a continuous medium. Let's write this equation and try to understand its structure in detail.

The first thing that we have to learn is that the form of the mass conservation equation (and many other *time-dependent* conservation equations) can be either *Eulerian* or *Lagrangian* depending on the nature of a geometrical point for which this equation is written. The Eulerian continuity equation is written for an *immobile*, or fixed point in space; it has the form:

$$\frac{\partial \rho}{\partial t} + \text{div}(\rho \vec{v}) = 0. \tag{1.1a}$$

Or, in a slightly different symbolic notation often used in continuum mechanics

$$\frac{\partial \rho}{\partial t} + \nabla \cdot (\rho \vec{v}) = 0, \tag{1.1b}$$

where $\frac{\partial}{\partial t}$ is the Eulerian time derivative; ρ is the local density, which characterises the amount of mass per unit volume (kg/m^3); $\vec{v}$ is local velocity (m/s) and div() or $\nabla \cdot$ denotes the divergence operator. The divergence is a scalar function of a vector field, and is defined as follows,

in one dimension (1D): $\qquad \text{div}(\vec{v}) = \dfrac{\partial v_x}{\partial x},$ $\qquad\qquad$ (1.2a)

in two dimensions (2D): $\qquad \text{div}(\vec{v}) = \dfrac{\partial v_x}{\partial x} + \dfrac{\partial v_y}{\partial y},$ $\qquad\qquad$ (1.2b)

in three dimensions (3D): $\qquad \text{div}(\vec{v}) = \dfrac{\partial v_x}{\partial x} + \dfrac{\partial v_y}{\partial y} + \dfrac{\partial v_z}{\partial z},$ $\qquad$ (1.2c)

where x, y and z are Cartesian coordinates and v_x, v_y and v_z are components parallel to the respective coordinate axes of the velocity vector $\vec{v}$ (Fig. 1.3). In simple words, the divergence of a vector at a given point is positive when the surrounding vector

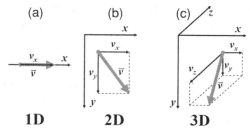

Fig. 1.3 Components of a local velocity vector $\vec{v}$ (grey arrow) in one (a) two (b) and three (c) dimensions.

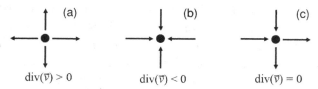

Fig. 1.4 Examples of divergent (a) convergent (b) and neutral (c) 2D velocity fields around a point.

field is directed outward from the point (divergent flow, Fig. 1.4(a)) and is negative when this field is directed towards the point (convergent flow, Fig. 1.4(b)).

The Lagrangian continuity equation is written for a *moving* point of reference; it has the form:

$$\frac{D\rho}{Dt} + \rho \,\mathrm{div}(\vec{v}) = 0, \tag{1.3a}$$

or

$$\frac{D\rho}{Dt} + \rho \nabla \cdot \vec{v} = 0, \tag{1.3b}$$

where $\dfrac{D}{Dt}$ is the Lagrangian time derivative and the other parameters were explained before (see Eq. (1.1)).

1.3 Eulerian and Lagrangian points – what is the difference?

Understanding the difference between Eulerian and Lagrangian points is *funda-mental* for continuum mechanics. A Lagrangian point is strictly connected to a single material point and is moving with this point. Therefore, the same material point is always found in a given Lagrangian point independent of the moment of time. For this reason, the Lagrangian time derivative of density $\dfrac{D\rho}{Dt}$ (i.e., changes in density with time at the Lagrangian point) is also called the *substantive* or *objective* time derivative. On the other hand, an Eulerian point is an immobile observation point, not related to any specific material point. Therefore, at different moments of time, different Lagrangian material points can be found at the same Eulerian point. In other words, different Lagrangian material points are *passing through* the same Eulerian observation point with time. Many equations of continuum mechanics containing time derivatives can be written in both Eulerian and Lagrangian forms which differ from each other (e.g. Eqs. (1.1) and (1.3)). Choosing either the Eulerian or Lagrangian form of an equation notably affects the way of representing *advective transport processes* (i.e. the movement of material with the flow) which will be discussed in detail in Chapter 8, together with the advantages and disadvantages of the two approaches.

1.4 Derivation of the Eulerian continuity equation

Let's now analyse the Eulerian continuity equation (Eq. (1.1)) which contains both vector (velocity) and scalar (density) variables. This equation establishes the balance of mass in an elementary observation volume. It implies, in particular, that if mass is leaving (fluxing out of) the volume (i.e., $\text{div}(\rho \vec{v}) > 0$), the local density (i.e., the amount of mass per unit volume) decreases with time (i.e., $\frac{\partial \rho}{\partial t} < 0$).

First we need to understand that $\rho \vec{v}$ is nothing else but the local *mass flux* vector

$$\rho \vec{v} = (\rho v_x, \rho v_y, \rho v_z), \tag{1.4}$$

which has the dimension of unit mass, fluxing through a unit surface, per unit time $\left(\frac{\text{kg}}{\text{m}^2\text{s}} \right)$. This definition follows from the fact that the velocity in a continuous medium can be considered as *material volume flux* (Fig. 1.5) i.e., unit volume fluxing through a unit surface per unit time $\left(\frac{\text{m}}{\text{s}} = \frac{\text{m}^3}{\text{m}^2\text{s}} \right)$. Therefore, velocity (i.e. volume flux) multiplied by the density (i.e. mass per unit volume) gives the mass flux.

The Eulerian continuity equation can be derived by analysing material fluxes through a small, immobile, rectangular Eulerian (observation) volume of constant dimensions Δx, Δy and Δz (Fig. 1.6(a)). Let's assume that the initial mass of fluid in this volume is m_0. Then, the initial average fluid density (ρ_0) within this volume is thus

$$\rho_0 = \frac{m_0}{\Delta x \, \Delta y \, \Delta z}. \tag{1.5}$$

Mass enters the volume through the boundaries A, C and E and leaves it through the opposite boundaries B, D and F. Material fluxes affect the mass of fluid in the

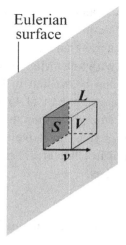

Fig. 1.5 Relationship between the flow velocity v and material volume V, passing through the element S of the immobile Eulerian surface (grey) during the time t. The relations $V = S\,L$ and $L = v\,t$ allows one to formulate velocity as the material volume flux $v = \dfrac{V}{S\,t}$.

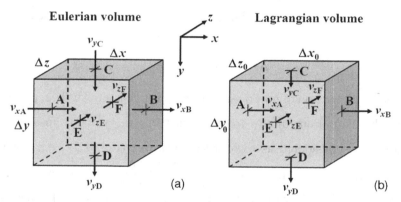

Fig. 1.6 Eulerian (a) and Lagrangian (b) elementary volumes considered for the derivation of the continuity equation. Arrows in (a) show the velocity components which are responsible for *material fluxes through* the respective boundaries (A, B, C, D, E and F). Arrows in (b) show the velocity components responsible for *moving* the respective boundaries.

observation volume and after a small period of time Δt (or *time increment*), this mass becomes m_1 and the average fluid density (ρ_1) changes to

$$\rho_1 = \frac{m_1}{\Delta x\,\Delta y\,\Delta z}. \tag{1.6}$$

The balance between the old (m_0) and new (m_1) mass results in the following relation

$$m_1 = m_0 + m_{in} - m_{out},$$
$$m_{in} = m_A + m_C + m_E,$$
$$m_{out} = m_B + m_D + m_F,$$
$$m_A = \rho_A v_{xA} \Delta y \Delta z \Delta t,$$
$$m_B = \rho_B v_{xB} \Delta y \Delta z \Delta t, \qquad (1.7)$$
$$m_C = \rho_C v_{yC} \Delta x \Delta z \Delta t,$$
$$m_D = \rho_D v_{yD} \Delta x \Delta z \Delta t,$$
$$m_E = \rho_E v_{zE} \Delta x \Delta y \Delta t,$$
$$m_F = \rho_F v_{zF} \Delta x \Delta y \Delta t,$$

where m_{in} and m_{out} are the incoming and outgoing mass, respectively; $m_A - m_F$ are masses that passed through the respective boundaries during the time Δt; $\rho_A - \rho_F$ is the density at the respective boundaries; $v_{xA} - v_{zD}$ are the velocity components responsible for material fluxes through the boundaries (Fig. 1.6(a)). If Δt is small, we can now write an approximate expression for the Eulerian time derivative of the average density in the volume as:

$$\frac{\partial \rho}{\partial t} \approx \frac{\Delta \rho}{\Delta t} = \frac{\rho_1 - \rho_0}{\Delta t} = \frac{m_1 - m_0}{\Delta x \, \Delta y \, \Delta z \, \Delta t}. \qquad (1.8)$$

By using Equation (1.7) the following expression can be further obtained (verify as an exercise)

$$\frac{\Delta \rho}{\Delta t} = -\frac{\rho_B v_{xB} - \rho_A v_{xA}}{\Delta x} - \frac{\rho_D v_{yD} - \rho_C v_{yC}}{\Delta y} - \frac{\rho_F v_{zF} - \rho_E v_{zE}}{\Delta z}, \qquad (1.8a)$$

or

$$\frac{\Delta \rho}{\Delta t} = -\frac{\Delta(\rho v_x)}{\Delta x} - \frac{\Delta(\rho v_y)}{\Delta y} - \frac{\Delta(\rho v_z)}{\Delta z}, \qquad (1.8b)$$

or

$$\frac{\Delta \rho}{\Delta t} + \frac{\Delta(\rho v_x)}{\Delta x} + \frac{\Delta(\rho v_y)}{\Delta y} + \frac{\Delta(\rho v_z)}{\Delta z} = 0,$$
$$\Delta(\rho v_x) = \rho_B v_{xB} - \rho_A v_{xA},$$
$$\Delta(\rho v_y) = \rho_D v_{yD} - \rho_C v_{yC}, \qquad (1.8c)$$
$$\Delta(\rho v_z) = \rho_F v_{zF} - \rho_E v_{zE},$$

where $\Delta(\rho v_x)$, $\Delta(\rho v_y)$ and $\Delta(\rho v_z)$ are differences in the mass fluxes in x, y and z directions respectively (i.e. through the respective pairs of boundaries, Fig. 1.6(a)).

Obviously, in such cases when Δt, Δx, Δy and Δz all tend to zero, the differences can be replaced by derivatives and we obtain the Eulerian continuity equation

$$\frac{\partial \rho}{\partial t} + \frac{\partial(\rho v_x)}{\partial x} + \frac{\partial(\rho v_y)}{\partial y} + \frac{\partial(\rho v_z)}{\partial z} = 0, \tag{1.9a}$$

or

$$\frac{\partial \rho}{\partial t} + \mathrm{div}(\rho \vec{v}) = 0. \tag{1.9b}$$

1.5 Derivation of the Lagrangian continuity equation

Similarly, the Lagrangian continuity equation can be derived by analysing the motion of a small, mobile, rectangular Lagrangian (material) volume of initial dimensions Δx_0, Δy_0 and Δz_0 (Fig. 1.6(b)). In contrast to the steady Eulerian volume, the amount of mass m in the moving Lagrangian volume remains constant (since this volume is always related to the same material points), but the dimensions of the volume may change due to *internal* expansion/contraction processes. The initial average fluid density (ρ_0), within the volume is given by

$$\rho_0 = \frac{m}{\Delta x_0 \, \Delta y_0 \, \Delta z_0}. \tag{1.10}$$

Internal expansion or contraction affects the dimensions of the Lagrangian volume, and after a small period of time Δt, these dimensions become Δx_1, Δy_1 and Δz_1 and the average fluid density (ρ_1) changes to

$$\rho_1 = \frac{m}{\Delta x_1 \, \Delta y_1 \, \Delta z_1}. \tag{1.11}$$

The relationship between the old (Δx_0, Δy_0, Δz_0) and the new (Δx_1, Δy_1, Δz_1) dimensions of the Lagrangian volume can be established on the basis of the relative movements of the boundaries of the volume (A, B, C, D, E, F, Fig. 1.6(b)) which leads to the following relations

$$\Delta x_1 = \Delta x_0 + \Delta t \, \Delta v_x, \tag{1.12}$$
$$\Delta v_x = v_{xB} - v_{xA}$$
$$\Delta y_1 = \Delta y_0 + \Delta t \, \Delta v_y, \tag{1.13}$$
$$\Delta v_y = v_{yD} - v_{yC}$$
$$\Delta z_1 = \Delta z_0 + \Delta t \, \Delta v_z, \tag{1.14}$$
$$\Delta v_z = v_{zF} - v_{zE}$$

where Δv_x, Δv_y and Δv_z are the differences in the velocity components that correspond to the relative movements of their respective pairs of boundaries (Fig. 1.6(b)). Taking Δt to be small, we can now write an approximate expression for the

Lagrangian time derivative of the average density in the volume as

$$\frac{D\rho}{Dt} \approx \frac{\Delta\rho}{\Delta t} = \frac{\rho_1 - \rho_0}{\Delta t} = \frac{m}{\Delta x_1\,\Delta y_1\,\Delta z_1\,\Delta t} - \frac{m}{\Delta x_0\,\Delta y_0\,\Delta z_0\,\Delta t}. \qquad (1.15)$$

By using the equations derived for Δx_1, Δy_1 and Δz_1 (Eqs. (1.12)–(1.14)) the following expression can be obtained (verify as an exercise)

$$\frac{\Delta\rho}{\Delta t} = \rho_0 \frac{1 - \left(1 + \Delta t\dfrac{\Delta v_x}{\Delta x_0}\right)\left(1 + \Delta t\dfrac{\Delta v_y}{\Delta y_0}\right)\left(1 + \Delta t\dfrac{\Delta v_z}{\Delta z_0}\right)}{\Delta t\left(1 + \Delta t\dfrac{\Delta v_x}{\Delta x_0}\right)\left(1 + \Delta t\dfrac{\Delta v_y}{\Delta y_0}\right)\left(1 + \Delta t\dfrac{\Delta v_z}{\Delta z_0}\right)}, \qquad (1.16a)$$

or

$$\frac{\Delta\rho}{\Delta t} = \rho_0 \frac{-\dfrac{\Delta v_x}{\Delta x_0} - \dfrac{\Delta v_y}{\Delta y_0} - \dfrac{\Delta v_z}{\Delta z_0} - \Delta t\left(\dfrac{\Delta v_x\,\Delta v_y}{\Delta x_0\,\Delta y_0} + \dfrac{\Delta v_x\,\Delta v_z}{\Delta x_0\,\Delta z_0} + \dfrac{\Delta v_y\,\Delta v_z}{\Delta y_0\,\Delta z_0} + \Delta t\dfrac{\Delta v_x\,\Delta v_y\,\Delta v_z}{\Delta x_0\,\Delta y_0\,\Delta z_0}\right)}{\left(1 + \Delta t\dfrac{\Delta v_x}{\Delta x_0}\right)\left(1 + \Delta t\dfrac{\Delta v_y}{\Delta y_0}\right)\left(1 + \Delta t\dfrac{\Delta v_z}{\Delta z_0}\right)}, \qquad (1.16b)$$

or

$$\frac{\Delta\rho}{\Delta t} + \rho_0 \frac{\dfrac{\Delta v_x}{\Delta x_0} + \dfrac{\Delta v_y}{\Delta y_0} + \dfrac{\Delta v_z}{\Delta z_0} + K_1}{K_2} = 0,$$

$$K_1 = \Delta t\left(\frac{\Delta v_x\,\Delta v_y}{\Delta x_0\,\Delta y_0} + \frac{\Delta v_x\,\Delta v_z}{\Delta x_0\,\Delta z_0} + \frac{\Delta v_y\,\Delta v_z}{\Delta y_0\,\Delta z_0} + \Delta t\frac{\Delta v_x\,\Delta v_y\,\Delta v_z}{\Delta x_0\,\Delta y_0\,\Delta z_0}\right), \qquad (1.16c)$$

$$K_2 = \left(1 + \Delta t\frac{\Delta v_x}{\Delta x_0}\right)\left(1 + \Delta t\frac{\Delta v_y}{\Delta y_0}\right)\left(1 + \Delta t\frac{\Delta v_z}{\Delta z_0}\right),$$

where K_1 and K_2 are coefficients which respectively tend to *zero* and *unity* when Δt tends to zero. Obviously in the case when Δt, Δx_0, Δy_0 and Δz_0 all tend towards zero, the differences in Eq. (1.16c) can be replaced by derivatives and taking $K_1 = 0$ and $K_2 = 1$ we obtain the Lagrangian continuity equation

$$\frac{D\rho}{Dt} + \rho\frac{\partial v_x}{\partial x} + \rho\frac{\partial v_y}{\partial y} + \rho\frac{\partial v_z}{\partial z} = 0, \qquad (1.17a)$$

or

$$\frac{D\rho}{Dt} + \rho\,\text{div}(\vec{v}) = 0. \qquad (1.17b)$$

1.6 Comparing Eulerian and Lagrangian continuity equations. Advective transport term

Let's now perform transformations of the Eulerian continuity equation (Eq. (1.1)) in order to decipher its structure and to establish a relationship with the Lagrangian continuity equation (Eq. (1.3)). It is convenient to decompose $\operatorname{div}(\rho\vec{v})$ using the standard *product rule* (also called *Leibniz's law*) $(u \cdot v)' = u' \cdot v + v' \cdot u$, or in an alternative notation $\dfrac{\partial}{\partial x}(uv)' = \dfrac{\partial u}{\partial x} \cdot v + \dfrac{\partial v}{\partial x} \cdot u$.

$$\operatorname{div}(\rho\vec{v}) = \rho \operatorname{div}(\vec{v}) + \vec{v}\operatorname{grad}(\rho), \tag{1.18a}$$

or in a different symbolic notation

$$\nabla \cdot (\rho\vec{v}) = \rho\nabla \cdot \vec{v} + \vec{v}\nabla\rho, \tag{1.18b}$$

or 'deciphering' what we actually are doing in three dimensions

$$\frac{\partial}{\partial x}(\rho v_x) + \frac{\partial}{\partial y}(\rho v_y) + \frac{\partial}{\partial z}(\rho v_z) = \left(\rho\frac{\partial v_x}{\partial x} + \rho\frac{\partial v_y}{\partial y} + \rho\frac{\partial v_z}{\partial z}\right)$$
$$+ \left(v_x\frac{\partial \rho}{\partial x} + v_y\frac{\partial \rho}{\partial y} + v_z\frac{\partial \rho}{\partial z}\right), \tag{1.18c}$$

where $\operatorname{grad}(\rho)$ or $\nabla\rho$ is the gradient of the density ρ. The gradient is a vector function of a scalar field defined as follows

in one dimension (1D):

$$\operatorname{grad}(\rho) = \frac{\partial \rho}{\partial x}, \tag{1.19a}$$

in two dimensions (2D):

$$\operatorname{grad}(\rho) = \left(\frac{\partial \rho}{\partial x}, \frac{\partial \rho}{\partial y}\right), \tag{1.19b}$$

in three dimensions (3D):

$$\operatorname{grad}(\rho) = \left(\frac{\partial \rho}{\partial x}, \frac{\partial \rho}{\partial y}, \frac{\partial \rho}{\partial z}\right). \tag{1.19c}$$

Therefore, both $\nabla\rho$ and $\vec{v}$ in Eq. (1.18) are vectors and *scalar multiplying* of these two vectors (1.18c) gives the following result

in one dimension (1D):

$$\vec{v}\operatorname{grad}(\rho) = v_x\frac{\partial \rho}{\partial x}, \tag{1.20a}$$

in two dimensions (2D):

$$\vec{v}\operatorname{grad}(\rho) = v_x\frac{\partial \rho}{\partial x} + v_y\frac{\partial \rho}{\partial y}, \tag{1.20b}$$

in three dimensions (3D):

$$\vec{v}\operatorname{grad}(\rho) = v_x\frac{\partial \rho}{\partial x} + v_y\frac{\partial \rho}{\partial y} + v_z\frac{\partial \rho}{\partial z}. \tag{1.20c}$$

Now by comparing Equations (1.1), (1.3) and (1.18) we can establish the relationship between the Eulerian ($\frac{\partial \rho}{\partial t}$) and Lagrangian ($\frac{D\rho}{Dt}$) time derivatives of density

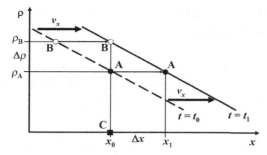

Fig. 1.7 Schematic representation of the advective transport in the case of uniform 1D movement of a fluid with a linear density distribution. The dashed and solid thick lines correspond to the density distribution for the moments of time t_0 and t_1, respectively. Circles A and B denote the density for two different Lagrangian points passing through an Eulerian observation point (solid rectangle C) at the different moments of time (t_0 and t_1, respectively).

as

$$\frac{D\rho}{Dt} = \frac{\partial \rho}{\partial t} + \vec{v}\,\text{grad}(\rho). \tag{1.21}$$

The extra term, $\vec{v}\,\text{grad}(\rho)$ in the Eulerian continuity equation is an *advective trans-port* term that reflects changes of density in an immobile (Eulerian) point, due to the movement of an inhomogeneous medium with existing density gradients relative to this point (Fig. 1.7). Obviously, the advective transport terms in the Eulerian continuity equation are only relevant (i.e. nonzero) in situations when both the velocity of the medium and density gradient are nonzero. On the other hand, *substantive* changes of density ($D\rho/Dt$) in the moving Lagrangian point do not depend on density gradients and the Lagrangian continuity equation (1.3) thus does not contain advective terms.

When the density in all moving material points does not change with time (i.e. $\dfrac{D\rho}{Dt} = 0$) the Eulerian continuity equation reduces to the Eulerian *advective transport equation*

$$\frac{\partial \rho}{\partial t} = -\vec{v}\,\text{grad}(\rho). \tag{1.22}$$

The minus sign in the right-hand side of equation (1.22) reflects the relation between the density gradient and the direction of motion (Fig. 1.7): if a medium is moving *in the direction* of decreasing density (i.e. $\vec{v}\,\text{grad}(\rho) < 0$), then the density in an immobile observation point *increases* with time (i.e., $\dfrac{\partial \rho}{\partial t} > 0$).

Let's now derive the advective transport relation (Eq. (1.22)) for the simple 1D case shown in Fig. 1.7. A fluid with a linear density distribution moves at a constant velocity v_x. At the moment of time t_0 the density (ρ_A) in the Eulerian observation point C (solid rectangle) corresponds to the Lagrangian point A (solid circle). Due to fluid motion, the density profile shifts to the right with time. Therefore, at a later moment of time t_1, the density (ρ_B) at the same Eulerian point C will correspond to another Lagrangian point B (open circle). Under the assumption that the difference in time between the two moments ($\Delta t = t_1 - t_0$) is small, we can approximate the time derivative of density at the Eulerian point C as

$$\frac{\partial \rho}{\partial t} \approx \frac{\Delta \rho_t}{\Delta t} = \frac{\rho_B - \rho_A}{t_1 - t_0}, \tag{1.23}$$

where $\Delta \rho_t$ corresponds to changes in density with time at the Eulerian point C. If on the other hand, the displacement of the density profile (i.e. the distance between two Lagrangian points A and B, $\Delta x = x_1 - x_0$) is also small, we can approximate the local derivative of density with respect to x, at the moment of time t_1 as

$$\frac{\partial \rho}{\partial x} \approx \frac{\Delta \rho_x}{\Delta x} = \frac{\rho_A - \rho_B}{x_1 - x_0}, \tag{1.24}$$

where $\Delta \rho_x$ corresponds to a change in density with coordinate in the proximity of the Eulerian point C at the instant when time equals t_1. Note that $\Delta \rho_t$ and $\Delta \rho_x$ are different quantities, and that in the case considered they are similar in absolute value but are different in sign (cf. Eq. (1.23) and (1.24)). Taking into account the relationships

$$\Delta x = v_x \Delta t$$

and

$$\Delta \rho_x = -\Delta \rho_t,$$

the following expression can be obtained (derive as an exercise)

$$\frac{\Delta \rho_t}{\Delta t} = -v_x \frac{\Delta \rho_x}{\Delta x}. \tag{1.25}$$

When Δt and respectively Δx tend to zero, the differences in Eq. (1.25) can be replaced by derivatives and we obtain the 1D advection equation

$$\frac{\partial \rho}{\partial t} = -v_x \frac{\partial \rho}{\partial x}. \tag{1.26}$$

The advective transport equation is very important for geodynamic modelling since it describes changes in the distribution of *transport properties* (such as density, temperature, composition, etc.) in non-homogeneous, deforming media. We will come back to this issue in Chapter 8.

1.7 Incompressible continuity equation

For many geological media (such as the Earth's crust and mantle), one may assume an incompressibility condition (i.e. density of material points does not change with time), which is written as:

$$\frac{D\rho}{Dt} = \frac{\partial \rho}{\partial t} + \vec{v}\,\mathrm{grad}\,(\rho) = 0. \tag{1.27}$$

It is valid in cases when pressure and temperature changes are not very large and no phase transformations leading to volume changes occur in the medium. In this situation, one can use an *incompressible continuity equation* which is the same in both Eulerian and Lagrangian forms

$$\mathrm{div}(\vec{v}) = 0. \tag{1.28}$$

The incompressible continuity equation is broadly used in numerical geodynamic modelling, although in many cases it is a rather big simplification (e.g. in the case of the whole Earth mantle convection, e.g., Tackley, 2008). Typical examples of geodynamic settings where deformations are strongly defined by the incompressibility condition are, for example, corner flow in the mantle wedge above subducting slabs (e.g. Turcotte and Schubert, 2002) and circulation of tectonic melanges in subduction channels (e.g. Cloos, 1982).

Analytical exercise

Exercise 1.1
In a region of the Earth's mantle, the velocity field is given by

$$v_x = 10^{-10} + x \cdot 10^{-13} + y \cdot 10^{-13} + z \cdot 10^{-13}$$
$$v_y = 10^{-10} - x \cdot 10^{-13} + y \cdot 2 \times 10^{-13} + z \cdot 3 \times 10^{-13}$$
$$v_z = 10^{-10} - x \cdot 10^{-13} - y \cdot 10^{-13} - z \cdot 2 \times 10^{-13}.$$

The mantle density field in the same region is given by

$$\rho = 3300 + x \cdot 0.001 - y \cdot 0.002 + z \cdot 0.001.$$

Calculate ρ, $\mathrm{div}(\vec{v})$, $\dfrac{\partial \rho}{\partial t}$ and $\dfrac{D\rho}{Dt}$ for the point with coordinates $x = 1000$, $y = 1000$, $z = 1000$.

Programming exercise and homework

Exercise 1.2

Write a MATLAB code for computing and visualising a 2D velocity field and its divergence. Model design: an area of the mantle (1000×1500 km) is convecting with one central upwelling in the middle of the model box and two downwellings at the sides. The velocity field is given by the following equations

$$v_x = -v_{x0} \sin\left(2\pi \frac{x}{W}\right) \cos\left(\pi \frac{y}{H}\right),$$

$$v_y = v_{y0} \cos\left(2\pi \frac{x}{W}\right) \sin\left(\pi \frac{y}{H}\right),$$

where x and y are respectively horizontal and vertical coordinates inside the box in m; $W = 1\,000\,000$ m and $H = 1\,500\,000$ m are the width and height of the model, respectively (i.e., 1000×1500 km); $v_{x0} = 10^{-9}$ m/s and $v_{y0} = 10^{-9}$ m/s are scaling values for respectively horizontal and vertical velocity components (10^{-9} m/s $\approx$ 3 cm/year). Compute (analytically) v_x, v_y, $\frac{\partial v_x}{\partial x}$, $\frac{\partial v_y}{\partial y}$ and $\mathrm{div}(\bar{v}) = \frac{\partial v_x}{\partial x} + \frac{\partial v_y}{\partial y}$ on a 2D grid of points (e.g. 31×31) which are regularly distributed inside the model and visualise these parameters separately as colourmaps (*pcolor*) in order to see how they are distributed relative to each other. Visualise the velocity as an arrow field (*quiver*). Try to tune the scaling velocity values v_{x0} and v_{y0}, such that the divergence of velocity goes to (almost) zero in the entire model (i.e. absolute values of $\mathrm{div}(\bar{v})$ should be many orders of magnitude less than that of $\frac{\partial v_x}{\partial x}$ and $\frac{\partial v_y}{\partial y}$). An example is in **Divergence.m**.

2

Density and gravity

Theory: Density of rocks and minerals. Thermal expansion and compressibility. Dependence of density on pressure and temperature. Equations of state. Poisson equation for gravitational potential and its derivation.

Exercises: Computing and visualising density, thermal expansion and compressibility.

2.1 Density of rocks and minerals. Equations of state

Many geodynamic processes are either directly or indirectly driven by the gravity force due to the spatial variation of rock density inside the Earth. The density of rocks (ρ) depends on pressure (P), temperature (T), chemical composition (C) and mineralogical composition (M)

$$\rho = f(P, T, C, M). \tag{2.1}$$

These factors are not fully independent. The mineralogical composition of a rock with a constant composition may for example change due to changes in pressure and temperature. *Easy-to-remember* densities of major rock types used in geodynamics are:

felsic rocks (e.g., granites) $\sim$2700 kg/m^3,
mafic rocks (e.g. basalts) $\sim$3000 kg/m^3,
ultramafic rocks (e.g., peridotites) $\sim$3300 kg/m^3.

Other relevant rock densities are given in Table 17.2 and in Appendix 2 of Turcotte and Schubert (2002).

Variations in the density of minerals and rocks with T and P are often characterised by thermal expansion (α) and compressibility (β)

$$\alpha = -\frac{1}{\rho}\frac{\partial\rho}{\partial T},\tag{2.2}$$

$$\beta = \frac{1}{\rho}\frac{\partial\rho}{\partial P}.\tag{2.3}$$

The thermal expansion of rocks and minerals is typically positive ($\alpha > 0$), i.e.
density decreases with increasing temperature (cf. minus in the right part of
Eq. (2.2)). There are, indeed, exceptions, for example, beta-quartz possesses a
density which remains almost constant with increasing temperature at room pres-
sure ($\alpha \approx 0$). Typical values of α are on the order of $n \times 10^{-5}$ K^{-1}. Compressibility
is always positive ($\beta > 0$) and density always increases with increasing pressure.
Typical values of β are on the order of $n \times 10^{-2}$ GPa^{-1} (or $n \times 10^{-11}$ Pa^{-1}). In
cases of constant α and β, integration of Equations (2.2) and (2.3) versus T and P
gives

$$\rho = \rho_r e^{\beta(P-P_r)-\alpha(T-T_r)},\tag{2.4a}$$

where ρ_r is the density of a given material at *reference* pressure P_r (typically 10^5
Pa $= 1$ bar) and temperature T_r (typically 298.15 K $= 25\,°$C).

Since both $\alpha(T - T_r)$ and $\beta(P - P_r)$ are typically very small (much less than
unity), the equations can be simplified either to (using the rules that $e^a \approx 1 + a$
and $e^{-a} \approx 1 - a$ when $a \ll 1$)

$$\rho = \rho_r\,[1 + \beta(P - P_r)] \times [1 - \alpha(T - T_r)],\tag{2.4b}$$

or (using the rules that $e^a \approx 1 + a$ and $e^{-a} = \dfrac{1}{e^a} \approx \dfrac{1}{1+a}$ when $a \ll 1$) to

$$\rho = \rho_r\frac{1 + \beta(P - P_r)}{1 + \alpha(T - T_r)}.\tag{2.4c}$$

Equations (2.4a)–(2.4c), however, do not account for changes in density due to
changes in mineralogical composition with changing pressure and temperature.
These *equations of state* (*EOS*) are much simplified since they are based on the
assumption that α and β remain constant with pressure and temperature. For real
rocks and minerals, these two parameters however strongly depend on both P and
T (Fig. 2.1) and therefore more realistic EOS must be used for describing the
density changes (e.g. Anderson, 1995). If compressibility of the mineral depends
on pressure, then the following well-known equations can be used:

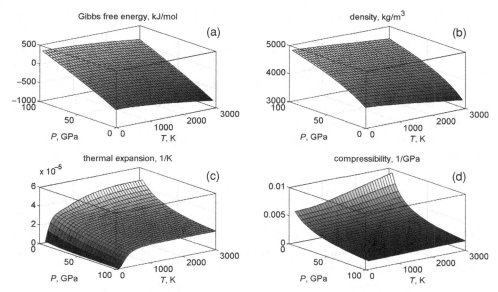

Fig. 2.1 Gibbs free energy (a), density (b), thermal expansion (c) and compressibility (d) of periclase (MgO) computed on the basis of the semi-empirical parameterisation of Gerya *et al.* (2004d) for a broad range of temperature and pressure values. Note that pressure axis in (c) and (d) is inverted compared to (a) and (b). Results are computed with the code **Periclase_EOS.m**.

Murnaghan equation of state (Murnaghan, 1944)

$$\rho = \rho_0 \left(1 + B_0' \frac{P}{B_0} \right)^{1/B_0'}, \tag{2.5}$$

Birch–Murnaghan equation of state (Birch, 1947)

$$P(\rho) = \frac{3B_0}{2} \left[\left(\frac{\rho}{\rho_0} \right)^{7/3} - \left(\frac{\rho}{\rho_0} \right)^{5/3} \right] \left\{ 1 + \frac{3}{4} (B_0' - 4) \left[\left(\frac{\rho}{\rho_0} \right)^{2/3} - 1 \right] \right\}, \tag{2.6}$$

where $B_0 = 1/\beta_0$ is the bulk modulus (this quantity will be also discussed in relation to elasticity in Chapter 12) at $P = 0$ (β_0 is the compressibility at $P = 0$),

$B_0' = \left(\dfrac{\partial B}{\partial P} \right)_{T=const}$ is the pressure derivative of the bulk modulus at a constant

temperature and ρ_0 is the density at $P = 0$. The Murnaghan equation is derived with an experimentally based assumption that the pressure derivative of the bulk modulus B_0' is independent of pressure (many substances have a fairly constant B_0' value of about 3.5). The Birch–Murnaghan equation (2.6) is also derived empirically based on measurements of volumes of solid substances, held at a constant temperature.

Obviously, this equation has to be solved iteratively to obtain a density at a given pressure.

Equations (2.5) and (2.6), however, do not establish the dependence of density upon temperature, and therefore, an even more complicated EOS must be used when the density description over a wide range of both pressure and temperature values is needed. For example, in mineralogy and petrology, the molar volume of minerals (V) is described in many cases by semi-empirical equations (e.g. the Murnaghan-like EOS) with additional temperature-dependent terms (Holland and Powell, 1998)

$$V = V_r \left[1 + a(T - T_r) + b\left(\sqrt{T} - \sqrt{T_r}\right)\right] \times \left\{1 - \frac{B'_r P}{B_r[1 - c(T - T_r)] + B'_r P}\right\}^{1/B'_r},$$

(2.7)

where V_r is molar volume at P_r and T_r (i.e. at a *reference PT-conditions*) and a, b and c are empirical parameters computed from experimental measurements of molar volume at elevated pressures and temperatures.

It should also be mentioned that in mineralogy and petrology, self-consistent descriptions of thermodynamic properties (including density) of minerals and fluids is often based on formulating the equations for their molar thermodynamic potentials (typically either for Gibbs $G_{(P, T)}$ or for Helmholz $F_{(V, T)}$ potential, e.g., Karpov *et al.*, 1976; Helgesson *et al.*, 1978; Dorogokupets and Karpov, 1984; Berman, 1988; Holland and Powell, 1990, 1998). Density as well as many other physical properties (such as heat capacity, thermal expansion, compressibility, etc.) are then computed from the potential (Fig. 2.1) using standard thermodynamic relations

$$V = \left(\frac{\partial G_{(P,T)}}{\partial P}\right)_{T=cont},$$

(2.8)

$$P = -\left(\frac{\partial F_{(V,T)}}{\partial V}\right)_{T=cont},$$

(2.9)

$$V = \frac{m}{\rho},$$

(2.10)

where m is molar mass, kg/mol.

Self-consistent thermodynamic databases (e.g., Karpov *et al.*, 1976; Helgesson *et al.*, 1978; Berman, 1988; Holland and Powell, 1990, 1998) are often based on the standard formulation of Gibbs free energy in the form:

$$G_{m(P,T)} = \Delta H_r - T \cdot S_r + \int_{T_r}^{T} [C_{Pr(T)}]dT - T \cdot \int_{T_r}^{T} [C_{Pr(T)}/T]dT + \int_{P_r}^{P} [V_{(P,T)}]dP,$$

(2.11)

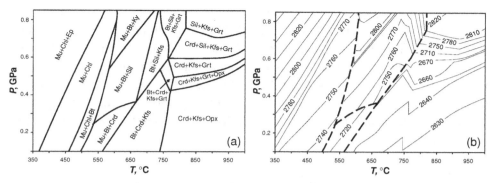

Fig. 2.2 Equilibrium mineral assemblages (a) and the corresponding density (kg/m^3) map (b) computed for a typical composition of metamorphosed aluminous sediment (high-grade metapelite) on the basis of Gibbs free energy minimisation (Gerya *et al.*, 2001). Quartz, plagioclase and Fe-Ti oxides are present in all mineral assemblages. Other minerals are: Bt = biotite, Chl = chlorite, Crd = cordierite, Ep = epidote, Grt = garnet, Kfs = K-feldspar, Ky = kyanite, Mu = muscovite, Opx = opthopyroxene, Sil = sillimanite. Heavy dashed lines in (b) indicate sharp changes in density related to changes in the mineral assemblages.

where $G_{m(P,\,T)}$ is the molar Gibbs free energy (i.e. Gibbs potential) at a given P and T; ΔH_r and S_r are the enthalpy of formation and entropy respectively, of a substance at standard pressure P_r and temperature T_r; $C_{Pr(T)}$ is the heat capacity as a function of temperature at a standard pressure P_r; $V_{(P,\,T)}$ is the molar volume of substance as a function of pressure and temperature defined by a semi-empirical EOS-function (such as, for example, Eq. 2.7).

Natural rocks typically contain several different minerals and therefore the density of a rock can be calculated from its mineralogical composition as follows

$$\rho_{rock} = \sum_{i=1}^{n} \rho_i X_i, \qquad (2.12)$$

where n is the number of different minerals in the rock, X_i is the *volumetric fraction* of the i-th mineral in the rock and ρ_i is the density of the i-th mineral as a function of P and T. At any given P, T and chemical composition of the rock, both the amount and the composition of the minerals can be computed using the concept of thermodynamic equilibrium. According to this concept, the Gibbs free energy of the rock in an equilibrium state corresponds to a global minimum. The amount and composition of minerals can then be obtained from internally consistent thermodynamic databases by using the so-called *Gibbs free energy minimisation approach* (e.g. Karpov *et al.*, 1976; Dorogokupets and Karpov, 1984, Connolly and Kerrick, 1987; de Capitani and Brown, 1987). In this case, the density of rocks in an equilibrium state (Sobolev and Babeyko, 1994; Petrini *et al.*, 2001; Gerya *et al.*,

2001; Kaus *et al.*, 2005) can be computed from Eqs. (2.8), (2.10), (2.11) and (2.12) by using the densities and volumetric fractions of minerals composing a computed equilibrium mineral assemblage (Fig. 2.2). Such an approach has recently been used to constrain coupled petrological-thermomechanical numerical geodynamic models (e.g. Gerya *et al.*, 2004c, 2006; Tackley, 2008; Mishin *et al.*, 2008) which take into account changes of rock density and thermal properties during the course of various geodynamic processes (we will discuss the details of this approach in Chapter 17).

2.2 Gravity and gravitational potential

The density distribution in a continuous medium is inherently related to the gravitational field in this medium. This can be formulated in the form of a so-called Poisson equation, which describes spatial changes in gravitational potential Φ inside a *self-gravitating* continuum

$$\frac{\partial^2 \Phi}{\partial x^2} + \frac{\partial^2 \Phi}{\partial y^2} + \frac{\partial^2 \Phi}{\partial z^2} = 4\pi G \rho_{(x,y,z)}, \tag{2.13a}$$

or in a different symbolic notation

$$\boldsymbol{\Delta}\Phi = 4\pi G \rho_{(x,y,z)}, \tag{2.13b}$$

or

$$\nabla^2 \Phi = 4\pi G \rho_{(x,y,z)}, \tag{2.13c}$$

where $G = 6.672 \times 10^{-11}$ (N · m^2)/kg^2 is the gravitational constant, $\rho_{(x,y,z)}$ is the spatially variable density and $\boldsymbol{\Delta}$ (big delta), or ∇^2 is the Laplace operator or Laplacian, which is a *differential operator* (like the divergence and gradient operator from Chapter 1) often used in continuum mechanics for representing the sum of second-order partial derivatives of a variable A

$$\text{In 1D:} \quad \boldsymbol{\Delta}A = \nabla^2 A = \frac{\partial^2 A}{\partial x^2}, \tag{2.14a}$$

$$\text{In 2D:} \quad \boldsymbol{\Delta}A = \nabla^2 A = \frac{\partial^2 A}{\partial x^2} + \frac{\partial^2 A}{\partial y^2}, \tag{2.14b}$$

$$\text{In 3D:} \quad \boldsymbol{\Delta}A = \nabla^2 A = \frac{\partial^2 A}{\partial x^2} + \frac{\partial^2 A}{\partial y^2} + \frac{\partial^2 A}{\partial z^2}. \tag{2.14c}$$

The gravitational potential Φ(J/kg), characterises the *amount of potential energy* per unit mass for a given location, related to the interaction of the local mass with all other surrounding masses. Another interpretation of Φ is the *amount of work* needed to be done to move a unit mass (i.e. overcoming its gravitational interactions with surrounding masses) from a given location to infinity (where no interactions with other masses occur). Since such work is always positive, the amount of potential energy and therefore the gravitational potential is maximal at infinity. The relative gravitational potential (i.e. changes in potential energy relative to infinity) will then always be negative.

The Poisson equation (2.13) can be derived from Newton's law of gravitation which quantifies the gravitational attraction force f_g acting between two bodies with masses m_1 and m_2, separated by a distance r

$$f_g = G\frac{m_1 m_2}{r^2}, \tag{2.15a}$$

or in 3-D vector notation

$$f_{x1} = G\frac{m_1 m_2}{r^2}\frac{(x_2 - x_1)}{r}, \tag{2.15b}$$

$$f_{y1} = G\frac{m_1 m_2}{r^2}\frac{(y_2 - y_1)}{r}, \tag{2.15c}$$

$$f_{z1} = G\frac{m_1 m_2}{r^2}\frac{(z_2 - z_1)}{r}, \tag{2.15d}$$

$$g_{x1} = G\frac{m_2}{r^2}\frac{(x_2 - x_1)}{r}, \tag{2.15e}$$

$$g_{y1} = G\frac{m_2}{r^2}\frac{(y_2 - y_1)}{r}, \tag{2.15f}$$

$$g_{z1} = G\frac{m_2}{r^2}\frac{(z_2 - z_1)}{r}, \tag{2.15g}$$

where f_{x1}, f_{y1} and f_{z1} are components of the gravitational force vector $\vec{f}_1$ acting on mass m_1 and g_{x1}, g_{y1} and g_{z1} are components of gravitational acceleration vector $\bar{g}_1$ felt by mass m_1 (in accordance with the second Newton's law of motion, acceleration can be defined as the amount of force per unit mass, $g_{i1} = \frac{f_{i1}}{m_1}$).

The simplified explanation of such a derivation is the following. Firstly, we should realize that local derivatives of gravitational potential by the spatial coordinates are equal to the respective components of the gravitational acceleration vector $\vec{g}$, taken with a negative sign

$$\frac{\partial \Phi}{\partial x} = -g_x, \tag{2.16a}$$

$$\frac{\partial \Phi}{\partial y} = -g_y, \tag{2.16b}$$

$$\frac{\partial \Phi}{\partial z} = -g_z. \tag{2.16c}$$

The minus sign in the right-hand side of equations (2.16a)–(2.16c) reflects the fact that the potential energy Φ should increase in the direction opposite to the direction of the local gravity force, i.e. work contributing to the increase in potential energy has to be done by applying a force $\vec{f}$ in the opposite direction compared to the local gravity force

$$d\Phi_x = f_x dx = -g_x dx, \tag{2.17a}$$

$$d\Phi_y = f_y dy = -g_y dy, \tag{2.17b}$$

$$d\Phi_z = f_z dz = -g_z dz, \tag{2.17c}$$

where $d\Phi_x$, $d\Phi_y$ and $d\Phi_z$ are increments in potential energy of a given unit mass due to small changes (dx, dy and dz) in the respective coordinates of this unit mass. According to Newton's law of gravitation (Eq. 2.15), the acceleration (i.e. gravitational force per unit mass) $\bar{g} = (g_x, g_y, g_z)$ felt at any given point within a continuum with coordinates x, y, z due to the gravitational attraction *from surrounding masses* is obtained by summing up (i.e. integrating) the accelerations exerted by each small mass element (δm_i), as follows

$$g_x = \sum_{i=1}^{\infty} G \frac{\delta m_i}{r_i^2} \frac{(x_i - x)}{r_i}, \tag{2.18a}$$

$$g_y = \sum_{i=1}^{\infty} G \frac{\delta m_i}{r_i^2} \frac{(y_i - y)}{r_i}, \tag{2.18b}$$

$$g_z = \sum_{i=1}^{\infty} G \frac{\delta m_i}{r_i^2} \frac{(z_i - z)}{r_i}, \tag{2.18c}$$

$$r_i = \sqrt{(x_i - x)^2 + (y_i - y)^2 + (z_i - z)^2}, \tag{2.18d}$$

where x_i, y_i, z_i are the coordinates of the i-th mass element and r_i is the distance between this element and the given point. The divergence of the integrated acceleration field at a given point can be then computed as follows (verify as an exercise)

$$-\Delta\Phi = \frac{\partial}{\partial x}\left(-\frac{\partial\Phi}{\partial x}\right) + \frac{\partial}{\partial y}\left(-\frac{\partial\Phi}{\partial y}\right) + \frac{\partial}{\partial z}\left(-\frac{\partial\Phi}{\partial z}\right)$$

$$= \text{div}(\bar{g}) = \frac{\partial g_x}{\partial x} + \frac{\partial g_y}{\partial y} + \frac{\partial g_z}{\partial z}, \tag{2.19a}$$

$$\text{div}(\bar{g}) = \sum_{i=1}^{\infty} G\delta m_i \left[\frac{\partial}{\partial x}\left(\frac{x - x_i}{r_i^3}\right) + \frac{\partial}{\partial y}\left(\frac{y - y_i}{r_i^3}\right) + \frac{\partial}{\partial z}\left(\frac{z - z_i}{r_i^3}\right)\right], \tag{2.19b}$$

$$\text{div}(\bar{g}) = \sum_{i=1}^{\infty} G\delta m_i \left[\left(\frac{3(x - x_i)^2}{r_i^5} - \frac{1}{r_i^3}\right) + \left(\frac{3(y - y_i)^2}{r_i^5} - \frac{1}{r_i^3}\right)\right.$$

$$\left. + \left(\frac{3(z - z_i)^2}{r_i^5} - \frac{1}{r_i^3}\right)\right], \tag{2.19c}$$

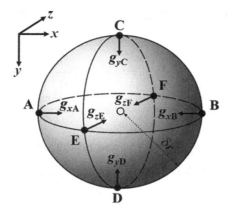

Fig. 2.3 Sphere of radius δr and mass δm considered for the derivation of the Poisson equation. Black arrows show respective components of gravitational acceleration vector felt at the surface of the sphere in six different points (black dots, A, B, C, D, E and F). Open circle shows the centre of the sphere in which the divergence of the gravitational acceleration is computed.

and finally by using Eq. (2.18)

$$\text{div}(\bar{g}) = \sum_{i=1}^{\infty} G\delta m_i \left[3\frac{r_i^2}{r_i^5} - 3\frac{1}{r_i^3} \right]. \tag{2.19d}$$

Note that differentiation is done with respect to the coordinates x, y, z of the given point and not by those of the surrounding masses (x_i, y_i, z_i) which are independent of x, y, z.

For all $r_i \neq 0$, Eq. (2.19d) is obviously *zero*. This means that the divergence of gravitational acceleration at a given point is independent of the surrounding masses and must come from the point itself (i.e. for $x = x_i$, $y = y_i$, $z = z_i$ and $r_i = 0$). Then one may restrict the volume of integration to a small sphere with mass δm and radius δr centred on this point (Fig. 2.3). The divergence of acceleration for this small sphere can be approximated by differences with the use of six points (A, B, C, D, E and F, Fig. 2.3) located on the sphere as follows

$$\text{div}(\vec{g}) = \frac{g_{xB} - g_{xA}}{2\delta r} + \frac{g_{yD} - g_{yC}}{2\delta r} + \frac{g_{zF} - g_{zE}}{2\delta r}, \tag{2.20}$$

where $g_{xA} - g_{zF}$ are the respective components of the acceleration vector at different points on the sphere's surface. Density inside the sphere can be considered constant since the mass inside the small sphere is uniform. Therefore, at the considered six points, the respective components obviously coincide with the acceleration vectors directed toward the centre of the sphere and can be computed from the common gravity formula for a sphere (see Turcotte and Schubert, 2002, Eqs. (5–1)–(5–15)) as

$$g_{xA} = g_{yC} = g_{zE} = G\frac{\delta m}{\delta r^2}, \tag{2.21a}$$

$$g_{xB} = g_{yD} = g_{zF} = -G\frac{\delta m}{\delta r^2}. \tag{2.21b}$$

Combining Equations (2.14) and (2.15) and taking into account that the average density inside the sphere with volume $V = \frac{4}{3}\pi\delta r^3$ is given by $\rho = \frac{\delta m}{V} = \frac{3\delta m}{4\pi\delta r^3}$, under the condition that δr tend to zero we obtain the Poisson equation (verify as an exercise)

$$\Delta\Phi = -\text{div}(\bar{g}) = 3G\frac{\delta m}{\delta r^3} = 4\pi G\rho. \tag{2.22}$$

It should be noted that our derivation is simplified and uses a standard expression (Eq. (2.21)) for the gravitational acceleration at the surface of the sphere, which can in turn be obtained by integrating Equation (2.18) for a point outside the sphere (see Turcotte and Schubert, 2002, Eqs. (5–1)–(5–15) for details of this derivation). It can also be shown on the basis of Gauss's theorem that one can consider not only a spherical geometry, but any arbitrary shape of the local mass and still obtain the same Poisson equation.

Analytical exercise

Exercise 2.1
Molar Gibbs potential of periclase (MgO, molar mass $m = 0.0403044$ kg/mol) is given by the following equation (Gerya et al., 2004d)

$$G_{m(P,T)} = H_r + V_r\Psi + \sum_{i=1}^{3} c_i[RT\ln(1 - e_i) - \Delta H_i e_{oi}/(1 - e_{oi})],$$

$$e_i = \exp[-(\Delta H_i + \Delta V_i\Psi)/RT],$$

$$e_{oi} = \exp(-\Delta H_i/RT_r),$$

$$\Psi = 5/4(P_r + \phi)^{1/5}[(P + \phi)^{4/5} - (P_r + \phi)^{4/5}],$$

where $R = 8.314$ J/mol is the gas constant, $P_r = 100\,000$ Pa and $T_r = 298.15$ K are the reference pressure and temperature respectively, and $H_r = -601\,500.00$ J, $V_r = 1.122\,28 \times 10^{-5}$ J/Pa, $\phi = 30\,179\,500\,000$ Pa, $c_1 = 1.966\,12$, $c_2 = 4.127\,56$,

$c_3 = 0.536\,90$, $\Delta H_1 = 2966.88$ J, $\Delta H_2 = 5621.69$ J, $\Delta H_3 = 27\,787.19$ J, $\Delta V_1 = \Delta V_2 = 3.52971 \times 10^{-8}$ J/Pa, $\Delta V_3 = 1.984\,956\,8 \times 10^{-6}$ J/Pa are empirical parameters.

Derive expressions for the density ρ, thermal expansion α and compressibility β as functions of pressure and temperature (Fig. 2.1) using the equations given in this chapter.

Programming exercises and homework

Exercise 2.2

Compute and visualise as colour maps with isolines (*pcolor, contour, contourf*) and surfaces (*surf, light, lighting*) the density ρ, thermal expansion α and compressibility β for periclase (Fig. 2.1) based on the equations derived for the analytical exercise. Take a temperature interval from 100 to 4000 K and a pressure interval from 10^9 to 10^{11} Pa (1 to 100 GPa). Try also to define the Gibbs free energy equation as an external function $G_{m(P,\,T)}$ and use differences instead of derivatives to compute $V_{(P,\,T)}$, $\rho_{(P,\,T)}$, $\alpha_{(P,\,T)}$ and $\beta_{(P,\,T)}$.

$$\rho_{(P,T)} = \frac{m}{V_{(P,T)}}, \text{ where}$$

$$V_{(P,T)} = \left(\frac{\partial G_{m(P,T)}}{\partial P}\right)_{T=const} \approx \frac{\Delta G_{m(P,T)}}{\Delta P} = \frac{G_{m(P+\Delta P,T)} - G_{m(P,T)}}{\Delta P}$$

$$\alpha_{(P,T)} = -\frac{1}{\rho_{(P,T)}} \times \frac{\partial \rho_{(P,T)}}{\partial T} \approx -\frac{1}{\rho_{(P,T)}} \times \frac{\Delta \rho_{(P,T)}}{\Delta T} = -\frac{1}{\rho_{(P,T)}} \times \frac{\rho_{(P,T+\Delta T)} - \rho_{(P,T)}}{\Delta T},$$

$$\beta_{(P,T)} = \frac{1}{\rho_{(P,T)}} \times \frac{\partial \rho_{(P,T)}}{\partial P} \approx \frac{1}{\rho_{(P,T)}} \times \frac{\Delta \rho_{(P,T)}}{\Delta P} = \frac{1}{\rho_{(P,T)}} \times \frac{\rho_{(P+\Delta P,T)} - \rho_{(P,T)}}{\Delta P},$$

where ΔP and ΔT are small increments in pressure and temperature, respectively.

Compare the results based on differences with your analytical solutions. An example is in the code **Periclase_EOS.m** which calls function **G_periclase.m**.

Exercise 2.3

Load from files (*fopen, fscanf, fclose*) and visualise (*pcolor*) density maps (Mishin *et al.*, 2008) corresponding to the phase diagrams computed for pyrolite (mantle, file m895_ro) and MORB (oceanic crust, file morn_ro) based on a Gibbs free energy minimisation approach. Compute and visualise the density difference between these two contrasting types of rocks at various P–T parameters. Check at which P and T this difference is maximal/minimal.

The first nine positions in the data files are as follows

pl8951 350 350 800.003 10001.5 9.16904 4269.33 T(K) P(bar) where *pl8951* denotes the rock identification (skip that); *350 and 350* are resolutions for T and P respectively; *800.003* and *10001.5* are starting value for T (K) and P (bar, 1 bar $= 10^5$ Pa), respectively; *9.16904* and *4269.33* are steps for T (K) and P (bar), respectively; T (K) and P (bar) are P and T identifications (to skip). Further data in the files are 350×350 maps of rock density (kg/m^3) at variable T (inner cycle) and P (outer cycle). An example is in **Density_map.m**.

3

Numerical solutions of partial differential equations

> **Theory:** Analytical and numerical methods for solving partial differential equations. Using finite differences to compute various derivatives. Eulerian and Lagrangian approaches. Transition from partial differential equations to systems of linear equations. Methods of solving large systems of linear equations: iterative methods (Jacobi iteration, Gauss–Seidel iteration), direct methods (Gaussian elimination). Indexing of unknowns in 1D and 2D.
> **Exercises:** Numerical solutions of Poisson equation in 1D and 2D.

3.1 Finite-difference method

Two principal methods are used for solving partial differential equations (PDEs) of continuum mechanics: *analytical* and *numerical*. Analytical methods are restricted to relatively simple problems and cannot be applied to a general case. This caveat is due, in particular, to the fact that it is sometimes impossible to analytically express the distribution of field variables (T, P, $\vec{v}$, η, ρ, etc.) in space and time. In effect, analytical methods are very useful for the general understanding of geodynamic processes (e.g. Turcotte and Schubert, 2002). In addition, analytical solutions are broadly used for benchmarking numerical codes (testing accuracy, Chapter 16).

Numerical methods for solving PDEs are universal and can be applied for both continuous and discontinuous distributions of field variables (e.g. Gustafsson, 2008). The following groups of numerical approaches are most used in geomodelling (e.g. Lynch, 2005; Zhong *et al.*, 2007):

(1) *finite-difference methods (FDM)*
(2) *finite-volume methods (FVM)*
(3) *finite-element methods (FEM)*
(4) *spectral methods*

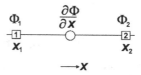

Fig. 3.1 1D numerical grid *stencil* (i.e. pattern of points) used for computing the first order derivative of the gravity potential $\dfrac{\partial \Phi}{\partial x}$ by using finite differences.

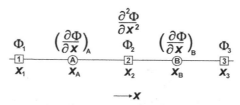

Fig. 3.2 1D numerical grid stencil used for computing the second derivative of the gravity potential $\dfrac{\partial^2 \Phi}{\partial x^2}$ by using finite differences.

In this book, we will concentrate on the finite-difference method (e.g. Patankar, 1980), which is the simplest of the four methods both for understanding and from a programming point of view. Finite differences are linear mathematical expressions which are used to represent derivatives to a certain degree of accuracy. For example, the first derivative of gravity potential by x-coordinate (Fig. 3.1) can be computed *within a certain degree of accuracy* (locally) by using finite differences as follows

$$\frac{\partial \Phi}{\partial x} = \frac{\Delta \Phi}{\Delta x} = \frac{\Phi_2 - \Phi_1}{x_2 - x_1}, \tag{3.1}$$

where $\Delta \Phi = \Phi_2 - \Phi_1$ and $\Delta x = x_2 - x_1$ are the *differences* in gravity potential and x-coordinate respectively between points 1 and 2. The smaller the distance Δx between points 1 and 2 becomes, the more accurate the computed derivative is.

By analogy, higher-order derivatives can be computed using lower-order derivatives. For example, the second derivative of gravity potential (Fig. 3.2) can be computed by repetitive use of finite-differences equation (3.1) as follows

$$\frac{\partial^2 \Phi}{\partial x^2} = \frac{\left(\dfrac{\partial \Phi}{\partial x}\right)_B - \left(\dfrac{\partial \Phi}{\partial x}\right)_A}{x_B - x_A}, \tag{3.2}$$

where

$$\left(\frac{\partial \Phi}{\partial x}\right)_A = \frac{\Phi_2 - \Phi_1}{x_2 - x_1},$$

$$\left(\frac{\partial \Phi}{\partial x}\right)_B = \frac{\Phi_3 - \Phi_2}{x_3 - x_2}.$$

Using a similar procedure we can formulate third-, fourth-, fifth- and higher-order derivatives as well.

As follows from the above examples, we need a *grid* of points representing the distribution of field variables in space (and time) to apply finite differences. This so called *numerical grid* is also often called a *numerical mesh*. Similarly, two types of geometrical points may exist, grid points can be either *Eulerian* or *Lagrangian*. Eulerian points have steady positions and an Eulerian grid does not deform with deformation of the medium. Lagrangian points move according to the local flow and a Lagrangian grid deforms with the deformation of medium. Time derivatives of field variables for Eulerian and Lagrangian points may differ from each other e.g.,

$$\frac{D\rho}{Dt} = \frac{\partial \rho}{\partial t} + \vec{v}\,\text{grad}(\rho), \tag{3.3}$$

where $\dfrac{D\rho}{Dt}$ is the substantive time derivative of density for a moving Lagrangian point and $\dfrac{\partial \rho}{\partial t}$ is the time derivative of density for an immobile Eulerian point *in the same location*. The main advantage of using an Eulerian grid is the possibility of having a relatively simple grid geometry that does not change during the model deformation; this simplifies the numerical formulation. The main disadvantage is the necessity to account for advective terms in time-dependent PDEs, which often causes numerical problems (e.g. *numerical diffusion*, Chapter 8). For a Lagrangian grid it is the contrary: a deforming grid ultimately produces numerical problems (and requires *re-gridding* or *re-meshing* when it is deformed too strongly) while the absence of advective terms in PDEs is an advantage. The use of either an Eulerian, or a Lagrangian grid depends on the partial differential equations to be solved as well as on the type of physical processes to be modelled. In geodynamic modelling, combinations of Eulerian and Lagrangian grids for different field variables are often used to explore advantages of both approaches (see e.g. Zhong *et al.*, 2007).

What are we actually gaining by approximating derivatives by finite differences? This is a very important issue at the 'core' of numerical modelling. The use of finite differences allows us to transform partial differential equations, which are

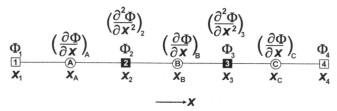

Fig. 3.3 1D numerical grid for solving 1D Poisson equation in form $\dfrac{\partial^2 \Phi}{\partial x^2} = 1$.

applicable to every geometrical point of a continuum (i.e. to an *infinite amount of points*), into *a system of finite amount of linear equations* formulated for a limited amount of grid points. The logical steps of applying finite differences are as follows:

(1) Replacing an infinite amount of geometrical points of a continuum within the model by a finite amount of grid points.
(2) Defining physical properties of the continuum at these points.
(3) Applying partial differential equations (including *boundary condition equations*) to the grid points and substituting them by linear equations expressed via finite differences. These linear equations relate the physical properties defined for different grid points.
(4) Solving the resulting system of linear equations and obtaining unknown values of the physical parameters for the grid points.

Let's consider an example of this procedure for solving the 1D Poisson equation of the form $\dfrac{\partial^2 \Phi}{\partial x^2} = 1$, with the boundary conditions $\Phi(x_1) = R_1$ and $\Phi(x_4) = R_4$, where x_1, x_4 are coordinates of the model boundaries and $\Phi(x_1)$, $\Phi(x_4)$ are the values of the gravity potential at these boundaries.

Step (1): defining a numerical grid. A 1D grid for this problem is shown in Fig. 3.3. Different symbols on this grid show different *grid points* (also called *nodal points* or *nodes*) where different parameters are defined and different equations are applied:

- nodes [1] and [4] (empty squares) are *basic boundary nodes* where only the gravitational potential Φ is defined and the boundary condition equation is applied

$$\Phi = R_1 \quad \text{when} \quad x = x_1,$$

$$\Phi = R_4 \quad \text{when} \quad x = x_4,$$

- nodes [2] and [3] (solid squares) are *basic internal nodes* where both gravitational potential Φ and its second derivative $\dfrac{\partial^2 \Phi}{\partial x^2}$ are defined and the Poisson equation is applied

$$\frac{\partial^2 \Phi}{\partial x^2} = 1.$$

- nodes [A], [B] and [C] (open circles) are *additional nodes* located exactly in the middle of the intervals between the basic nodes. At these additional nodes, the first derivative of gravitational potential $\dfrac{\partial \Phi}{\partial x}$ is defined, however no equations are applied. We only need these additional nodes to formulate second derivatives in the basic internal nodes by using finite differences.

Step (2): applying equations for the nodes and converting them to linear equations.
The following equations are applied for different nodes

node [1]: $\qquad\qquad\qquad \Phi = R_1$

node [2]: $\qquad\qquad\qquad \left(\dfrac{\partial^2 \Phi}{\partial x^2}\right)_2 = 1$

node [3]: $\qquad\qquad\qquad \left(\dfrac{\partial^2 \Phi}{\partial x^2}\right)_3 = 1$

node [4]: $\qquad\qquad\qquad \Phi = R_4.$

To transform these equations, we first define a way of computing the derivatives via finite differences for nodes [2] and [3]

node [2]: $\qquad\qquad \left(\dfrac{\partial^2 \Phi}{\partial x^2}\right)_2 = \dfrac{\left(\dfrac{\partial \Phi}{\partial x}\right)_B - \left(\dfrac{\partial \Phi}{\partial x}\right)_A}{x_B - x_A}$

where

$$\left(\frac{\partial \Phi}{\partial x}\right)_A = \frac{\Phi_2 - \Phi_1}{x_2 - x_1}$$

$$\left(\frac{\partial \Phi}{\partial x}\right)_B = \frac{\Phi_3 - \Phi_2}{x_3 - x_2}$$

node [3]: $\qquad\qquad \left(\dfrac{\partial^2 \Phi}{\partial x^2}\right)_3 = \dfrac{\left(\dfrac{\partial \Phi}{\partial x}\right)_C - \left(\dfrac{\partial \Phi}{\partial x}\right)_B}{x_C - x_B}$

where

$$\left(\frac{\partial \Phi}{\partial x}\right)_B = \frac{\Phi_3 - \Phi_2}{x_3 - x_2}$$

$$\left(\frac{\partial \Phi}{\partial x}\right)_C = \frac{\Phi_4 - \Phi_3}{x_4 - x_3}.$$

Applying these finite differences results in the following system of four equations

$$\Phi_1 = R_1$$

$$\frac{(\Phi_3 - \Phi_2)/(x_3 - x_2) - (\Phi_2 - \Phi_1)/(x_2 - x_1)}{(x_3 - x_1)/2} = 1$$

$$\frac{(\Phi_4 - \Phi_3)/(x_4 - x_3) - (\Phi_3 - \Phi_2)/(x_3 - x_2)}{(x_4 - x_2)/2} = 1$$

$$\Phi_4 = R_4.$$

Assembling coefficients for each unknown gives a final system of four linear equations

$$1 \times \Phi_1 = R_1 \qquad\qquad\qquad (3.4)$$

$$\frac{2/(x_2 - x_1)}{(x_3 - x_1)} \times \Phi_1 + \frac{-2/(x_2 - x_1) - 2/(x_3 - x_2)}{(x_3 - x_1)} \times \Phi_2 + \frac{2/(x_3 - x_2)}{(x_3 - x_1)} \times \Phi_3 = 1$$
$$(3.5)$$

$$\frac{2/(x_3 - x_2)}{(x_4 - x_2)} \times \Phi_2 + \frac{-2/(x_3 - x_2) - 2/(x_4 - x_3)}{(x_4 - x_2)} \times \Phi_3 + \frac{2/(x_4 - x_3)}{(x_4 - x_2)} \times \Phi_4 = 1$$
$$(3.6)$$

$$1 \times \Phi_4 = R_4. \qquad\qquad\qquad (3.7)$$

In order to be able to find a solution, *the number of independent linear equations must be equal to the number of unknown parameters.* In this example, we know the coordinates x_1, x_2, x_3 and x_4 as well as coefficients R_1 and R_4 for the boundary equations. Unknown parameters are the discrete values of gravitational potential Φ in the basic grid points, i.e. Φ_1, Φ_2, Φ_3 and Φ_4. Therefore, the number of equations (four) is equal to the number of unknowns (four) and these equations are *linear with respect to the unknowns* and, thus, can be easily solved.

Step (3): solving the system of linear equations. Solving the above system of equations is trivial since we have only four equations with only four unknowns.

As an exercise solve this problem manually for the following values of coordinates and coefficients.

$$x_1 = 0, x_2 = 1, x_3 = 2, x_4 = 3, R_1 = 0, R_4 = 0.$$

This example looks trivial. Why should we spend so much time discussing it in so many details that are apparently so obvious? This is just to be finally convinced that the mathematical basis of numerical modelling *is exactly trivial*. What one needs to know are *derivatives* and *linear algebra* as was postulated in the Introduction (Golden Rule 1).

3.2 Solving linear equations

What is going to change if, instead of four basic nodes we have one thousand? Not much . . . We will merely have to formulate and solve thousands of linear equations to obtain values of thousands of unknowns. Of course, our numerical solution will be more accurate and it is not necessary to solve thousands of equations manually. One can obtain much larger systems of equations trying to numerically solve geodynamic problems in 2D and 3D. Such systems typically contain thousands, millions and even billions of linear equations (such as (3.4)–(3.7)) having the following general form

$$L_{1,1}S_1 + L_{1,2}S_2 + L_{1,3}S_3 + \cdots + L_{1,n-1}S_{n-1} + L_{1,n}S_n = R_1$$
$$L_{2,1}S_1 + L_{2,2}S_2 + L_{2,3}S_3 + \cdots + L_{2,n-1}S_{n-1} + L_{2,n}S_n = R_2$$
$$\cdots \tag{3.8}$$
$$L_{n-1,1}S_1 + L_{n-1,2}S_2 + L_{n-1,3}S_3 + \cdots + L_{n-1,n-1}S_{n-1} + L_{n-1,n}S_n = R_{n-1}$$
$$L_{n,1}S_1 + L_{n,2}S_2 + L_{n,3}S_3 + \cdots + L_{n,n-1}S_{n-1} + L_{n,n}S_n = R_n$$

where S_l are unknowns, which are components of an n-dimensional vector $\{S\}$ (i.e. $n \times 1$ array, line), $L_{k,l}$ are coefficients, which are elements of an $n \times n$ *square matrix* $\{L\}$, R_k are the right-hand sides which are components of an n-dimensional vector $\{R\}$ (i.e. $1 \times n$ array, column). Obviously, the system of equations can only be solved when the number of linearly independent equations in the system is equal to the number of unknowns. In the case of numerical modelling with finite differences, most of the coefficients $L_{k,l}$ in the equations are equal to zero, i.e. the $\{L\}$ matrix of coefficients is *sparse*. This is because linear equations formulated with finite differences for a node only contain parameters from the *nearest* nodes and not from all nodes of the grid (e.g. Eq. (3.5)). The methods of solving large systems of equations are subdivided into iterative (e.g., *Jacobi iteration, Gauss–Seidel iteration*) and direct (e.g. *Gaussian elimination*) methods.

In order to apply iterative methods, *initial guesses* for the unknown variables are first defined, which represent a current approximation of the solution $S_1^{current}, S_2^{current}, \ldots, S_n^{current}$ (all unknowns can for example initially be set to zeros). Then *residuals* (i.e. errors) for each equation can be computed as follows

$$\Delta R_1 = R_1 - L_{1,1} S_1^{current} - L_{1,2} S_2^{current} - L_{1,3} S_3^{current} - \cdots$$
$$- L_{1,n-1} S_{n-1}^{current} - L_{1,n} S_n^{current}$$

$$\Delta R_2 = R_2 - L_{2,1} S_1^{current} - L_{2,2} S_2^{current} - L_{2,3} S_3^{current} - \cdots$$
$$- L_{2,n-1} S_{n-1}^{current} - L_{2,n} S_n^{current}$$

$$\cdots \qquad (3.9)$$

$$\Delta R_{n-1} = R_{n-1} - L_{n-1,1} S_1^{current} - L_{n-1,2} S_2^{current} - L_{n-1,3} S_3^{current} - \cdots$$
$$- L_{n-1,n-1} S_{n-1}^{current} - L_{n-1,n} S_n^{current}$$

$$\Delta R_n = R_n - L_{n,1} S_1^{current} - L_{n,2} S_2^{current} - L_{n,3} S_3^{current} - \cdots$$
$$- L_{n,n-1} S_{n-1}^{current} - L_{n,n} S_n^{current}.$$

The values of the residuals can then be used to obtain new, more accurate, values of the unknowns

$$S_1^{new} = S_1^{current} + \theta_1 \frac{\Delta R_1}{L_{1,1}}$$

$$S_2^{new} = S_2^{current} + \theta_2 \frac{\Delta R_2}{L_{2,2}}$$

$$\cdots \qquad (3.10)$$

$$S_{n-1}^{new} = S_{n-1}^{current} + \theta_{n-1} \frac{\Delta R_{n-1}}{L_{n-1,n-1}}$$

$$S_n^{new} = S_n^{current} + \theta_n \frac{\Delta R_n}{L_{n,n}},$$

where $\theta_1, \theta_2, \ldots, \theta_n$ are *relaxation parameters* defining how strongly the computed residuals $\Delta R_1, \Delta R_2, \ldots, \Delta R_n$ will contribute to the changes in the current values of the unknown. Relaxation parameters are typically taken within the range between 0 and 1. In the case when the relaxation parameters are equal to 1, a new solution can be obtained with the use of simplified formulas (please derive)

$$S_1^{new} = \frac{R_1 - L_{1,2}S_2^{current} - L_{1,3}S_3^{current} - \cdots - L_{1,n-1}S_{n-1}^{current} - L_{1,n}S_n^{current}}{L_{1,1}}$$

$$S_2^{new} = \frac{R_2 - L_{2,1}S_1^{current} - L_{2,3}S_3^{current} - \cdots - L_{2,n-1}S_{n-1}^{current} - L_{2,n}S_n^{current}}{L_{2,2}}$$

$$\cdots \qquad (3.11)$$

$$S_{n-1}^{new} = \frac{R_{n-1} - L_{n-1,1}S_1^{current} - L_{n-1,2}S_2^{current} - L_{n-1,3}S_3^{current} - \cdots - L_{n-1,n}S_n^{current}}{L_{n-1,n-1}}$$

$$S_n^{new} = \frac{R_n - L_{n,1}S_1^{current} - L_{n,2}S_2^{current} - L_{n,3}S_3^{current} - \cdots - L_{n,n-1}S_{n-1}^{current}}{L_{n,n}}.$$

New values of the unknowns are then used for the next iteration to obtain the next (more accurate) solution. Iteration cycles are repeated several times to reach a certain level of accuracy which is defined by values of residuals.

There are several methods of doing such iterations. In the *Jacobi iteration*, new values are assigned to all unknowns simultaneously *after* finishing one cycle of iterations for all n equations. In *Gauss–Seidel iterations*, new values are assigned to each unknown *during* the cycle of iterations, immediately after obtaining the updated value from the respective equation. Residuals for the following equations in the cycle are then computed using the new values of unknowns obtained from the previous equations.

Computational advantages of iterative methods are (i) small amount of consumed memory, typically proportional to the number of unknowns and (ii) small amount of operations, also typically proportional to the number of unknowns per solution cycle. Therefore, *iterative solvers* are frequently used in 3D when the number of equations is large. Among the disadvantages are (i) lowered accuracy of solution and (ii) problems of convergence to an accurate solution, which are especially relevant for solving problems with large variations in material properties (e.g., mechanical problems with strong and sharp variations in viscosity). Indeed, there are various possibilities for notable improvement of the iterative methods discussed here based for example on the *multigrid* approach, which will be discussed in Chapter 14.

Direct methods of solutions do not require an initial guess and are based on mathematical transformations of the matrix $\{L\}$ and vector $\{R\}$, which allows one to compute the vector $\{S\}$. One of the best known direct methods is *Gaussian elimination*, which produces a diagonal matrix $\{L_{diagonal}\}$ and a corresponding new vector $\{R_{diagonal}\}$ that can be then used to directly compute vector $\{S\}$. The procedure of Gaussian elimination is as follows

(a) Divide all Equations (3.8) by their respective (non-zero) coefficients at S_1 to obtain a new system with new coefficients at the unknowns and new right-hand sides

$$S_1 + \frac{L_{1,2}}{L_{1,1}} S_2 + \frac{L_{1,3}}{L_{1,1}} S_3 + \cdots + \frac{L_{1,n-1}}{L_{1,1}} S_{n-1} + \frac{L_{1,n}}{L_{1,1}} S_n = \frac{R_1}{L_{1,1}}$$

$$S_1 + \frac{L_{2,2}}{L_{2,1}} S_2 + \frac{L_{2,3}}{L_{2,1}} S_3 + \cdots + \frac{L_{2,n-1}}{L_{2,1}} S_{n-1} + \frac{L_{2,n}}{L_{2,1}} S_n = \frac{R_2}{L_{2,1}}$$

$$\cdots \tag{3.12}$$

$$S_1 + \frac{L_{n-1,2}}{L_{n-1,1}} S_2 + \frac{L_{n-1,3}}{L_{n-1,1}} S_3 + \cdots + \frac{L_{n-1,n-1}}{L_{n-1,1}} S_{n-1} + \frac{L_{n-1,n}}{L_{n-1,1}} S_n = \frac{R_{n-1}}{L_{n-1,1}}$$

$$S_1 + \frac{L_{n,2}}{L_{n,1}} S_2 + \frac{L_{n,3}}{L_{n,1}} S_3 + \cdots + \frac{L_{n,n-1}}{L_{n,1}} S_{n-1} + \frac{L_{n,n}}{L_{n,1}} S_n = \frac{R_n}{L_{n,1}}.$$

(b) Subtract the first equation from all other equations (starting from the second one) to obtain a subsystem of $n - 1$ equations with $n - 1$ unknowns (since S_1 will be *eliminated* from all equations starting from the second one), i.e. new matrix $\{L^1\}$ and vector $\{R^1\}$

$$\left(\frac{L_{2,2}}{L_{2,1}} - \frac{L_{1,2}}{L_{1,1}}\right) S_2 + \left(\frac{L_{2,3}}{L_{2,1}} - \frac{L_{1,3}}{L_{1,1}}\right) S_3 + \cdots + \left(\frac{L_{2,n-1}}{L_{2,1}} - \frac{L_{1,n-1}}{L_{1,1}}\right)$$

$$\times S_{n-1} + \left(\frac{L_{2,n}}{L_{2,1}} - \frac{L_{1,n}}{L_{1,1}}\right) S_n = \frac{R_2}{L_{2,1}} - \frac{R_1}{L_{1,1}}$$

$$\cdots$$

$$\left(\frac{L_{n-1,2}}{L_{n-1,1}} - \frac{L_{1,2}}{L_{1,1}}\right) S_2 + \left(\frac{L_{n-1,3}}{L_{n-1,1}} - \frac{L_{1,3}}{L_{1,1}}\right) S_3 + \cdots + \left(\frac{L_{n-1,n-1}}{L_{n-1,1}} - \frac{L_{1,n-1}}{L_{1,1}}\right) \tag{3.13}$$

$$\times S_{n-1} + \left(\frac{L_{n-1,n}}{L_{n-1,1}} - \frac{L_{1,n}}{L_{1,1}}\right) S_n = \frac{R_{n-1}}{L_{n-1,1}} - \frac{R_1}{L_{1,1}}$$

$$\left(\frac{L_{n,2}}{L_{n,1}} - \frac{L_{1,2}}{L_{1,1}}\right) S_2 + \left(\frac{L_{n,3}}{L_{n,1}} - \frac{L_{1,3}}{L_{1,1}}\right) S_3 + \cdots + \left(\frac{L_{n,n-1}}{L_{n,1}} - \frac{L_{1,n-1}}{L_{1,1}}\right)$$

$$\times S_{n-1} + \left(\frac{L_{n,n}}{L_{n,1}} - \frac{L_{1,n}}{L_{1,1}}\right) S_n = \frac{R_n}{L_{n,1}} - \frac{R_1}{L_{1,1}},$$

or by using new notations for the coefficients in the left-hand side $L_{k,l}^1 = \frac{L_{k,l}}{L_{k,1}} - \frac{L_{1,l}}{L_{1,1}}$ and for the right-hand side $R_k^1 = \frac{R_k}{L_{k,1}} - \frac{R_1}{L_{1,1}}$ we write

$$L_{2,2}^1 S_2 + L_{2,3}^1 S_3 + \cdots + L_{2,n-1}^1 S_{n-1} + L_{2,n}^1 S_n = R_2^1$$

$$\cdots \tag{3.14}$$

$$L_{n-1,2}^1 S_2 + L_{n-1,3}^1 S_3 + \cdots + L_{n-1,n-1}^1 S_{n-1} + L_{n-1,n}^1 S_n = R_{n-1}^1$$

$$L_{n,2}^1 S_2 + L_{n,3}^1 S_3 + \cdots + L_{n,n-1}^1 x_{n-1} + C_{n,n}^1 x_n = R_n^1.$$

(c) Repeat (a) and (b) on this new subsystem to obtain a subsystem of $n - 2$ equations with $n - 2$ unknowns.

(d) Repeat (a) and (b) on the new subsystem (in total $n - 1$ *elimination* cycles are needed).

(e) The last equation in the final system (diagonal matrix) will have the form

$$L_{n,n}^{n-1} S_n = R_n^{n-1},$$ (3.15)

i.e., will only contain only one coefficient $L_{n,n}^{n-1}$ for the unknown x_n and right-hand side R_n^{n-1}. Then S_n can be directly calculated as

$$S_n = \frac{R_n^{n-1}}{L_{n,n}^{n-1}}.$$ (3.16)

(f) The penultimate equation will have the form

$$L_{n-1,n-1}^{n-2} S_{n-1} + L_{n-1,n}^{n-2} S_n = R_{n-1}^{n-2},$$ (3.17)

and S_{n-1} can be calculated with the already known S_n as follows

$$S_{n-1} = \frac{R_{n-1}^{n-2} - L_{n-1,n}^{n-2} S_n}{L_{n-1,n-1}^{n-2}}.$$ (3.18)

(g) Repeat for all other unknowns from S_{n-2} to S_1.

The main advantage of direct methods is that the solution can be done to computer accuracy and no iterations are needed. Among their disadvantages are (i) large amounts of consumed memory, typically proportional to the square of the number of unknowns and (ii) large amounts of operations, typically proportional to the square or even to the cube of the number of unknowns. Due to limitations in computer power, *direct solvers* are more often used in 1D and 2D modelling, particularly for solving numerical problems, where *iterative solvers* are inefficient.

3.3 Geometrical and global indexing of unknowns

In composing the system of Equations (3.4)–(3.7), we indexed our unknown parameters Φ_1, Φ_2, Φ_3 and Φ_4 in 1D using the principle of growing index of the parameter with x-coordinate of the respective geometrical point to which this unknown is assigned (cf. points 1, 2, 3 and 4 in Fig. 3.3). We may then have an impression that a general system of equations (3.8) can only be applicable for 1D problems since in 2D unknown parameters should have both horizontal and vertical indices, for example $\Phi_{i,j}$ and respectively $S_{i,j}$. This is not correct and it is a small, but important point to understand. *Geometrical indexing* of unknowns $\Phi_{i,j}$ in a 2D grid is different from overall *global indexing* of these unknowns given by S_l and used in the system of equations (3.8). Global indexing of unknowns (Fig. 3.4) is always needed when direct methods are used for solving the equations as one has to compose the matrix $\{L\}$ and vector $\{R\}$. In the case of iterative solutions, one can formulate equations (such as (3.4)–(3.7)) using a geometrical indexing.

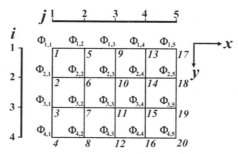

Fig. 3.4 Geometrical indexing of gravity potential values $\Phi_{i,j}$ assigned to the nodes of 2D grid and global indexing (italic numbers) of these parameters in the vector $\{S\}$. The global indexing is done by columns of nodal points.

Programming exercises and homework

Exercise 3.1

Solve the 1D Poisson equation, written in form $\dfrac{\partial^2 \Phi}{\partial x^2} = 1$, on a regular grid of 1000 points with finite differences and visualise the solution. The model length is 1000 km. Use sparse initialisation for the matrix of coefficients $\{L\}$, i.e. $L = \text{sparse}(1000, 1000)$. Compose a matrix $\{L\}$ and a right-hand side vector $\{R\}$ (use Eqs. (3.4)–(3.7) as an example) and obtain the solution vector $\{S\}$ with a direct solver (in MATLAB with the command $S = L\backslash R$). Use $\Phi = 0$ as the boundary condition for the two external nodes of the grid (e.g. Fig. 3.3). An example is in **Poisson1D.m**.

Exercise 3.2

Solve the 2D Poisson equation with finite differences and visualise the solution. The governing equation is given by

$$\frac{\partial^2 \Phi}{\partial x^2} + \frac{\partial^2 \Phi}{\partial y^2} = 1, \tag{3.19}$$

on a regular grid of 31×41 points. The model size is 1000×1500 km. Use the principle of global indexing in 2D as shown in Fig. 3.4. A finite-difference representation of the Poisson equation in 2D can be derived from Eq. (3.19) by analogy with Eq. (3.2), but applied separately for the x and y directions (Fig. 3.5)

$$\frac{\Phi_{i,j-1} - 2\Phi_{i,j} + \Phi_{i,j+1}}{\Delta x^2} + \frac{\Phi_{i-1,j} - 2\Phi_{i,j} + \Phi_{i+1,j}}{\Delta y^2} = 1, \tag{3.20}$$

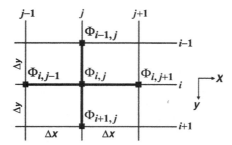

Fig. 3.5 2D numerical grid stencil (5-point cross) used for formulating the Poisson equation by using finite differences on a regular rectangular grid.

or by assembling coefficients at each unknown

$$\left(\frac{1}{\Delta x^2}\right)\Phi_{i,j-1} + \left(\frac{1}{\Delta x^2}\right)\Phi_{i,j+1} + \left(\frac{-2}{\Delta x^2} + \frac{-2}{\Delta y^2}\right)\Phi_{i,j}$$
$$+ \left(\frac{1}{\Delta y^2}\right)\Phi_{i-1,j} + \left(\frac{1}{\Delta y^2}\right)\Phi_{i+1,j} = 1. \tag{3.21}$$

Use $\Phi = 0$ as the boundary conditions for all external nodes of the grid. Compute the global index of unknown k, based on geometrical indices i and j (Fig. 3.5) as

$$k = N_y \times (j - 1) + i, \tag{3.22}$$

where N_y is vertical resolution. An example is in **Poisson2D_direct.m**.

Exercise 3.3
Solve the same 2D problem using Gauss–Seidel iterations. Use Eq. (3.9) to compute residuals. Use $\theta = 1.5$ as a relaxation factor for all points (Eq. (3.10)). Plot the gravitational potential and residuals every 10 iterations. An example is in **Poisson2D_Gauss_Seidel.m**.

Exercise 3.4
Solve the same 2D problem using Jacobi iterations. Use $\theta = 1.0$ as relaxation factor for all points (Eq. (3.10)). An example is in **Poisson2D_ Jacobi.m**.

4

Stress and strain

Theory: Deformation and stresses. Definition of stress, strain and strain-rate tensors. Deviatoric stresses. Mean stress as a dynamic (non-lithostatic) pressure. Symmetry of stress tensor. Stress and strain rate invariants.

Exercises: Computing the strain rate tensor components in 2D from the material velocity fields.

4.1 Stress

Tensors are field variables which characterise the internal state of a continuum and are, perhaps, the most difficult quantities to intuitively understand. Indeed, at least three of them have to be used in the following and these are the *stress, strain* and *strain rate* tensors.

Stress is the *internal distribution and intensity of force* acting at any point within a continuum in response to various internal and external loads applied to the continuum. Stress is defined as a *force per unit area* and we can easily 'apprehend' its effect by pressing two fingers against each other – equal *force is applied from both sides and therefore nothing moves*, but we have a *feeling of pressure* between the fingers, which is a sign of the presence of stress. This stress is directly proportional to the applied force – the stronger we press the stronger the feeling is. On the other hand, the stress is inversely proportional to the contact surface between the fingers – if we press one finger with the nail of the other the feeling is much stronger because the same force is applied to a much smaller area. This is why pricking a finger with a needle is so painful – the force applied to the needle is not big but the contact surface of the needle with the finger is very small and the resulting stress is consequently very big.

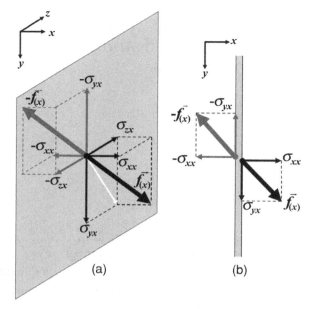

Fig. 4.1 Components of stress tensor defined from the force balance on a surface. (a) relationship between the stress components (thin arrows σ_{xx}, σ_{yx}, σ_{zx} and $-\sigma_{xx}$, $-\sigma_{yx}$, $-\sigma_{zx}$) and force vectors (thick arrows $\vec{f}_{(x)}$ and $-\vec{f}_{(x)}$) acting on the two sides of the unit element (grey) of a Lagrangian surface orthogonal to x-axis (i.e. x-surface). White arrow in (a) shows the direction of shear along the surface. (b) physical analogy: normal and shear stress components acting on a thin plate (cross-section of the plate in x-y-plane is shown).

In order to characterise the stress tensor, let us consider the force $\vec{f}_{(x)}$ acting on a unit element of a Lagrangian x-surface (i.e. surface orthogonal to the x-axis) (Fig. 4.1(a)). First of all, we need to understand that the force vector $\vec{f}_{(x)}$ acting on one side of the surface element is balanced by the counterforce vector $-\vec{f}_{(x)}$ which acts on the other side, and therefore this *stressed surface element* does not move. Thus, in order to characterise the *force balance state* of the stressed surface element, one needs to characterise the magnitude and direction of the force (balanced by the counterforce) acting on this element. Let us adopt a convention that the characterisation will be based on the force vector $\vec{f}_{(x)}$, applied to the side of the x-surface *from which the x-axis is exiting*. As we will see in the following, according to this convention, *extensional stresses are positive* as is usually assumed in continuum mechanics (e.g., Ranalli, 1995). Notice that this *usual continuum mechanics convention* is opposite to that used in the book of Turcotte and Schubert (2002), where stresses are taken positive under compression (which geoscientists find more intuitive since pressure is also positive under compression).

The force vector $\vec{f}_{(x)}$ can obviously be decomposed into three components (σ_{xx}, σ_{xy}, σ_{xz}) parallel to each coordinate axis (Fig. 4.1(a)). These are the components of the stress tensor since force $\vec{f}_{(x)}$ is acting on the unit element. According to the common continuum mechanics convention (e.g. Ranalli, 1995), which is again opposite to that used in the book of Turcotte and Schubert (2002), the first index (*i*) of a stress component σ_{ij} denotes the axis along which this stress component is taken (i.e. $i = z$ for the component parallel to the *z* axis) and the second index (*j*) indicates the surface on which force balance is considered (i.e. $j = x$ for the surface orthogonal to *x* axis). It should be pointed out that our 'hard choice' of a stress definition and notation is, indeed, very convenient for formulating several crucial equations, such as the momentum equation and the rheological constitutive relationships, which is the main reason why we deviated from the 'geological convention'. On the other hand, our vertical axis *y*, is always pointing down, thus preserving common 'geological logic' that the vertical coordinate is depth (and not height as in continuum mechanics) and increases downward rather than upward. A stress component that is orthogonal to the surface (cf. σ_{xx} in Fig. 4.1(a)) is called a *normal stress component* and the components which are parallel to the surface are called *shear stress components* (cf. σ_{yx} and σ_{zx} in Fig. 4.1(a)). The normal stress component characterises the magnitude of extension/compression *across the surface*. The two shear stress components characterise the magnitude and direction (cf. white arrow in Fig. 4.1) of shearing applied *along the considered surface*. A useful physical analogy (Fig. 4.1(b)) – if one imagines that the force and counterforce are applied on two sides of a very thin plate, then the normal component defines how strongly two opposite surfaces of the plate are forced to be shifted from/toward each other and the shear stress components define where and how strong these surfaces are forced to be shifted parallel to each other.

In order to fully characterise the force balance at a point (a small material volume), it is convenient to represent the stress tensor as a $N \times N$ matrix where N is the dimension of the problem such that in one, two and three dimensions we will have one, four and nine stress components respectively (Fig. 4.2)

1D stress tensor, $N = 1$ (Fig. 4.2(a)): $\sigma_{ij} = (\sigma_{xx})$,

2D stress tensor, $N = 2$ (Fig. 4.2(b)): $\sigma_{ij} = \begin{pmatrix} \sigma_{xx} & \sigma_{xy} \\ \sigma_{yx} & \sigma_{yy} \end{pmatrix}$,

3D stress tensor, $N = 3$ (Fig. 4.2(c)): $\sigma_{ij} = \begin{pmatrix} \sigma_{xx} & \sigma_{xy} & \sigma_{xz} \\ \sigma_{yx} & \sigma_{yy} & \sigma_{yz} \\ \sigma_{zx} & \sigma_{zy} & \sigma_{zz} \end{pmatrix}$,

where *i* and *j* are symbolic coordinate indices (*x*, *y*, *z*) which vary in vertical and horizontal directions, respectively. In continuum mechanics books a numerical

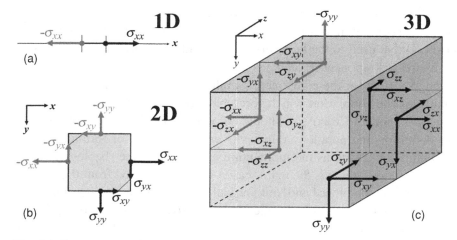

Fig. 4.2 Components of the stress tensor (black and grey arrows) defined in one-
(a) two- (b) and three- (c) dimensions on faces of a small interval, square and cube,
respectively. The faces are always oriented orthogonal to the main axis. Thin lines
in (b) and (c) connect pairs of shear stress components which should be equal to
each other in the absence of internal sources of angular momentum.

(1, 2, 3) notation for the coordinate indices i and j and stresses (σ_{11}, σ_{12}, σ_{32}, etc.)
is commonly used as well (e.g. Ranalli, 1995). *Note that i and j are indices and not
spatial coordinates of geometric points.*

Normal stresses are always located on the main diagonal of the matrix. Due
to the *condition of force balance in the absence of internal sources of angular
momentum*, this matrix is symmetric relative to the main diagonal so that

$$\sigma_{ij} = \sigma_{ji},$$

i.e. (Fig. 4.2(b), (c))

$$\sigma_{xy} = \sigma_{yx},$$
$$\sigma_{xz} = \sigma_{zx},$$
$$\sigma_{yz} = \sigma_{zy}.$$

Like components of a vector, components of the stress tensor at a point depend on
the orientation of the coordinate system. We will discuss this in more detail later
in relation to elasticity (Chapter 12).

In continuum mechanics, *pressure* is defined as the mean normal stress:

$$P = -(\sigma_{xx} + \sigma_{yy} + \sigma_{zz})/3 \qquad (4.1)$$

where the negative sign on the right-hand side of Eq. (4.1) reflects another convention according to which pressure is positive under compression. Pressure is an

invariant and, thus, *does not change with changing the coordinate system.* In the case of a *hydrostatic stress state* (which is the state of a fluid *at rest*) all shear stresses are zero and all normal stresses are equal to each other

$$\sigma_{xy} = \sigma_{yx} = \sigma_{xz} = \sigma_{zx} = \sigma_{yz} = \sigma_{zy} = 0 \qquad (4.2a)$$

$$\sigma_{xx} = \sigma_{yy} = \sigma_{zz} = -P. \qquad (4.2b)$$

In geosciences, pressure is often considered as corresponding to the hydrostatic condition everywhere and it is computed as a function of depth y and vertical density profile $\rho(y)$

$$P(y) = P_0 + g \int_0^y \rho(y)dy, \qquad (4.3)$$

where $P_0 = 0.1$ MPa is pressure on the Earth's surface and g is the gravitational acceleration.

This simplification does not hold when deformations of geological media occur and real *dynamic* pressure may notably deviate from the lithostatic value given by Eq. (4.3).

It is often convenient to define the *deviatoric* stresses σ'_{ij}, which are deviations of stresses from the hydrostatic stress state (i.e., deviations from conditions (4.2))

$$\sigma'_{ij} = \sigma_{ij} + P\delta_{ij}, \qquad (4.4)$$

where δ_{ij} is the *Kronecker delta*: $\delta_{ij} = 1$ when $i = j$ and $\delta_{ij} = 0$ when $i \neq j$, i and j are coordinate indices (x, y, z). The Kronecker delta is a peculiar abbreviation used in the mechanics of continuum. It only takes values of either 1 or 0 and is analogous to the logical operator 'if' used in many programming languages. Any equation with δ_{ij} represents a *group of equations*. For example, Eq. (4.4) in 3D represents the following equations:

Normal deviatoric stresses

$$\sigma'_{xx} = \sigma_{xx} + P,$$

$$\sigma'_{yy} = \sigma_{yy} + P,$$

$$\sigma'_{zz} = \sigma_{zz} + P,$$

and shear stresses which are entirely deviatoric

$$\sigma'_{xy} = \sigma'_{yx} = \sigma_{xy} = \sigma_{yx},$$

$$\sigma'_{xz} = \sigma'_{zx} = \sigma_{xz} = \sigma_{zx},$$

$$\sigma'_{yz} = \sigma'_{zy} = \sigma_{yz} = \sigma_{zy}.$$

It is worth mentioning that the sum of the normal deviatoric stresses is zero by definition (Eq. (4.4))

$$\sigma'_{xx} + \sigma'_{yy} + \sigma'_{zz} = 0,$$

since

$$\sigma_{xx} + \sigma_{yy} + \sigma_{zz} = -3P.$$

The *second invariant* of the deviatoric stress tensor can be calculated as follows:

$$\sigma_{II} = \sqrt{\tfrac{1}{2}\sigma'^2_{ij}}, \tag{4.5}$$

where the indices *ij* imply a *summation*! This is another abbreviation that is commonly used in continuum mechanics and makes equations shorter (but, indeed, not easier to understand for inexperienced readers). The spelled-out form of Eq. (4.5) is much longer

$$\sigma_{II} = \sqrt{\tfrac{1}{2}\left(\sigma'^2_{xx} + \sigma'^2_{yy} + \sigma'^2_{zz} + \sigma^2_{xy} + \sigma^2_{yx} + \sigma^2_{xz} + \sigma^2_{zx} + \sigma^2_{yz} + \sigma^2_{zy}\right)}, \tag{4.6a}$$

or, using the condition of force balance $\sigma_{ij} = \sigma_{ji}$

$$\sigma_{II} = \sqrt{\tfrac{1}{2}\left(\sigma'^2_{xx} + \sigma'^2_{yy} + \sigma'^2_{zz}\right) + \sigma^2_{xy} + \sigma^2_{xz} + \sigma^2_{yz}}. \tag{4.6b}$$

The second stress invariant σ_{II} does not depend on the coordinate system and characterises the local deviation of stresses in the medium from the hydrostatic state.

4.2 Strain and strain rate

Another important quantity is the *strain* γ, that characterises the amount of deformation. Strain is dimensionless and is computed as the ratio of displacement ΔL to the initial length of deforming body L (Fig. 4.3)

$$\gamma = \frac{\Delta L}{L}. \tag{4.7}$$

By analogy with stress, one can discriminate normal and shear strain corresponding to axial and shear deformation, respectively (Fig. 4.3(a) and (b)).

The definition of strain given by Eq. (4.7) can only be applied in cases of relatively simple axial and shear deformations. In case of more complex deformation, the *strain tensor* ε_{ij}, is defined as

$$\varepsilon_{ij} = \frac{1}{2}\left(\frac{\partial u_i}{\partial x_j} + \frac{\partial u_j}{\partial x_i}\right), \tag{4.8}$$

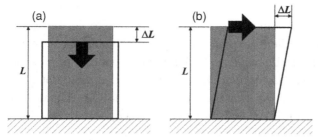

Fig. 4.3 Axial (a) and shear (b) deformation corresponding to normal and shear strain components. The strain in both cases is estimated as $\gamma = \dfrac{\Delta L}{L}$. Note that in case of shear deformation length L is measured orthogonal to the displacement direction.

where i and j *are coordinate indices* (x, y, z) and x_i and x_j *are spatial coordinates* (i.e., x_x, x_y and x_z are x-, y- and z-coordinates respectively). *Note that in contrast to symbolic i- and j-indices* x_i *and* x_j *are physical coordinates of geometrical points.* Do not confuse them with each other! In 3D, we can define nine tensor components:

three normal strain components

$$\varepsilon_{xx} = \frac{1}{2}\left(\frac{\partial u_x}{\partial x} + \frac{\partial u_x}{\partial x}\right) = \frac{\partial u_x}{\partial x},$$

$$\varepsilon_{yy} = \frac{1}{2}\left(\frac{\partial u_y}{\partial y} + \frac{\partial u_y}{\partial y}\right) = \frac{\partial u_y}{\partial y},$$

$$\varepsilon_{zz} = \frac{1}{2}\left(\frac{\partial u_z}{\partial z} + \frac{\partial u_z}{\partial z}\right) = \frac{\partial u_z}{\partial z}$$

and six shear strain components

$$\varepsilon_{xy} = \varepsilon_{yx} = \frac{1}{2}\left(\frac{\partial u_x}{\partial y} + \frac{\partial u_y}{\partial x}\right),$$

$$\varepsilon_{xz} = \varepsilon_{zx} = \frac{1}{2}\left(\frac{\partial u_x}{\partial z} + \frac{\partial u_z}{\partial x}\right),$$

$$\varepsilon_{yz} = \varepsilon_{zy} = \frac{1}{2}\left(\frac{\partial u_z}{\partial y} + \frac{\partial u_y}{\partial z}\right).$$

Note that stress and strain tensors are very different physical quantities (although they can be strongly correlated in case of reversible elastic deformation, Chapter 12): stress characterises the distribution of forces acting in a continuum at a given moment of time, while strain quantifies in an integrated way the entire *deformation history* of the continuum from the initial state, up until this given moment

(Fig. 4.3). The symmetric form of the strain tensor subtracts the rotational component of the velocity field which does not contribute to the deformation (rotation of
a rigid body has gradients in material displacement, but does not produce any internal deformation). In Eq. (4.8), u_i and u_j are components of *material displacement*
vector, $\vec{u} = (u_x, u_y, u_z)$ which characterise the displacement of a material point
relative to its original position (i.e. before deformation). The time derivative of the
displacement vector is the velocity vector $\vec{v} = (v_x, v_y, v_z)$ so that

$$v_i = \frac{Du_i}{Dt} \tag{4.9}$$

and, in 3D deformation

$$v_x = \frac{Du_x}{Dt},$$
$$v_y = \frac{Du_y}{Dt},$$
$$v_z = \frac{Du_z}{Dt}.$$

The strain tensor is widely used when elastic deformation is considered
(Chapter 12).

In numerical geodynamic modelling, it is convenient to use the *strain rate*, which
characterises the dynamics of changes in the internal deformation rather than the
strain which characterises the total amount of deformation compared to the initial
state. The strain rate tensor $\dot{\varepsilon}_{ij}$ is the time derivative (indicated by the dot on top of
the strain symbol) of the strain tensor ε_{ij}. Components of the strain rate tensor are
defined via spatial gradients of the velocity as follows

$$\dot{\varepsilon}_{ij} = \frac{1}{2} \left(\frac{\partial v_i}{\partial x_j} + \frac{\partial v_j}{\partial x_i} \right), \tag{4.10}$$

where i and j are coordinate indices and x_i and x_j are spatial coordinates such that
in 3D we can define nine tensor components:

three normal strain rate components

$$\dot{\varepsilon}_{xx} = \frac{1}{2} \left(\frac{\partial v_x}{\partial x} + \frac{\partial v_x}{\partial x} \right) = \frac{\partial v_x}{\partial x},$$
$$\dot{\varepsilon}_{yy} = \frac{1}{2} \left(\frac{\partial v_y}{\partial y} + \frac{\partial v_y}{\partial y} \right) = \frac{\partial v_y}{\partial y},$$
$$\dot{\varepsilon}_{zz} = \frac{1}{2} \left(\frac{\partial v_z}{\partial z} + \frac{\partial v_z}{\partial z} \right) = \frac{\partial v_z}{\partial z}$$

and six shear strain rate components

$$\dot{\varepsilon}_{xy} = \dot{\varepsilon}_{yx} = \frac{1}{2}\left(\frac{\partial v_x}{\partial y} + \frac{\partial v_y}{\partial x}\right),$$

$$\dot{\varepsilon}_{xz} = \dot{\varepsilon}_{zx} = \frac{1}{2}\left(\frac{\partial v_x}{\partial z} + \frac{\partial v_z}{\partial x}\right),$$

$$\dot{\varepsilon}_{yz} = \dot{\varepsilon}_{zy} = \frac{1}{2}\left(\frac{\partial v_y}{\partial z} + \frac{\partial v_z}{\partial y}\right).$$

Similarly to the strain tensor, the symmetric form of the strain rate tensor is obtained by subtracting the rotational component of the velocity field: it is easy to check that rigid body rotation in 2D has gradients in the velocity field which do not produce any internal deformation, i.e.

$$\dot{\varepsilon}_{xy} = \frac{1}{2}\left(\frac{\partial v_x}{\partial y} + \frac{\partial v_y}{\partial x}\right) = 0.$$

By analogy to the stress tensor, the strain rate tensor can also be subdivided to isotropic $\dot{\varepsilon}_{kk}$ (which is an invariant) and deviatoric $\dot{\varepsilon}'_{ij}$ components

$$\dot{\varepsilon}_{kk} = \dot{\varepsilon}_{xx} + \dot{\varepsilon}_{yy} + \dot{\varepsilon}_{zz} = \frac{\partial v_x}{\partial x} + \frac{\partial v_y}{\partial y} + \frac{\partial v_z}{\partial z} = \mathrm{div}(\bar{v}), \quad (4.11)$$

$$\dot{\varepsilon}'_{ij} = \dot{\varepsilon}_{ij} - \delta_{ij}\frac{1}{3}\dot{\varepsilon}_{kk}, \quad (4.12)$$

where i, j and k are coordinate indices.

According to Eqs. (4.11), (4.12) the sum of normal deviatoric strain rate components is zero

$$\dot{\varepsilon}'_{xx} + \dot{\varepsilon}'_{yy} + \dot{\varepsilon}'_{zz} = 0. \quad (4.13)$$

Like the second stress invariant, the second invariant of the deviatoric strain rate tensor is calculated as follows

$$\dot{\varepsilon}_{\mathrm{II}} = \sqrt{\frac{1}{2}\dot{\varepsilon}'^{2}_{ij}}. \quad (4.14)$$

Analytical exercise

Exercise 4.1

Show that the symmetric form of the strain rate tensor satisfies the condition

$$\dot{\varepsilon}_{xy} = \frac{1}{2}\left(\frac{\partial v_x}{\partial y} + \frac{\partial v_y}{\partial x}\right) = 0,$$

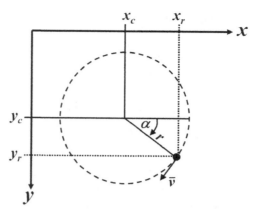

Fig. 4.4 Geometrical relationship in the case of 2D rigid body rotation.

in the case of rigid body rotation with constant angular velocity. Use the fact that coordinates of a rotating point in 2D are given by (Fig. 4.4)

$$x_r = x_c + r\cos(\alpha), \quad y_r = y_c + r\sin(\alpha),$$

where r is the distance to the centre of rotation, x_c and y_c are the coordinates of the centre and α is the clockwise rotation angle taken from the horizontal axis.

Programming exercise and homework

Exercise 4.2

Compute and visualise the deviatoric strain rate tensor components and invariants for the model described in the Exercise 1.2. An example is in **Strain_rate.m**.

5

The momentum equation

Theory: Momentum equation. Viscosity and Newtonian law of viscous friction. Navier–Stokes equation for the motion of a viscous fluid. Stokes equation of slow laminar flow of highly viscous incompressible fluid and its application to geodynamics. Simplification of the Stokes equation in case of constant viscosity and its relation to the Poisson equation. Analytical example for channel flow. Stream function approach.
Exercises: Solving continuity and momentum equations for the case of constant viscosity with a stream function approach.

5.1 Momentum equation

The deformation of continuous media always results from the balance of various internal and external forces that act on these media. In order to relate forces and deformation, an equation of motion should be used. This is the so-called *momentum equation*, which describes the *conservation of momentum* for a continuous medium in the gravity field:

$$\text{Eulerian form:} \quad \frac{\partial \sigma_{ij}}{\partial x_j} + \rho g_i = \rho \left(\frac{\partial v_i}{\partial t} + v_j \frac{\partial v_i}{\partial x_j} \right), \tag{5.1a}$$

$$\text{Lagrangian form:} \quad \frac{\partial \sigma_{ij}}{\partial x_j} + \rho g_i = \rho \frac{D v_i}{Dt}. \tag{5.1b}$$

The momentum equation is a differential equivalent to the famous Newton's second law of motion, which describes changes in velocity of an object with mass m according to

$$f = ma, \tag{5.2}$$

61

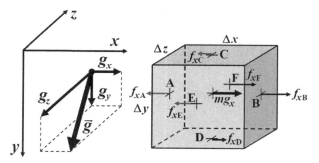

Fig. 5.1 Lagrangian elementary volume considered for the derivation of the respective form of x-momentum equation. Thin arrows show x-components of stress-related forces acting *from the outside of the volume* on respective boundaries (A, B, C, D, E and F). The thick arrow *inside the volume* shows the x-component of the gravity force (mg_x) which is proportional to the mass (m) embedded inside the volume. Orientation and components of gravity vector $\vec{g} = (g_x, g_y, g_z)$ are displayed outside of the volume.

where f is a net force acting on the object and $a = \dfrac{Dv}{Dt}$, is the acceleration of the object. This law can be written in vector form:

$$\vec{f} = m\vec{a} \quad \text{or} \quad f_i = ma_i, \tag{5.3a}$$

which gives three equations for 3D considerations

$$f_x = ma_x, \tag{5.3b}$$
$$f_y = ma_y, \tag{5.3c}$$
$$f_z = ma_z, \tag{5.3d}$$

where i denotes the coordinate index and $a_i = \dfrac{Dv_i}{Dt}$ are the components of the acceleration vector $\vec{a}$. Equation (5.1b) can in fact be derived from Equation (5.3a) while considering each material point of the continuum as a very small Lagrangian volume, for which the net acting force can be computed locally.

   Let us make this derivation on an intuitive basis by considering forces that act on a small Lagrangian volume with dimensions Δx, Δy and Δz (Fig. 5.1). From Equation (5.3b), one first obtains the expression for the x-momentum equation. The net force f_x acting in the x-direction on the Lagrangian volume in a gravity field can be represented as a sum of seven elementary forces

$$f_x = f_{xA} + f_{xB} + f_{xC} + f_{xD} + f_{xE} + f_{xF} + mg_x, \tag{5.4}$$

where $f_{xA} - f_{xF}$ are stress-related forces *acting from the outside* of the volume on the respective boundaries (A–F) and mg_x is the gravity force, which is proportional to the mass embedded in the volume. Stress-related forces are proportional to the surfaces of the respective boundaries and can be computed as follows

$$f_{xA} = -\sigma_{xxA}\Delta y\Delta z, \tag{5.5a}$$

$$f_{xB} = +\sigma_{xxB}\Delta y\Delta z, \tag{5.5b}$$

$$f_{xC} = -\sigma_{xyC}\Delta x\Delta z, \tag{5.5c}$$

$$f_{xD} = +\sigma_{xyD}\Delta x\Delta z, \tag{5.5d}$$

$$f_{xE} = -\sigma_{xzE}\Delta x\Delta y, \tag{5.5e}$$

$$f_{xF} = +\sigma_{xzF}\Delta x\Delta y, \tag{5.5f}$$

where σ_{xxA}, σ_{xxB} and σ_{xyC}, σ_{xyD}, σ_{xzE}, σ_{xzF} are the normal and shear stress components defined at respective boundaries. Note that the sign in front of the stress components on the right-hand side of Equations (5.5a)−(5.5f) solely depends on whether the force (positive sign, boundaries B, D and F) or the counterforce (negative sign, boundaries A, C and E) is acting on the respective boundary from outside of the volume (see Fig. 4.1 for the stress convention). On the other hand, the stress components themselves can also be either positive or negative in sign. By combining Eqs. (5.3b) and (5.4)−(5.5), one can now write Newton's second law of motion for the considered Lagrangian volume (verify as an exercise)

$$f_{xA} + f_{xB} + f_{xC} + f_{xD} + f_{xE} + f_{xF} + mg_x = ma_x, \tag{5.6a}$$

or

$$(\sigma_{xxB} - \sigma_{xxA})\Delta y\Delta z + (\sigma_{xyD} - \sigma_{xyC})\Delta x\Delta z + (\sigma_{xzF} - \sigma_{xzE})\Delta x\Delta y + mg_x = ma_x. \tag{5.6b}$$

Normalising both sides of Eq. (5.6b) by the considered Lagrangian volume

$$V = \Delta x\Delta y\Delta z, \tag{5.7}$$

we can now obtain the x-momentum equation in the differences representation (verify as an exercise)

$$\frac{(\sigma_{xxB} - \sigma_{xxA})\Delta y\Delta z}{V} + \frac{(\sigma_{xyD} - \sigma_{xyC})\Delta x\Delta z}{V} + \frac{(\sigma_{xzF} - \sigma_{xzE})\Delta x\Delta y}{V} + \frac{m}{V}g_x = \frac{m}{V}a_x, \tag{5.8a}$$

or

$$\frac{(\sigma_{xxB} - \sigma_{xxA})}{\Delta x} + \frac{(\sigma_{xyD} - \sigma_{xyC})}{\Delta y} + \frac{(\sigma_{xzF} - \sigma_{xzE})}{\Delta z} + \rho g_x = \rho a_x, \tag{5.8b}$$

or

$$\frac{\Delta\sigma_{xx}}{\Delta x} + \frac{\Delta\sigma_{xy}}{\Delta y} + \frac{\Delta\sigma_{xz}}{\Delta z} + \rho g_x = \rho a_x, \tag{5.8c}$$

$$\rho = \frac{m}{V}, \tag{5.8d}$$

$$\Delta\sigma_{xx} = \sigma_{xxB} - \sigma_{xxA}, \tag{5.8e}$$

$$\Delta\sigma_{xy} = \sigma_{xyD} - \sigma_{xyC}, \tag{5.8f}$$

$$\Delta\sigma_{xz} = \sigma_{xzF} - \sigma_{xzE}, \tag{5.8g}$$

where ρ is the average material density in the Lagrangian volume V, and $\Delta\sigma_{xx}$, $\Delta\sigma_{xy}$, $\Delta\sigma_{xz}$ are differences of the respective stress components taken in x- y- and z-directions (i.e. between respective pairs of boundaries), respectively.

When Δx, Δy and Δz all tend towards zero, the differences in Eq. 5.8c can be replaced by derivatives and we obtain the Lagrangian x-momentum equation

$$\frac{\partial\sigma_{xx}}{\partial x} + \frac{\partial\sigma_{xy}}{\partial y} + \frac{\partial\sigma_{xz}}{\partial z} + \rho g_x = \rho a_x, \tag{5.9a}$$

or

$$\frac{\partial\sigma_{xj}}{\partial x_j} + \rho g_x = \rho a_x. \tag{5.9b}$$

In this differential equation, written for an infinitely small Lagrangian volume, the density ρ corresponds to the local density at a point. Obviously, based on similar considerations y- and z-momentum equations can also be derived (verify as an exercise).

5.2 Newtonian law of viscous friction

As we discussed in the Introduction, rocks often behave on a geological time scale as highly viscous fluids. For this reason, the viscous *rheological* relationship between stress and strain rate known as *Newtonian law of viscous friction* is widely used in geodynamic modelling. The Newtonian law of viscous friction relates the shear stress τ (Pa) with the shear strain rate $\dfrac{\partial v}{\partial x}$ (1/s) according to

$$\tau = \eta\frac{\partial v}{\partial x}, \tag{5.10}$$

where η (Pa s), the *viscosity*, characterises the degree of resistance a material has to shear deformation. Viscosity is generally different for different materials and may also depend on pressure (P), temperature (T), strain rate and some other parameters. The viscosity of rocks is typically greater than 10^{17} Pa s: the viscosity of the asthenospheric upper mantle, for example, is about 10^{21} Pa s (Turcotte and Schubert, 2002). We will discuss this issue in more detail in the next chapter.

In 3D, the law of viscous friction is formulated with the components of both the deviatoric stress (σ'_{ij}) and the deviatoric strain rate ($\dot{\varepsilon}'_{ij}$) tensors in form of *viscous constitutive relationship* as follows:

$$\sigma'_{ij} = 2\eta\dot{\varepsilon}'_{ij} + \delta_{ij}\eta_{bulk}\dot{\varepsilon}_{kk}, \qquad (5.11a)$$

or

$$\sigma'_{ij} = 2\eta(\dot{\varepsilon}_{ij} - \tfrac{1}{3}\delta_{ij}\dot{\varepsilon}_{kk}) + \delta_{ij}\eta_{bulk}\dot{\varepsilon}_{kk}, \qquad (5.11b)$$

where η and η_{bulk} are the *shear viscosity* and *bulk viscosity*, respectively; $\dot{\varepsilon}_{kk}$ is the bulk strain rate (Eq. 4.11) in response to *irreversible inelastic volume changes* (such as due to phase transformation or compaction).

In the absence of mineralogical phase transformations, rocks exhibit relatively small density variations (see Chapter 2). Therefore, the *incompressible fluid approximation* ($\rho = \text{const}$, $\dfrac{D\rho}{Dt} = 0$, see Chapter 1) is generally valid. In this case, $\dot{\varepsilon}_{kk} = div(\vec{v}) = 0$, $\dot{\varepsilon}'_{ij} = \dot{\varepsilon}_{ij}$ and the law of viscous friction can be simplified to:

$$\sigma'_{ij} = 2\eta\dot{\varepsilon}_{ij}. \qquad (5.12)$$

5.3 Navier–Stokes equation

Using the momentum equation (5.1b) and the relationship between the total (σ_{ij}) and deviatoric (σ'_{ij}) stresses (Eq. 4.4), we can introduce pressure into the momentum equation (5.1b) and obtain the *Navier–Stokes equation of motion*, which describes the conservation of momentum for a fluid in the gravity field:

$$\frac{\partial\sigma'_{ij}}{\partial x_j} - \frac{\partial P}{\partial x_i} + \rho g_i = \rho\frac{Dv_i}{Dt}, \qquad (5.13)$$

where i and j are coordinate indices; x_i and x_j are spatial coordinates; g_i is the i-th component of the gravity vector $\vec{g} = (g_x, g_y, g_z)$; $\dfrac{Dv_i}{Dt}$ is the substantive time derivative of i-th component of the velocity vector (i.e., acceleration vector). By analogy to other substantive time derivatives, it can be related to the Eulerian time derivative as:

$$\frac{Dv_i}{Dt} = \frac{\partial v_i}{\partial t} + \vec{v} \cdot grad(v_i), \qquad (5.14)$$

or in 3D via

$$\frac{Dv_x}{Dt} = \frac{\partial v_x}{\partial t} + v_x \frac{\partial v_x}{\partial x} + v_y \frac{\partial v_x}{\partial y} + v_z \frac{\partial v_x}{\partial z},$$

$$\frac{Dv_y}{Dt} = \frac{\partial v_y}{\partial t} + v_x \frac{\partial v_y}{\partial x} + v_y \frac{\partial v_y}{\partial y} + v_z \frac{\partial v_y}{\partial z},$$

$$\frac{Dv_z}{Dt} = \frac{\partial v_z}{\partial t} + v_x \frac{\partial v_z}{\partial x} + v_y \frac{\partial v_z}{\partial y} + v_z \frac{\partial v_z}{\partial z}.$$

In the case of 3D deformation, the Navier–Stokes equation of motion corresponds to a system of three partial differential equations (with $\sigma'_{ij} = \sigma_{ij}$ for $i \neq j$):

x-Navier–Stokes equation $\dfrac{\partial \sigma'_{xx}}{\partial x} + \dfrac{\partial \sigma_{xy}}{\partial y} + \dfrac{\partial \sigma_{xz}}{\partial z} - \dfrac{\partial P}{\partial x} + \rho g_x = \rho \dfrac{Dv_x}{Dt},$ (5.15)

y-Navier–Stokes equation $\dfrac{\partial \sigma'_{yy}}{\partial y} + \dfrac{\partial \sigma_{yx}}{\partial x} + \dfrac{\partial \sigma_{yz}}{\partial z} - \dfrac{\partial P}{\partial y} + \rho g_y = \rho \dfrac{Dv_y}{Dt},$ (5.16)

z-Navier–Stokes equation $\dfrac{\partial \sigma'_{zz}}{\partial z} + \dfrac{\partial \sigma_{zx}}{\partial x} + \dfrac{\partial \sigma_{zy}}{\partial y} - \dfrac{\partial P}{\partial z} + \rho g_z = \rho \dfrac{Dv_z}{Dt}.$ (5.17)

In highly viscous flows, the inertial forces $(\rho \dfrac{Dv_i}{Dt})$ are negligible with respect to viscous resistance and gravitational forces. For example, a typical plate velocity in geodynamics is on the order of several cm/year ($n \times 10^{-9}$ m/s) and it may notably change only within millions of years ($n \times 10^{13}$ s). Consequently, the typical magnitude of mantle flow 'accelerations' will be on the order of $\dfrac{Dv_i}{Dt} \approx \dfrac{\Delta v}{\Delta t} = \dfrac{n \times 10^{-9}}{n \times 10^{13}} = n \times 10^{-22}$ m/s². This makes the right-hand side of the Navier–Stokes equation $\rho \dfrac{Dv_i}{Dt}$ negligible compared to the ρg_i term in the left-hand side of the equation, since the magnitude of the gravitational acceleration g_i is on the order of 10 m/s², i.e. 10^{23} times bigger than the mantle flow accelerations. Under such circumstances deformation of highly viscous flows can be accurately described by the *Stokes equation of slow flow* (with $\sigma'_{ij} = \sigma_{ij}$ for $i \neq j$):

$$\frac{\partial \sigma'_{ij}}{\partial x_j} - \frac{\partial P}{\partial x_i} + \rho g_i = 0, \tag{5.18}$$

or

x-Stokes equation $\dfrac{\partial \sigma'_{xx}}{\partial x} + \dfrac{\partial \sigma_{xy}}{\partial y} + \dfrac{\partial \sigma_{xz}}{\partial z} - \dfrac{\partial P}{\partial x} + \rho g_x = 0,$ (5.19)

y-Stokes equation $\dfrac{\partial \sigma'_{yy}}{\partial y} + \dfrac{\partial \sigma_{yx}}{\partial x} + \dfrac{\partial \sigma_{yz}}{\partial z} - \dfrac{\partial P}{\partial y} + \rho g_y = 0,$ (5.20)

z-Stokes equation $\dfrac{\partial \sigma'_{zz}}{\partial z} + \dfrac{\partial \sigma_{zx}}{\partial x} + \dfrac{\partial \sigma_{zy}}{\partial y} - \dfrac{\partial P}{\partial z} + \rho g_z = 0.$ (5.21)

The equations can be further simplified if the viscosity is constant and the fluid is incompressible. In this case, the Stokes equations simplify to:

$$\eta \frac{\partial^2 v_i}{\partial x_j^2} - \frac{\partial P}{\partial x_i} + \rho g_i = 0. \tag{5.22}$$

Let us go through the logic of this simplification for the x-Stokes equation (5.19).
First, applying Eq. (5.12) we obtain

$$\frac{\partial}{\partial x}(2\eta\, \dot{\varepsilon}_{xx}) + \frac{\partial}{\partial y}(2\eta\, \dot{\varepsilon}_{xy}) + \frac{\partial}{\partial z}(2\eta\, \dot{\varepsilon}_{xz}) - \frac{\partial P}{\partial x} + \rho g_x = 0. \tag{5.23}$$

Then using Eq. (4.10) for strain rate components, we get

$$\eta\left[\frac{\partial}{\partial x}\left(\frac{\partial v_x}{\partial x} + \frac{\partial v_x}{\partial x}\right) + \frac{\partial}{\partial y}\left(\frac{\partial v_x}{\partial y} + \frac{\partial v_y}{\partial x}\right) + \frac{\partial}{\partial z}\left(\frac{\partial v_x}{\partial z} + \frac{\partial v_z}{\partial x}\right)\right] - \frac{\partial P}{\partial x} + \rho g_x = 0. \tag{5.24a}$$

Further regrouping of the above equation gives

$$\eta\left[\left(\frac{\partial^2 v_x}{\partial x^2} + \frac{\partial^2 v_x}{\partial y^2} + \frac{\partial^2 v_x}{\partial z^2}\right) + \left(\frac{\partial^2 v_x}{\partial x \partial x} + \frac{\partial^2 v_y}{\partial x \partial y} + \frac{\partial^2 v_z}{\partial x \partial z}\right)\right] - \frac{\partial P}{\partial x} + \rho g_x = 0 \tag{5.24b}$$

$$\eta\left[\left(\frac{\partial^2 v_x}{\partial x^2} + \frac{\partial^2 v_x}{\partial y^2} + \frac{\partial^2 v_x}{\partial z^2}\right) + \frac{\partial}{\partial x}\left(\frac{\partial v_x}{\partial x} + \frac{\partial v_y}{\partial y} + \frac{\partial v_z}{\partial z}\right)\right] - \frac{\partial P}{\partial x} + \rho g_x = 0. \tag{5.24c}$$

And finally using the fact that for the incompressible fluid $\dfrac{\partial v_x}{\partial x} + \dfrac{\partial v_y}{\partial y} + $

$\dfrac{\partial v_z}{\partial z} = \text{div}(\bar{v}) = 0$, we obtain

$$\eta\left(\frac{\partial^2 v_x}{\partial x^2} + \frac{\partial^2 v_x}{\partial y^2} + \frac{\partial^2 v_x}{\partial z^2}\right) - \frac{\partial P}{\partial x} + \rho g_x = 0, \tag{5.24d}$$

which is analogous to Eq. (5.22). Obviously, one can get similar expressions for both the y-Stokes and z-Stokes equations (derive as an exercise).
As usual, for the case of 3D deformation we have three equations;

$$\eta\left(\frac{\partial^2 v_x}{\partial x^2} + \frac{\partial^2 v_x}{\partial y^2} + \frac{\partial^2 v_x}{\partial z^2}\right) - \frac{\partial P}{\partial x} + \rho g_x = 0 \quad \text{or} \quad \eta \mathbf{\Delta} v_x = \frac{\partial P}{\partial x} - \rho g_x,$$

$$\eta\left(\frac{\partial^2 v_y}{\partial x^2} + \frac{\partial^2 v_y}{\partial y^2} + \frac{\partial^2 v_y}{\partial z^2}\right) - \frac{\partial P}{\partial y} + \rho g_y = 0 \quad \text{or} \quad \eta\, \mathbf{\Delta} v_y = \frac{\partial P}{\partial y} - \rho g_y,$$

$$\eta\left(\frac{\partial^2 v_z}{\partial x^2} + \frac{\partial^2 v_z}{\partial y^2} + \frac{\partial^2 v_z}{\partial z^2}\right) - \frac{\partial P}{\partial z} + \rho g_z = 0 \quad \text{or} \quad \eta\, \mathbf{\Delta} v_z = \frac{\partial P}{\partial z} - \rho g_z,$$

where $\mathbf{\Delta} = \dfrac{\partial^2}{\partial x^2} + \dfrac{\partial^2}{\partial y^2} + \dfrac{\partial^2}{\partial z^2}$ is the *Laplace operator*.

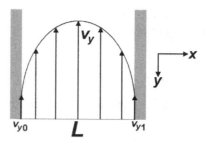

Fig. 5.2 Vertical laminar flow of a viscous fluid in a planar channel.

5.4 Poisson equation

If the pressure gradients are constant, Eq. (5.24) can be written in the form of the *Poisson equation*

$$\boldsymbol{\Delta} v_i = const_i \quad \text{where} \quad const_i = \frac{1}{\eta} \left(\frac{\partial P}{\partial x_i} - \rho g_i \right). \tag{5.25}$$

In 3D it is always worth writing out the system of three equations explicitly:

$$x\text{-Poisson equation} \quad \boldsymbol{\Delta} v_x = const_x, \quad \text{where} \quad const_x = \frac{1}{\eta} \left(\frac{\partial P}{\partial x} - \rho g_x \right),$$

$$y\text{-Poisson equation} \quad \boldsymbol{\Delta} v_y = const_y, \quad \text{where} \quad const_y = \frac{1}{\eta} \left(\frac{\partial P}{\partial y} - \rho g_y \right),$$

$$z\text{-Poisson equation} \quad \boldsymbol{\Delta} v_z = const_z, \quad \text{where} \quad const_z = \frac{1}{\eta} \left(\frac{\partial P}{\partial z} - \rho g_z \right).$$

Despite its simplicity, the Poisson equation is valid for several important geodynamic problems, for example in uniaxial (e.g. purely vertical) flow of a fluid in a channel (e.g. magma flow in the magmatic channel, rock flow in the subduction channel etc.).

In a vertical planar channel of width L (Fig. 5.2), the vertical velocity depends on the x-coordinate only, $v_x = 0$, $v_z = 0$, $\frac{\partial v_y}{\partial y} = 0$, $\frac{\partial v_y}{\partial z} = 0$ and the system of three Poisson equations (5.25) reduces to

$$\frac{\partial^2 v_y}{\partial x^2} = \frac{1}{\eta} \left(\frac{\partial P}{\partial y} - \rho g_y \right). \tag{5.26}$$

This can be solved analytically by a two-fold integration process;

$$\frac{\partial v_y}{\partial x} = \int \frac{\partial^2 v_y}{\partial x^2} dx = \int \frac{1}{\eta} \left(\frac{\partial P}{\partial y} - \rho g_y \right) dx = \frac{1}{\eta} \left(\frac{\partial P}{\partial y} - \rho g_y \right) x + C_1, \quad (5.27)$$

$$v_y = \int \frac{\partial v_y}{\partial x} dx = \int \left[\frac{1}{\eta} \left(\frac{\partial P}{\partial y} - \rho g_y \right) x + C_1 \right] dx$$

$$= \frac{1}{2\eta} \left(\frac{\partial P}{\partial y} - \rho g_y \right) x^2 + C_1 x + C_2, \quad (5.28)$$

where C_1 and C_2 are integration constants which can be defined from the boundary conditions:

left boundary, $v_y = v_{y0}$ when $x = 0$ and then,

$$v_{y0} = \frac{1}{2\eta} \left(\frac{\partial P}{\partial y} - \rho g_y \right) x^2 + C_1 x + C_2 = C_2$$

so that

$$C_2 = v_{y0};$$

right boundary, $v_y = v_{y1}$ when $x = L$. So then,

$$v_{y1} = \frac{1}{2\eta} \left(\frac{\partial P}{\partial y} - \rho g_y \right) x^2 + C_1 x + C_2 = \frac{1}{2\eta} \left(\frac{\partial P}{\partial y} - \rho g_y \right) L^2 + C_1 L + C_2 \text{ so that}$$

$$C_1 = \frac{v_{y1} - v_{y0}}{L} - \frac{1}{2\eta} \left(\frac{\partial P}{\partial y} - \rho g_y \right) L,$$

where v_{y0} and v_{y1} is vertical velocity of the left and right wall, respectively. The final expression for the vertical velocity profile is then

$$v_y = \frac{1}{2\eta} \left(\frac{\partial P}{\partial y} - \rho g_y \right) (x^2 - Lx) + v_{y0} + (v_{y1} - v_{y0}) \frac{x}{L}. \quad (5.29)$$

If the walls are immobile, $v_{y0} = v_{y1} = 0$ and Eq. (5.29) simplifies to

$$v_y = \frac{1}{2\eta} \left(\frac{\partial P}{\partial y} - \rho g_y \right) (x^2 - Lx). \quad (5.30)$$

Analytical exercises with the Poisson equation are very useful as analytical solutions are often used for *benchmarking* (i.e., testing the accuracy of) numerical solutions (Chapter 16).

5.5 Stream function approach

The stream function approach is a way to formulate and solve the coupled momentum and continuity equations for 2D incompressible flow with only one scalar variable – *the stream function* (Ψ) – whose derivatives are given by

$$\frac{\partial \Psi}{\partial y} = v_x, \tag{5.31}$$

$$\frac{\partial \Psi}{\partial x} = -v_y. \tag{5.32}$$

Consequently, the 2D incompressibility condition is automatically satisfied

$$\text{div}(\bar{v}) = \frac{\partial v_x}{\partial x} + \frac{\partial v_y}{\partial y} = \frac{\partial}{\partial x}\left(\frac{\partial \Psi}{\partial y}\right) + \frac{\partial}{\partial y}\left(-\frac{\partial \Psi}{\partial x}\right) = \frac{\partial^2 \Psi}{\partial x \partial y} - \frac{\partial^2 \Psi}{\partial y \partial x} = 0.$$

Also, the 2D Stokes equations for an incompressible fluid are first modified as

x-Stokes equation $$\frac{\partial}{\partial y}\left(\frac{\partial \sigma'_{xx}}{\partial x} + \frac{\partial \sigma_{xy}}{\partial y} - \frac{\partial P}{\partial x} + \rho g_x\right) = 0, \tag{5.33}$$

y-Stokes equation $$\frac{\partial}{\partial x}\left(\frac{\partial \sigma'_{yy}}{\partial y} + \frac{\partial \sigma_{yx}}{\partial x} - \frac{\partial P}{\partial y} + \rho g_y\right) = 0, \tag{5.34}$$

and then Eq. (5.34) is subtracted from Eq. (5.33) in order to eliminate pressure

$$\left(\frac{\partial^2 \sigma'_{xx}}{\partial x \partial y} + \frac{\partial^2 \sigma_{xy}}{\partial y^2}\right) - \left(\frac{\partial^2 \sigma'_{yy}}{\partial x \partial y} + \frac{\partial^2 \sigma_{yx}}{\partial x^2}\right) = \frac{\partial \rho}{\partial x} g_y - \frac{\partial \rho}{\partial y} g_x. \tag{5.35}$$

Using the *constitutive relation* (5.12) and Eqs. (5.31), (5.32) we can now reformulate Eq. (5.35) using the stream function only

$$\frac{\partial^2}{\partial x \partial y}(2\eta \dot{\varepsilon}_{xx}) + \frac{\partial^2}{\partial y^2}(2\eta \dot{\varepsilon}_{xy}) - \frac{\partial^2}{\partial x \partial y}(2\eta \dot{\varepsilon}_{yy}) - \frac{\partial^2}{\partial x^2}(2\eta \dot{\varepsilon}_{xy}) = \frac{\partial \rho}{\partial x} g_y - \frac{\partial \rho}{\partial y} g_x, \tag{5.36a}$$

$$\frac{\partial^2}{\partial x \partial y}\left(2\eta \frac{\partial v_x}{\partial x} - 2\eta \frac{\partial v_y}{\partial y}\right) + \frac{\partial^2}{\partial y^2}\left(\eta \frac{\partial v_x}{\partial y} + \eta \frac{\partial v_y}{\partial x}\right) - \frac{\partial^2}{\partial x^2}\left(\eta \frac{\partial v_x}{\partial y} + \eta \frac{\partial v_y}{\partial x}\right)$$
$$= \frac{\partial \rho}{\partial x} g_y - \frac{\partial \rho}{\partial y} g_x, \tag{5.36b}$$

$$\frac{\partial^2}{\partial x \partial y}\left(4\eta \frac{\partial^2 \Psi}{\partial x \partial y}\right) + \frac{\partial^2}{\partial y^2}\left(\eta \frac{\partial^2 \Psi}{\partial y^2} - \eta \frac{\partial^2 \Psi}{\partial x^2}\right) - \frac{\partial^2}{\partial x^2}\left(\eta \frac{\partial^2 \Psi}{\partial y^2} - \eta \frac{\partial^2 \Psi}{\partial x^2}\right)$$
$$= \frac{\partial \rho}{\partial x} g_y - \frac{\partial \rho}{\partial y} g_x. \tag{5.36c}$$

In the case of constant viscosity η, Eq. (5.36c) can be further simplified to

$$\eta \left[4\frac{\partial^4 \Psi}{\partial x^2 \partial y^2} + \frac{\partial^4 \Psi}{\partial y^2 \partial y^2} - \frac{\partial^4 \Psi}{\partial x^2 \partial y^2} - \frac{\partial^4 \Psi}{\partial x^2 \partial y^2} + \frac{\partial^4 \Psi}{\partial x^2 \partial x^2} \right] = \frac{\partial \rho}{\partial x} g_y - \frac{\partial \rho}{\partial y} g_x,$$

(5.37a)

$$\eta \left[\frac{\partial^2}{\partial y^2}\left(\frac{\partial^2 \Psi}{\partial x^2} + \frac{\partial^2 \Psi}{\partial y^2} \right) + \frac{\partial^2}{\partial x^2}\left(\frac{\partial^2 \Psi}{\partial x^2} + \frac{\partial^2 \Psi}{\partial y^2} \right) \right] = \frac{\partial \rho}{\partial x} g_y - \frac{\partial \rho}{\partial y} g_x, \quad (5.37b)$$

to obtain the so called *vorticity formulation*

$$\frac{\partial^2 \omega}{\partial y^2} + \frac{\partial^2 \omega}{\partial x^2} = \frac{1}{\eta}\left(\frac{\partial \rho}{\partial x} g_y - \frac{\partial \rho}{\partial y} g_x \right), \quad (5.38)$$

where ω is vorticity defined as

$$\frac{\partial^2 \Psi}{\partial x^2} + \frac{\partial^2 \Psi}{\partial y^2} = \omega. \quad (5.39)$$

As one can see, Eqs. (5.39) and (5.38) are both Poisson equations which can be solved subsequently, such that the solution of Eq. (5.38) goes into the right-hand side of Eq. (5.39). Obviously, boundary conditions should be formulated for both the vorticity and the stream function.

Analytical exercise

Exercise 5.1
Calculate the vertical velocity of a magma flow (v_y) in the middle of the vertical planar channel (Fig. 5.2) of width $L = 100$ m, for magma with a viscosity $\eta = 10^{16}$ Pa s and a density $\rho = 2800$ kg/m^3. The pressure gradient along the channel is $\frac{\partial P}{\partial y} = 25000$ Pa/m. Gravitational acceleration is $g_y = 10$ m/s^2, directed downward. The velocity on the channel boundaries is zero. Also calculate the solution when the right boundary is moving upward at a velocity $v_{y1} = 10^{-3}$ m/s.

Programming exercise and homework

Exercise 5.2
Solve the momentum and continuity equations with finite differences using the stream function – vorticity formulation (Eqs. 5.38, 5.39) on a regular grid (Fig. 5.3) of 51 × 41 points. Program a numerical model for buoyancy driven flow in a vertical gravity field ($g_x = 0$, $g_y = 10$ m/s^2 in Eq. 5.38) for a density structure with two vertical layers (3200 kg/m^3 and 3300 kg/m^3 for the left and right layer, respectively). The model size is 1000 × 1500 km (i.e. 1 000 000 × 1 500 000 m).

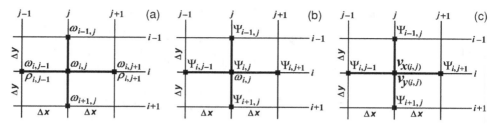

Fig. 5.3 Numerical grid stencils used for solving the coupled momentum and continuity equations in 2D with the stream function – vorticity formulation via finite differences, on a regular rectangular grid, in the case of constant viscosity and purely vertical gravity. (a), (b) stencils for formulating the Poisson equation for vorticity (Eq. 5.38) (a) and stream function (Eq. 5.39) (b). (c) stencil for computing the velocity components from the stream function.

Use a constant viscosity $\eta = 10^{21}$ Pa s for the entire model. Compose a matrix of coefficients $\{L\}$ and a right-hand side vector $\{R\}$ and obtain the solution vector $\{S\}$ with direct solver ($S = L\backslash R$) for the two Poisson equations (first for Eq. 5.38 and then for Eq. 5.39). A finite-difference representation of the Poisson equations in 2D can be formulated by analogy with Eq. (3.20) as follows (Fig. 5.3(a)(b))

$$\frac{\omega_{i,j-1} - 2\omega_{i,j} + \omega_{i,j+1}}{\Delta x^2} + \frac{\omega_{i-1,j} - 2\omega_{i,j} + \omega_{i+1,j}}{\Delta y^2} = g_y \frac{\rho_{i,j+1} - \rho_{i,j-1}}{\eta 2\Delta x}, \quad (5.40)$$

$$\frac{\Psi_{i,j-1} - 2\Psi_{i,j} + \Psi_{i,j+1}}{\Delta x^2} + \frac{\Psi_{i-1,j} - 2\Psi_{i,j} + \Psi_{i+1,j}}{\Delta y^2} = \omega_{i,j}. \quad (5.41)$$

Use $\omega = 0$ and $\Psi = 0$ as boundary conditions for all external nodes of the grid. Use the principle of global indexing of ω and Ψ in 2D shown in Fig. 3.4. Compute the global index k based on geometrical indices i and j (Fig. 5.3(a)(b)) according to Eq. (3.22).

After obtaining the solution for the stream function, convert it into the velocity field using finite differences (Fig. 5.3(c)) based on Eqs. (5.31), (5.32)

$$v_{x(i,j)} = \frac{\Psi_{i+1,j} - \Psi_{i-1,j}}{2\Delta y}, \quad (5.42)$$

$$v_{y(i,j)} = -\frac{\Psi_{i,j+1} - \Psi_{i,j-1}}{2\Delta x}. \quad (5.43)$$

Note that velocity components should only be computed for the *internal* nodes of the grid (i.e. for 49×39 points).

Plot the results for vorticity (*pcolor*), stream function (*contour*) and velocity (*quiver*) on the same diagram with the vertical axis directed downward (*axis ij*). An example is in **Streamfunction2D.m**.

6

Viscous rheology of rocks

Theory: Solid-state creep of minerals and rocks as the major mechanism of deformation of the Earth's interior. Dislocation and diffusion creep mechanisms. Rheological equations for minerals and rocks. Effective viscosity and its dependence on temperature, pressure and strain rate. Formulation of the effective viscosity from empirical flow laws.
Exercises: Programming viscous rheological equations for computing effective viscosities from empirical flow laws.

6.1 Rock rheology

Rheology is the physical property characterising flow/deformation behaviour of a material. Here we will discuss in more detail the meaning of viscosity and, generally, the rheology of rocks reflecting peculiarities of solid-state creep, which is the major mechanism of rock deformation. Solid-state creep is the ability of crystalline substances to deform irreversibly under applied stresses. Solid-state creep is the major deformation mechanism of the Earth's crust and mantle. Two major types of creep are known: *diffusion creep* and *dislocation creep*.

Diffusion creep is typically dominant at relatively low stresses and results from the diffusion of atoms through the interior (Herring–Nabarro creep) and along the boundaries (Coble creep) of crystalline grains subjected to stresses. As a result of this diffusion, grain deformation leads to bulk rock deformation. Diffusion creep is characterised by a linear (*Newtonian*) relationship between the strain rate $\dot{\gamma}$ and an applied shear stress τ

$$\dot{\gamma} = A_{diff}\tau,\tag{6.1}$$

where A_{diff} is a proportionality coefficient which is independent of stress, but depends on grain size, pressure, temperature, oxygen and water fugacity.

Dislocation creep is dominant at higher stresses and results from migration of dislocations (imperfections in the crystalline lattice structure). Dislocation density strongly depends on stresses, and therefore dislocation creep results in a non-linear (*non-Newtonian*) relationship between the strain rate and deviatoric stress

$$\dot{\gamma} = A_{disl}\tau^n, \tag{6.2}$$

where A_{disl} is a proportionality coefficient which is independent of stress and grain size, but depends on pressure, temperature, oxygen and water fugacity, and $n > 1$ is the stress exponent.

Both diffusion and dislocation creep rheologies are often calibrated from experimental data using a simple parameterised relationship (also called *flow law*) between the applied *differential stress* σ_d (the difference between maximal and minimal applied stress) and the resulting *ordinary strain rate* $\dot{\gamma}$

$$\dot{\gamma} = A_D h^m (\sigma_d)^n \exp\left(-\frac{E_a + V_a P}{RT}\right), \tag{6.3}$$

where P is pressure (Pa), T is temperature (K), R is the gas constant (8.314 J/K/mol), h is grain size (m) and A_D, n, m, E_a and V_a are experimentally determined *rheological parameters*: A_D is the material constant ($Pa^{-n}s^{-1}m^{-m}$), n is the stress exponent ($n = 1$ for diffusion creep and $n > 1$ in case of dislocation creep), m is the grain size exponent, E_a is activation energy (J/mol) and V_a is activation volume (J/Pa). Dislocation creep is grain size independent and therefore $m = 0$ and $h^m = 1$. In contrast, diffusion creep notably depends on grain size and the grain size exponent m is negative (i.e. strain rate increases with decreasing grain size).

6.2 Effective viscosity

In order to use the experimentally parameterised Equation (6.3) in numerical modelling, one needs to reformulate it in terms of an effective viscosity (η_{eff}), written as a function of the second invariant of either the deviatoric stress (σ_{II}), or strain rate ($\dot{\varepsilon}_{II}$). The following general relation which is valid for isotropic, incompressible materials can be used

$$\sigma_{II} = 2\eta_{eff}\dot{\varepsilon}_{II}, \tag{6.4a}$$

or

$$\eta_{eff} = \frac{\sigma_{II}}{2\dot{\varepsilon}_{II}}. \tag{6.4b}$$

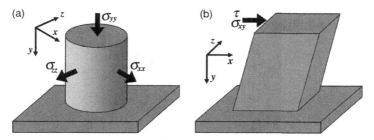

Fig. 6.1 Schematic geometrical relations for an axial compression (a) and a simple shear (b) experiment.

The reformulation should also take into account the type of performed rheological experiments in order to establish the proper relations between σ_{II} and σ_d, as well as between $\dot{\varepsilon}_{II}$ and $\dot{\gamma}$. The resulting expressions for the effective viscosity are as follows

$$\eta_{eff} = F_1 \frac{1}{A_D h^m (\sigma_{II})^{(n-1)}} \exp\left(\frac{E_a + V_a P}{RT}\right), \qquad (6.5a)$$

or

$$\eta_{eff} = F_2 \frac{1}{(A_D)^{1/n} h^{m/n} (\dot{\varepsilon}_{II})^{(n-1)/n}} \exp\left(\frac{E_a + V_a P}{nRT}\right), \qquad (6.5b)$$

where dimensionless coefficients F_1 and F_2 depend on the type of experiments used for calibration of Equation (6.3). Strain-rate-based formulations (6.5b) are more suitable for numerical modelling with viscous (visco-plastic) rheology, while a stress-based formulation is appropriate for visco-elastic (visco-elasto-plastic) problems (Chapter 12, 13). Below, two principal types of rheological experiments are considered in order to derive F_1 and F_2: axial compression (Fig. 6.1(a)) and simple shear (Fig. 6.1(b)). Note that in our consideration we will always use the incompressibility assumption $\mathrm{div}(\bar{v}) = 0$.

In case of an axial compression experiment (Fig. 6.1(a)) the F_1 and F_2 coefficients are obtained as follows (under compression oriented along y axis):

1. Establish the relationship between $\dot{\varepsilon}_{II}$ and $\dot{\gamma}$:

$$\dot{\gamma} = -\frac{\partial v_y}{\partial y} = -\dot{\varepsilon}_{yy},$$

$$\dot{\varepsilon}_{yy} = -\dot{\gamma},$$

$$\dot{\varepsilon}_{xx} = \dot{\varepsilon}_{zz} = -\frac{1}{2}\dot{\varepsilon}_{yy},$$

$$\dot{\varepsilon}_{II} = \sqrt{\frac{1}{2}\dot{\varepsilon}_{xx}^2 + \frac{1}{2}\dot{\varepsilon}_{yy}^2 + \frac{1}{2}\dot{\varepsilon}_{zz}^2} = \sqrt{\frac{3}{4}\dot{\varepsilon}_{yy}^2} = \frac{\sqrt{3}}{2}\dot{\gamma},$$

$$\dot{\gamma} = \frac{2}{\sqrt{3}}\dot{\varepsilon}_{II}. \qquad (6.6)$$

2. Establish the relationship between σ_{II} and σ_d:

$$\sigma'_{xx} = \sigma'_{zz} = -\frac{1}{2}\sigma'_{yy},$$

$$\sigma_d = \sigma_{max} - \sigma_{min} = \sigma_{xx} - \sigma_{yy} = \sigma'_{xx} - \sigma'_{yy} = -\frac{3}{2}\sigma'_{yy},$$

$$\sigma'_{yy} = -\frac{2}{3}\sigma_d,$$

$$\sigma_{II} = \sqrt{\frac{1}{2}\sigma'^2_{xx} + \frac{1}{2}\sigma'^2_{yy} + \frac{1}{2}\sigma'^2_{zz}} = \sqrt{\frac{3}{4}\sigma'^2_{yy}} = \frac{1}{\sqrt{3}}\sigma_d,$$

$$\sigma_d = \sqrt{3}\sigma_{II}. \tag{6.7}$$

3. Rewrite Equation (6.3) in term of $\dot{\varepsilon}_{II}$ and σ_{II} using Equations (6.6) and (6.7)

$$\frac{2}{\sqrt{3}}\dot{\varepsilon}_{II} = A_D h^m (\sqrt{3}\sigma_{II})^n \exp\left(-\frac{E_a + V_a P}{RT}\right), \tag{6.8a}$$

or

$$\dot{\varepsilon}_{II} = \frac{3^{(n+1)/2}}{2} A_D h^m (\sigma_{II})^n \exp\left(-\frac{E_a + V_a P}{RT}\right), \tag{6.8b}$$

or

$$\sigma_{II} = \frac{2^{1/n}}{3^{(n+1)/2n}} (\dot{\varepsilon}_{II})^{1/n} \frac{1}{(A_D)^{1/n} h^{m/n}} \exp\left(\frac{E_a + V_a P}{nRT}\right). \tag{6.8c}$$

4. Write an expression for η_{eff} as a function of the second invariant of deviatoric stress σ_{II} using Equations (6.4b) and (6.8b). Define F_1 by comparing with Equation (6.5a)

$$\eta_{eff} = \frac{\sigma_{II}}{2\dot{\varepsilon}_{II}} = \frac{\sigma_{II}}{2\left[\frac{3^{(n+1)/2}}{2} A_D h^m (\sigma_{II})^n \exp\left(-\frac{E_a + V_a P}{RT}\right)\right]},$$

$$\eta_{eff} = \frac{1}{3^{(n+1)/2}} \times \frac{1}{A_D h^m (\sigma_{II})^{(n-1)}} \exp\left(\frac{E_a + V_a P}{RT}\right),$$

$$F_1 = \frac{1}{3^{(n+1)/2}}. \tag{6.9}$$

5. Write an expression for η_{eff} as a function of the second strain rate invariant $\dot{\varepsilon}_{II}$ using Equations (6.4b) and (6.8c). Define F_2 by comparing with Equation (6.5b)

$$\eta_{eff} = \frac{\sigma_{II}}{2\dot{\varepsilon}_{II}} = \frac{\left[\dfrac{2^{1/n}}{3^{(n+1)/2n}}(\dot{\varepsilon}_{II})^{1/n}\dfrac{1}{(A_D)^{1/n}h^{m/n}}\exp\left(\dfrac{E_a + V_a P}{nRT}\right)\right]}{2\dot{\varepsilon}_{II}},$$

$$\eta_{eff} = \frac{1}{2^{(n-1)/n}3^{(n+1)/2n}} \times \frac{1}{(A_D)^{1/n}h^{m/n}(\dot{\varepsilon}_{II})^{(n-1)/n}}\exp\left(\frac{E_a + V_a P}{nRT}\right),$$

$$F_2 = \frac{1}{2^{(n-1)/n}3^{(n+1)/2n}}. \tag{6.10}$$

In case of simple shear experiments (Fig. 6.1(b)), under applied shear stress $\tau = \frac{1}{2}\sigma_d$ (by the way, τ may also be sometimes used in Eq. (6.3) instead of σ_d for calibrating experiments). In the case of an *xy*-shear $\tau = |\sigma_{xy}|$, the coefficients F_1 and F_2 are obtained as follows:

1. Establish a relationship between $\dot{\varepsilon}_{II}$ and $\dot{\gamma}$

$$\dot{\varepsilon}_{xx} = \dot{\varepsilon}_{yy} = \dot{\varepsilon}_{zz} = \dot{\varepsilon}_{xz} = \dot{\varepsilon}_{yz} = 0,$$

$$\dot{\gamma} = -\frac{\partial v_x}{\partial y} = -2\dot{\varepsilon}_{xy},$$

$$\dot{\varepsilon}_{II} = \sqrt{\dot{\varepsilon}_{xy}^2} = \frac{1}{2}\dot{\gamma},$$

$$\dot{\gamma} = 2\dot{\varepsilon}_{II}. \tag{6.11}$$

2. Establish a relationship between σ_{II} and σ_d

$$\sigma'_{xx} = \sigma'_{yy} = \sigma'_{zz} = \sigma_{xz} = \sigma_{yz} = 0,$$

$$\sigma_d = 2\tau = -2\sigma_{xy},$$

$$\sigma_{II} = \sqrt{\sigma_{xy}^2} = \frac{1}{2}\sigma_d,$$

$$\sigma_d = 2\sigma_{II}. \tag{6.12}$$

3. Rewrite Equation (6.3) in terms of $\dot{\varepsilon}_{II}$ and σ_{II} using Equations (6.11) and (6.12)

$$2\dot{\varepsilon}_{II} = A_D h^m (2\sigma_{II})^n \exp\left(-\frac{E_a + V_a P}{RT}\right), \tag{6.13a}$$

or

$$\dot{\varepsilon}_{II} = 2^{n-1} A_D h^m (\sigma_{II})^n \exp\left(-\frac{E_a + V_a P}{RT}\right), \tag{6.13b}$$

or

$$\sigma_{II} = \frac{1}{2^{(n-1)/n}}(\dot{\varepsilon}_{II})^{1/n}\frac{1}{(A_D)^{1/n}h^{m/n}}\exp\left(\frac{E_a + V_a P}{nRT}\right). \tag{6.13c}$$

4. Write an expression for η_{eff} as a function of the second invariant of deviatoric stress σ_{II} by using Equations (6.4b) and (6.13b). Define F_1 by comparing with Equation (6.5a)

$$\eta_{eff} = \frac{\sigma_{II}}{2\dot{\varepsilon}_{II}} = \frac{\sigma_{II}}{2\left[2^{(n-1)}A_D h^m (\sigma_{II})^n \exp\left(-\dfrac{E_a + V_a P}{RT}\right)\right]},$$

$$\eta_{eff} = \frac{1}{2^n} \times \frac{1}{A_D h^m (\sigma_{II})^{(n-1)}} \exp\left(\frac{E_a + V_a P}{RT}\right),$$

$$F_1 = \frac{1}{2^n}. \tag{6.14}$$

5. Write an expression for η_{eff} as a function of the second strain rate invariant $\dot{\varepsilon}_{II}$ using Equations (6.4b) and (6.13c). Defining F_2 by comparing with Equation (6.5b)

$$\eta_{eff} = \frac{\sigma_{II}}{2\dot{\varepsilon}_{II}} = \frac{\left[\dfrac{1}{2^{(n-1)/n}} (\dot{\varepsilon}_{II})^{1/n} \dfrac{1}{(A_D)^{1/n} h^{m/n}} \exp\left(\dfrac{E_a + V_a P}{nRT}\right)\right]}{2\dot{\varepsilon}_{II}},$$

$$\eta_{eff} = \frac{1}{2^{(2n-1)/n}} \times \frac{1}{(A_D)^{1/n} h^{m/n} (\dot{\varepsilon}_{II})^{(n-1)/n}} \exp\left(\frac{E_a + V_a P}{nRT}\right),$$

$$F_2 = \frac{1}{2^{(2n-1)/n}}. \tag{6.15}$$

It is also important to mention that in rocks and mineral aggregates, both dislocation and diffusion creep occur simultaneously under applied stress, which can be expressed in the following relation for the effective viscosity

$$\frac{1}{\eta_{eff}} = \frac{1}{\eta_{diff}} + \frac{1}{\eta_{disl}}, \tag{6.16}$$

where η_{diff} and η_{disl} is the viscosity for diffusion and dislocation creep respectively, which are defined by Eq. (6.5). This relation follows from the assumption that under condition of constant applied deviatoric stress σ'_{ij}, the strain rate $\dot{\varepsilon}_{ij}$ can be decomposed to the contribution of dislocation ($\dot{\varepsilon}_{ij(disl)}$) and diffusion ($\dot{\varepsilon}_{ij(diff)}$) creep

$$\dot{\varepsilon}_{ij} = \dot{\varepsilon}_{ij(disl)} + \dot{\varepsilon}_{ij(diff)}. \tag{6.17}$$

where

$$\dot{\varepsilon}_{ij} = \frac{\sigma'_{ij}}{2\eta_{eff}}, \quad \dot{\varepsilon}_{ij(disl)} = \frac{\sigma'_{ij}}{2\eta_{disl}}, \quad \dot{\varepsilon}_{ij(diff)} = \frac{\sigma'_{ij}}{2\eta_{diff}}. \tag{6.18}$$

Substituting relations (6.18) into Eq. (6.17) gives formula (6.16) (verify as an exercise).

6.3 Non-Newtonian channel flow

One important consequence of having a fluid with a *non-Newtonian* rheology is that the Poisson equation is *no longer valid* for channel flow (see Chapter 5) and the Stokes equation should be applied instead. Let's analyse such a non-Newtonian channel flow. In the case of a planar channel (Fig. 5.2), the Stokes equation reduces to

$$\frac{\partial \sigma_{xy}}{\partial x} = \frac{\partial P}{\partial y} - \rho g_y.$$

Under the assumption of a non-linear relationship between the stress and strain rate of the form $\dot{\varepsilon}_{xy} = A\sigma_{xy}^n$, the Stokes equation can be solved as follows. First we integrate Eq. (6.19) to obtain stress profile

$$\sigma_{xy} = \int \left(\frac{\partial \sigma_{xy}}{\partial x} \right) dx = \left(\frac{\partial P}{\partial y} - \rho g_y \right) x + C_1, \tag{6.19}$$

where C_1 is our first integration constant. Then we represent the stress as a function of strain rate

$$\sigma_{xy} = \left(\frac{\dot{\varepsilon}_{xy}}{A} \right)^{1/n}, \tag{6.20}$$

and modify Eq. (6.19) using Eq. (6.20) and the relation $\dot{\varepsilon}_{xy} = \frac{1}{2} \frac{\partial v_y}{\partial x}$ to give

$$\left(\frac{\dot{\varepsilon}_{xy}}{A} \right)^{1/n} = \left(\frac{\partial P}{\partial y} - \rho g_y \right) x + C_1,$$

$$\dot{\varepsilon}_{xy} = A \left[\left(\frac{\partial P}{\partial y} - \rho g_y \right) x + C_1 \right]^n,$$

$$\frac{\partial v_y}{\partial x} = 2A \left[\left(\frac{\partial P}{\partial y} - \rho g_y \right) x + C_1 \right]^n. \tag{6.21}$$

Integrating Eq. (6.21) we obtain an expression for vertical velocity profile

$$v_y = \int \left(\frac{\partial v_y}{\partial x} \right) dx = \frac{2A/(n+1)}{\partial P/\partial y - \rho g_y} \left[\left(\frac{\partial P}{\partial y} - \rho g_y \right) x + C_1 \right]^{n+1} + C_2, \tag{6.22}$$

where C_1 and C_2 are integration constants which can be defined from the boundary conditions.

For example, let us assume the following boundary conditions:

$$v_y = v_{y0} \quad \text{when} \quad x = 0,$$

and

$$\sigma_{xy} = 0 \quad \text{when} \quad x = L/2.$$

The latter condition is in fact an equivalent of the condition of symmetric velocity profile

$$v_y = v_{y0} = v_{y1} \quad \text{when} \quad x = L.$$

Then our integration constants become as follows (derive as an exercise based on Eqs. 6.19, 6.22)

$$C_1 = -\left(\frac{\partial P}{\partial y} - \rho g_y\right) \frac{L}{2}, \tag{6.23}$$

$$C_2 = v_{y0} - \frac{2A/(n+1)}{\partial P/\partial y - \rho g_y} C_1^{n+1}. \tag{6.24}$$

Programming exercises and homework

Exercise 6.1
Calculate and visualise a logarithmic viscosity profile across a 100 000 m (100 km) thick mantle lithosphere with a temperature which linearly increases from 400 °C at the top, to 1200 °C at the bottom. Apply the conditions of constant strain rate $\dot{\varepsilon}_{II} = 10^{-14} \mathrm{s}^{-1}$. Take the following mantle rheological parameters: $A_D = 2.5 \times 10^{-17}$ 1/Pa^n/s, $n = 3.5$, $E_a = 532\,000$ J/mol, $m = 0$; $V_a = 0$ (representative for dry olivine under upper mantle conditions). Take into account that all parameters are based on axial compression experiments (Fig. 6.1(a)). An example is in **Viscosity_profile.m**.

Exercise 6.2
Use the flow law from the previous exercise to compute and visualise a logarithmic viscosity map for the mantle in coordinates of temperature (400–1400 °C) and log stress (10^3–10^9 Pa). An example is in **Viscosity_map.m**.

Exercise 6.3
Repeat the previous exercise for an effective mantle viscosity based on Eqs. (6.4) and (6.16) using the flow laws for diffusion and dislocation creep written in the following form (Karato and Wu, 1993).

$$\dot{\varepsilon}_{II} = A \left(\frac{h}{b}\right)^m \left(\frac{\sigma_{II}}{\mu}\right)^n \exp\left(-\frac{E_a + V_a P}{RT}\right), \tag{6.25}$$

Table 6.1 *Flow parameters for olivine (Karato and Wu, 1993).*

Mechanism	Dry	Wet
Dislocation creep		
A, 1/s	3.5×10^{22}	2.0×10^{18}
n	3.5	3
m	0	0
E_a, J/mol	540 000	430 000
V_a, m^3/mol	15×10^{-6} to 25×10^{-6}	10×10^{-6} to 20×10^{-6}
Diffusion creep		
A, 1/s	8.7×10^{15}	5.3×10^{15}
n	1	1
m	−2.5	−2.5
E_a, J/mol	300 000	240 000
V_a, m^3/mol	6×10^{-6}	5×10^{-6}

where $b = 5 \times 10^{-10}$ m is the Burgers vector, $\mu = 8 \times 10^{10}$ Pa, $R = 8.314$ J/K/mol. The other flow law parameters are given in Table 6.1. In your calculations, assume that the grain size, $h = 0.001$ m (1 mm) and $P = 0$. Produce and compare log viscosity plots characteristic for dry and wet conditions. An example is in **Viscosity_comparison.m**.

7

Numerical solutions of the momentum and continuity equations

> **Theory:** Types of numerical grids and their applicability for different differential equations. Staggered, half-staggered and non-staggered grids in one, two and three dimensions. Discretisation of the continuity and Stokes equations on a rectangular grid. Conservative and non-conservative discretisation schemes for Stokes equations. Mechanical boundary conditions and their numerical implementation. No slip and free slip conditions.
>
> **Exercises:** Programming different mechanical boundary conditions. Solving continuity and momentum equations for the case of variable viscosity.

7.1 Grids

As we already discussed, the numerical solution of partial differential equations (PDEs) requires the definition of a grid of nodal points within the numerical model. The choice of this grid depends strongly on the type of equations to be solved. *Discretisation schemes* for these equations will also change with the changing types of numerical grids. The following types of numerical grid exist in numerical geodynamic modelling:

- Depending on the dimension of the problem, the numerical grid can be one, two and three dimensional (1D, 2D, 3D) (Fig. 7.1).
- Depending on the shape of the basic elements, the grid can be rectangular and triangular (Fig. 7.2).
- Depending on the distribution of nodal points, the grid can be regular and non-regular (irregular) (Fig. 7.3).
- Depending on the distribution of different variables within the grid, it can be non-staggered or staggered (Fig. 7.4).

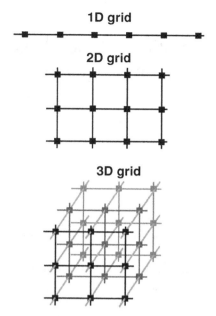

Fig. 7.1 Examples of 1D, 2D and 3D numerical grids.

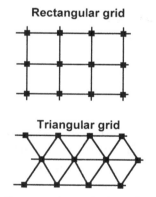

Fig. 7.2 Examples of rectangular and triangular 2D numerical grids.

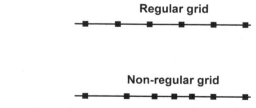

Fig. 7.3 Examples of regular and non-regular 1D numerical grids.

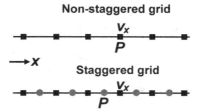

Fig. 7.4 Examples of non-staggered and staggered 1D numerical grids.

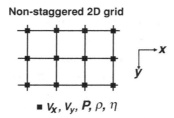

Fig. 7.5 Example of a non-staggered 2D numerical grid.

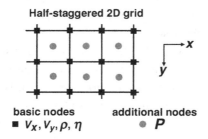

Fig. 7.6 Example of a half-staggered 2D numerical grid.

The simplest grids are non-staggered. All variables are defined at the same nodal points (Fig. 7.5). When using finite differences (FD) with such a grid, all equations are formulated at the same nodal points. Non-staggered grids are the natural choice for solving the Poisson, heat conservation and advective transport equations.

In the staggered grid, several types of nodal points exist (Fig. 7.6) at which different variables are defined. A half-staggered grid in two dimensions (Fig. 7.6) is a combination of a basic non-staggered grid, with an additional set of points defined at the centres of cells formed by the basic grid. Part of the variables is then defined at these additional nodes and not at the basic nodal points. A half-staggered grid is convenient for solving the combination of Stokes and continuity equations

Fully staggered 2D grid

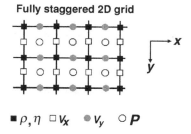

$\blacksquare \rho, \eta \quad \square v_x \quad \bullet v_y \quad \circ P$

Fig. 7.7 Example of a fully staggered 2D numerical grid.

in case of *constant viscosity* when unknown parameters are components of velocity (v_x, v_y) defined at the basic nodes, and pressure (P) defined at the additional nodes (Fig. 7.6). Accordingly, x- and y-Stokes equations (Eq. 5.22) are formulated at basic nodes, whilst the continuity equation is formulated at the additional nodes. Yet, half-staggered grids are less natural for mechanical and thermomechanical problems with *variable viscosity*.

Fully staggered grids are applied in two and three dimensions and consist of a combination of several types of nodal points having different geometrical positions (Fig. 7.7). Different variables are then defined at different nodal points. Different equations are also formulated at different nodal points. Despite the apparent geometrical complexity, fully staggered grids are the most convenient choice for thermomechanical numerical problems with variable viscosity when finite differences are used for solving the continuity, Stokes and temperature equations. Discretisation of thermomechanical equations on the fully staggered grid is very natural and gives simple FD formulas. Also, the accuracy of a numerical solution obtained on a fully staggered grid is notably (up to four times, Fornberg, 1995) higher than that on a non-staggered grid. Therefore: 'think in a fully staggered way!' (I've done this since 2002).

7.2 Discretisation of the equations

The discretisation of the equations depends on the type of numerical grid that is employed. The following discretisation schemes can for example be used to describe the 2D incompressible continuity equation $\dfrac{\partial v_x}{\partial x} + \dfrac{\partial v_y}{\partial y} = 0$ in the case of non-staggered (half-staggered) and a fully staggered 2D grid.

Non-staggered (half-staggered) grid (Fig. 7.8):

$$\frac{v_{x3} - v_{x1} + v_{x4} - v_{x2}}{2\Delta x} + \frac{v_{y2} - v_{y1} + v_{y4} - v_{y3}}{2\Delta y} = 0. \tag{7.1}$$

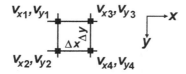

Fig. 7.8 Stencil used for the discretisation of the continuity equation for a 2D non-staggered grid.

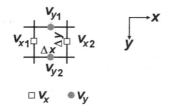

Fig. 7.9 Stencil used for the discretisation of the continuity equation for a 2D fully staggered grid.

Fully-staggered grid (Fig. 7.9):

$$\frac{v_{x2} - v_{x1}}{\Delta x} + \frac{v_{y2} - v_{y1}}{\Delta y} = 0. \tag{7.2}$$

In both examples, the continuity equation is formulated at the centre of a cell that forms an elementary volume of the numerical grid. However, in the case of a fully staggered grid, the *stencil* (i.e. pattern of points) used to represent the equation on the grid is more natural for the continuity equation (cf. Figs. 7.8 and 7.9) and therefore the resulting finite-difference formula involves fewer unknowns (cf. Eqs. 7.1 and 7.2).

7.3 Conservative finite differences

In order to discretise the Stokes equations in the case of variable viscosity, *conservative finite differences* should be used. Such finite differences provide conservation of stresses between nodal points and thus allow a correct numerical solution. Below, examples of non-conservative and conservative finite differences for the 1D incompressible Stokes equation ($\frac{\partial \sigma'_{xx}}{\partial x} - \frac{\partial P}{\partial x} = 0$, where $\sigma'_{xx} = 2\eta \frac{\partial v_x}{\partial x}$) are compared for 1D staggered grid (Fig. 7.10).

An **erroneous non-conservative** FD formulation of Stokes equation for two basic nodes 2 and 3 can be constructed, for example, (we can 'arrive' at this formula assuming erroneously that all we need to do is to use Eq. 5.24 and use a

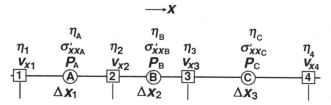

Fig. 7.10 Example of a 1D staggered grid used for the discretisation of Stokes equation with a variable viscosity. 1, 2, 3, 4 are basic nodes of the grid where Stokes equations are formulated. A, B, C, are additional (stress) nodes of the grid where the stresses are formulated.

different viscosity for each of the different nodes)

$$\text{node 2:} \quad 2\eta_2 \frac{(v_{x3} - v_{x2})/\Delta x_2 - (v_{x2} - v_{x1})/\Delta x_1}{(\Delta x_1 + \Delta x_2)/2} - \frac{P_B - P_A}{(\Delta x_1 + \Delta x_2)/2} = 0,$$

(7.3a)

$$\text{node 3:} \quad 2\eta_3 \frac{(v_{x4} - v_{x3})/\Delta x_3 - (v_{x3} - v_{x2})/\Delta x_2}{(\Delta x_2 + \Delta x_3)/2} - \frac{P_C - P_B}{(\Delta x_2 + \Delta x_3)/2} = 0,$$

(7.3b)

which implicitly means that formulation of the deviatoric stress σ'_{xxB} in Stokes equation written for nodes 2 and 3 are different due to different viscosities η_2 and η_3:

$$\text{node 2:} \quad \sigma'_{xxB} = 2\eta_2 \frac{v_{x3} - v_{x2}}{\Delta x_2},$$

$$\text{node 3:} \quad \sigma'_{xxB} = 2\eta_3 \frac{v_{x3} - v_{x2}}{\Delta x_2}.$$

This implies that stress is *not conserved* and artificially 'jumps' between the two nodes in response to the changing viscosity.

On the other hand a ***proper conservative*** FD formulation of Stokes equation is

$$\text{node 2:} \quad \frac{2\eta_B (v_{x3} - v_{x2})/\Delta x_2 - 2\eta_A (v_{x2} - v_{x1})/\Delta x_1}{(\Delta x_1 + \Delta x_2)/2} - \frac{P_B - P_A}{(\Delta x_1 + \Delta x_2)/2} = 0,$$

(7.4a)

$$\text{node 3:} \quad \frac{2\eta_C (v_{x4} - v_{x3})/\Delta x_3 - 2\eta_B (v_{x3} - v_{x2})/\Delta x_2}{(\Delta x_2 + \Delta x_3)/2} - \frac{P_C - P_B}{(\Delta x_2 + \Delta x_3)/2} = 0,$$

(7.4b)

which means that formulations of the deviatoric stress σ'_{xxB}, used in the Stokes equation written for nodes 2 and 3 are the same:

$$\sigma'_{xxB} = 2\eta_B \frac{v_{x3} - v_{x2}}{\Delta x_2},$$

implying that stress is *conserved* between the two nodes.

Thus, the conservative FD formulation is based on the following three formal rules:

(1) The Stokes equation is initially discretised in terms of stress components for the *basic nodes* of the grid (cf. nodes 2, 3, Fig. 7.10),

$$\text{node 2:} \quad \frac{\sigma'_{xxB} - \sigma'_{xxA}}{(\Delta x_1 + \Delta x_2)/2} - \frac{P_B - P_A}{(\Delta x_1 + \Delta x_2)/2} = 0,$$

$$\text{node 3:} \quad \frac{\sigma'_{xxC} - \sigma'_{xxB}}{(\Delta x_2 + \Delta x_3)/2} - \frac{P_C - P_B}{(\Delta x_2 + \Delta x_3)/2} = 0.$$

(2) These stress components are formulated at the *additional (stress) nodes* of the grid (cf. nodes A, B, C, Fig. 7.10)

$$\text{node A:} \quad \sigma'_{xxA} = 2\eta_A \frac{v_{x2} - v_{x1}}{\Delta x_1},$$

$$\text{node B:} \quad \sigma'_{xxB} = 2\eta_B \frac{v_{x3} - v_{x2}}{\Delta x_2},$$

$$\text{node C:} \quad \sigma'_{xxC} = 2\eta_C \frac{v_{x4} - v_{x3}}{\Delta x_3}.$$

Note that we have to use viscosity values η_A, η_B and η_C for the additional nodes (A, B, C) where the stress components are defined. If these values are not directly available at these locations, they can be computed by e.g. arithmetic averaging of the known viscosity values from the basic nodes (1, 2, 3, 4)

$$\eta_A = \frac{\eta_1 + \eta_2}{2},$$

$$\eta_B = \frac{\eta_2 + \eta_3}{2},$$

$$\eta_C = \frac{\eta_3 + \eta_4}{2}.$$

(3) Identical formulations of the stress components are used for the Stokes equation at the different basic nodes.

Applying these rules in 2D, the following conservative formulations can be derived for *x*- and *y*-Stokes equations (derive them as an exercise based on above principles)

x-Stokes equation (Fig. 7.11):

$$2\frac{\sigma'_{xxB} - \sigma'_{xxA}}{\Delta x_1 + \Delta x_2} + \frac{\sigma_{xy2} - \sigma_{xy1}}{\Delta y_2} - 2\frac{P_B - P_A}{\Delta x_1 + \Delta x_2} = -\frac{\rho_1 + \rho_2}{2}g_x \quad (7.5)$$

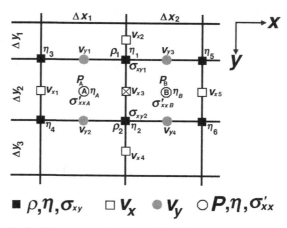

Fig. 7.11 Stencil of a 2D staggered grid used for discretisation of x-Stokes equation with a variable viscosity. The crossed square corresponds to the node at which the x-Stokes equation is formulated.

where

$$\sigma_{xy1} = 2\eta_1 \left(\frac{v_{x3} - v_{x2}}{\Delta y_1 + \Delta y_2} + \frac{v_{y3} - v_{y1}}{\Delta x_1 + \Delta x_2} \right),$$

$$\sigma_{xy2} = 2\eta_2 \left(\frac{v_{x4} - v_{x3}}{\Delta y_2 + \Delta y_3} + \frac{v_{y4} - v_{y2}}{\Delta x_1 + \Delta x_2} \right),$$

$$\sigma'_{xxA} = 2\eta_A \frac{v_{x3} - v_{x1}}{\Delta x_1},$$

$$\sigma'_{xxB} = 2\eta_B \frac{v_{x5} - v_{x3}}{\Delta x_2}.$$

If values of viscosity for the centres of cells (A, B) are not known, they can be computed by e.g. arithmetic averaging of the known viscosity values from the basic nodes (cf. black squares in Fig. 7.11)

$$\eta_A = \frac{\eta_1 + \eta_2 + \eta_3 + \eta_4}{4},$$

$$\eta_B = \frac{\eta_1 + \eta_2 + \eta_5 + \eta_6}{4}.$$

y-Stokes equation (Fig. 7.12):

$$2\frac{\sigma'_{yyB} - \sigma'_{yyA}}{\Delta y_1 + \Delta y_2} + \frac{\sigma_{yx2} - \sigma_{yx1}}{\Delta x_2} - 2\frac{P_B - P_A}{\Delta y_1 + \Delta y_2} = -\frac{\rho_1 + \rho_2}{2}g_y \qquad (7.6)$$

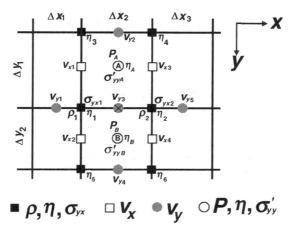

Fig. 7.12 Stencil of a 2D staggered grid used for the discretisation of the y-Stokes equation with a variable viscosity. The crossed circle corresponds to the node at which the y-Stokes equation is formulated.

where

$$\sigma_{yx1} = 2\eta_1 \left(\frac{v_{y3} - v_{y1}}{\Delta x_1 + \Delta x_2} + \frac{v_{x2} - v_{x1}}{\Delta y_1 + \Delta y_2} \right),$$

$$\sigma_{yx2} = 2\eta_2 \left(\frac{v_{y5} - v_{y3}}{\Delta x_2 + \Delta x_3} + \frac{v_{x4} - v_{x3}}{\Delta y_1 + \Delta y_2} \right),$$

$$\sigma'_{yyA} = 2\eta_A \frac{v_{y3} - v_{y2}}{\Delta y_1},$$

$$\sigma'_{yyB} = 2\eta_B \frac{v_{y4} - v_{y3}}{\Delta y_2}.$$

If values of viscosity for the centres of cells (A, B) are not known, they can be computed by e.g. arithmetic averaging of the known viscosity values from the basic nodes (cf. black squares in Fig. 7.12)

$$\eta_A = \frac{\eta_1 + \eta_2 + \eta_3 + \eta_4}{4},$$

$$\eta_B = \frac{\eta_1 + \eta_2 + \eta_5 + \eta_6}{4}.$$

Note that in all the examples above, the medium is assumed to be incompressible and that the deviatoric stress components σ'_{ij}, are formulated via the strain rate tensor components $\dot{\varepsilon}_{ij}$, and viscosity according to the Equations (4.10) and (5.12) as follows

$$\sigma'_{ij} = 2\eta \, \dot{\varepsilon}_{ij},$$

where $\dot{\varepsilon}_{ij} = \dfrac{1}{2}\left(\dfrac{\partial v_i}{\partial x_j} + \dfrac{\partial v_j}{\partial x_i}\right)$, i and j are coordinate indices and x_i and x_j are spatial coordinates.

7.4 Boundary conditions

As we have discussed before, in order to obtain numerical solutions, boundary conditions have to be defined. Mechanical boundary conditions depend on the type of numerical problem which is studied. The following boundary conditions are often used in geomodelling:

(1) free slip
(2) no slip
(3) free surface
(4) fast erosion
(5) infinity-like (external free slip, external no slip, Winkler basement)
(6) prescribed velocity (moving boundary)
(7) periodic
(8) combined conditions

(1) A free slip condition requires that the normal velocity component on the boundary is zero and the two other components do not change across the boundary (this condition also implies *zero shear strain rates and stresses* along the boundary). For example, for the boundary orthogonal to the x axis, the free slip condition is formulated as follows

$$v_x = 0,\tag{7.7a}$$

$$\frac{\partial v_y}{\partial x} = \frac{\partial v_z}{\partial x} = 0.\tag{7.7b}$$

(2) A no slip condition requires all velocity components on the boundary to be zero

$$v_x = v_y = v_z = 0.\tag{7.8}$$

(3) A free surface condition requires both shear and normal stresses at the boundary to be zero

$$\sigma'_{ij} = 0.\tag{7.9}$$

This condition allows the surface to be deformed. Numerical implementation of this condition requires programming either a deformable grid following the deforming surface, or the introduction of a low viscosity layer above the free surface (e.g. Schmeling *et al.*, 2008). This will be further discussed in Chapters 11 and 17.

(4) A fast erosion condition requires that all velocity components do not change across the boundary. For example, for the upper model boundary which is orthogonal to the y

axis, fast erosion condition is formulated as follows

$$\frac{\partial v_x}{\partial y} = \frac{\partial v_y}{\partial y} = \frac{\partial v_z}{\partial y} = 0. \tag{7.10}$$

This condition corresponds to an infinitely fast erosion/deposition at the upper free surface. This surface is always kept horizontal since erosion is so fast that all mountains are instantaneously scraped off and are deposited in valleys. This condition also ensures that the mass in the model is conserved.

(5) An infinity-like condition either mimics the absence of a boundary, or implies that this boundary is located very far away. External free slip (Burg and Gerya, 2005; Gerya *et al.*, 2008b) implies that conditions (7.7a) and (7.7b) are satisfied at a parallel boundary located at the distance ΔL from the actual boundary of the model and the velocity gradient between these two boundaries is constant. For example, the external free slip condition applied to the lower boundary of the model which is orthogonal to the y-axis is:

$$\frac{\partial v_y}{\partial y} \Delta L + v_y = 0, \tag{7.11a}$$

$$\frac{\partial v_x}{\partial y} = 0. \tag{7.11b}$$

$$\frac{\partial v_z}{\partial y} = 0. \tag{7.11c}$$

By analogy, the external no slip condition at the same boundary is formulated as

$$\frac{\partial v_x}{\partial x} \Delta L + v_x = 0, \tag{7.12a}$$

$$\frac{\partial v_y}{\partial y} \Delta L + v_y = 0, \tag{7.12b}$$

$$\frac{\partial v_z}{\partial z} \Delta L + v_z = 0. \tag{7.12c}$$

Note that relations (7.11) and (7.12) insure global conservation of mass in the computational domain, despite the presence of an 'open' boundary.

The Winkler's pliable basement (e.g. Burov *et al.*, 2001; Yamato *et al.*, 2008) assumes isostatic equilibrium at the model bottom and implies that the model overlies an infinite space filled with an inviscid fluid having a small density contrast (e.g. 10 kg/m^3) within the lower part of the model. This is a sort of free surface condition, applied at the lower boundary of the model, which is typically placed in the mantle asthenosphere. It assumes that the material underneath the boundary has zero viscosity and moves infinitely fast.

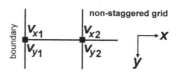

Fig. 7.13 Stencil of a 2D non-staggered grid used for the formulation of no slip and free slip boundary conditions.

(6) The prescribed velocity condition implies non-zero velocity at a model boundary. When velocity is prescribed orthogonal to the boundary (inward/outward flow), then a compensating outward/inward velocity should be prescribed on the other model boundary(ies) in order to insure mass conservation in the model. In this case, the model boundaries can also be displaced with time in response to the material movement (moving boundary condition, Chapter 17).

(7) Periodic boundary conditions are typically established for paired parallel lateral boundaries of a model and prescribe that all material properties as well as pressure and velocity fields at both sides of each boundary are identical. From a physical point of view, this implies that these two boundaries are open and that flow leaving the model through one boundary immediately re-enters through the opposite side. This condition is often used in mantle convection modelling to simulate part of a spherical/cylindrical shell with a convecting mantle (or mimic it, in Cartesian coordinates).

(8) Combined conditions represent a mixture between several types of boundary conditions.

All of the described boundary conditions can be time dependent. This could particularly imply that the physical location of the boundary condition may be a function of time (Chapter 17). Boundary conditions can also be applied *inside the model*.

We will now concentrate on the numerical implementation of the most common, and most simple, free slip and no slip conditions (we will discuss several examples of more complex conditions in Chapters 16 and 17). The numerical implementation of boundary conditions depends on the type of grid.

Non-staggered grid (Fig. 7.13):

$$\text{free slip,} \quad v_{x1} = 0, \quad v_{y1} = v_{y2}, \tag{7.13}$$

$$\text{no slip,} \quad v_{x1} = 0, \quad v_{y1} = 0. \tag{7.14}$$

Staggered grid (Fig. 7.14):

$$\text{free slip,} \quad v_{x1} = 0, v_{y1} = v_{y2}, \tag{7.15}$$

$$\text{no slip,} \quad v_{x_1} = 0, v_{y1} = v_{y2} \frac{\Delta x_1}{2\Delta x_1 + \Delta x_2}. \tag{7.16}$$

Condition for the vertical velocity v_{y1} implies that zero vertical velocity on the boundary $v_{yb} = 0$ (Fig. 7.14) is linearly extrapolated from vertical velocities in two

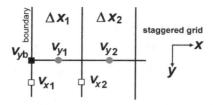

Fig. 7.14 Stencil of a 2D staggered grid used for the formulation of non slip and free slip boundary conditions.

internal nodes as

$$v_{yb} = v_{y1} + (v_{y1} - v_{y2})\frac{\Delta x_1}{\Delta x_1 + \Delta x_2} = 0.$$

7.5 Indexing of unknowns

Another very important issue, in relation to solving the Stokes and continuity equations on a fully staggered grid, is the indexing of the unknowns. This is particularly relevant when the system of linear equations (global matrix) formulated with finite differences is solved with Gaussian elimination as discussed in Chapter 3. This is a somewhat boring subject but it is extremely important to understand it properly. Remember, *90% of the bugs in your code are made with the indexing* (Bug Rule 5 in the Introduction). Both the possibility of obtaining the solution and the amount of computational work will strongly depend on the method used to index the unknowns (P, v_x and v_y) on the staggered grid. One of the optimal ways of numbering is illustrated in Fig. 7.15. The following rules are used for this indexing:

1. Staggered nodes of the grid are related in a uniform manner (cf. dashed arrows in Fig. 7.15) to the basic nodes of the grid formed by the intersections of the horizontal and vertical gridlines. The same amount of unknowns should be related to each basic node (P, v_x and v_y give us three unknowns per basic node, Fig. 7.15). If no staggered node with respective unknowns can be found inside the grid then 'ghost unknowns' are added (e.g., outside the grid) *for the uniformity of numbering*. Such 'ghost unknowns' are not used in the numerical solution and are set to zeros (cf. nodes with over-lined indices in Fig. 7.15).
2. Basic nodes of the grid are indexed (cf. indices in italics in Fig. 7.15) from 1 to $N_x \times N_y$, where N_x and N_y is the basic grid resolution (i.e. number of gridlines) in the horizontal and vertical direction, respectively (6 and 5, respectively in Fig. 7.15). The index numbering increase in the direction of the smallest amount of gridlines (i.e. in the vertical direction in Fig. 7.15) to ensure a minimal amount of computational work for

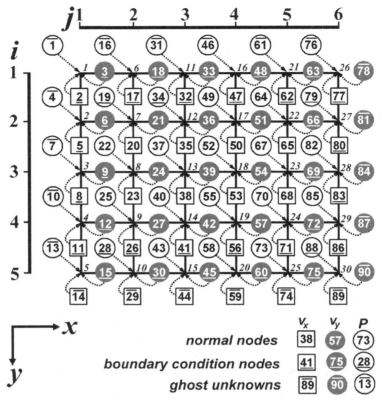

Fig. 7.15 Indexing of unknowns (P, v_x and v_y) for a 6×5 2D staggered grid. Dotted arrows show relations of staggered nodes to the basic nodes of the grid. Underlined indices denote parameters for which boundary conditions are defined. Over-lined indices correspond to the 'ghost unknowns' introduced for the uniformity of numbering and are not used in the numerical solution (these unknowns are set to zero). Indices in italics correspond to numbering of basic nodes. Indices i and j correspond to numbering of gridlines in the vertical and horizontal directions, respectively.

global matrix inversion

$$in_{node} = (j - 1) \times N_y + i, \tag{7.17}$$

where in_{node} is the index for the given node, i and j are the indices of the vertical and horizontal gridlines which are intersecting at the node (Fig. 7.15).

3. Unknown parameters attached to each basic node are indexed from 1 to $N_x \times N_y \times 3$, according to the increasing basic node index

$$in_P = 3 \times in_{node} - 2, \tag{7.17a}$$

$$in_{v_x} = 3 \times in_{node} - 1, \tag{7.17b}$$

$$in_{v_y} = 3 \times in_{node}, \tag{7.17c}$$

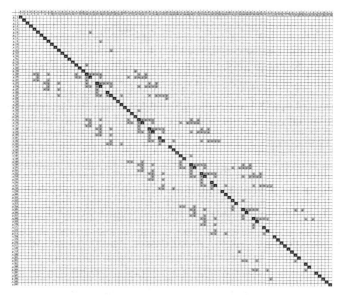

Fig. 7.16 Sparse tri-diagonal global 90 × 90 matrix corresponding to the staggered grid shown in Fig. 7.15. Coloured rectangles correspond to non-zero coefficients. Black rectangles are located on the main diagonal of the matrix. Note the absence of coefficients on the main diagonal when the continuity equation is solved in order to obtain pressure.

where in_P, in_{v_x} and in_{v_y} are indices for P, v_x and v_y related to the given basic node in_{node}.

As one can see on Fig. 7.15, index in_P located in the centre of a grid cell is always bigger than the indices in_{v_x} and in_{v_y} for the velocity nodes surrounding this cell. This is an important condition since pressure is obtained by solving the continuity equation. An incompressible continuity equation does not initially contain pressure and the solution is guaranteed by the order of processing during the inversion of the global matrix (Fig. 7.16).

The continuity equation (Eq. (1.28)) for pressure in a given cell (cf. Fig. 7.9) is processed after all Stokes equations (Eq. 5.18) for all surrounding v_x- and v_y-nodes (cf. Figs. 7.11, 7.12) which contain pressure in the cell. For this reason, the pressure values in the four corners of the grid (cf. pressure nodes 19, 28, 79 and 88 in Fig. 7.15) cannot be computed from the continuity equation since these pressure nodes are not used in the formulation of the momentum equation and are surrounded by v_x and v_y nodes for which boundary conditions are formulated (cf. nodes with underlined indices in Fig. 7.15). For these pressure nodes, some boundary conditions (e.g. horizontal symmetry) should be used instead. Additionally, the pressure value should be given at one of the remaining pressure nodes (cf. P-node

34 in Fig. 7.15) of the grid such that absolute pressure values can be defined through the pressure gradients present in the momentum equation. We thus need to formulate boundary conditions for five pressure nodes.

The idea about obtaining a solution for pressure by formulating an equation that does not contain pressure may sound bizarre. Let us just convince ourselves on a simple example. We apply Gaussian elimination (Eqs. 3.12–3.18) to the analogue system of three equations with variables v_x, v_y and P having the global indices 1, 2 and 3 respectively

$$\text{equation A1 (formulated for } v_x): \quad 2v_x + 4v_y + 2P = 10,$$
$$\text{equation A2 (formulated for } v_y): \quad 3v_x + 9v_y + 6P = 21,$$
$$\text{equation A3 (formulated for } P): \quad v_x + 3v_y = 5.$$

Dividing all equations by a number to normalise the coefficient of v_x, we get

$$\text{equation B1:} \quad v_x + 2v_y + P = 5,$$
$$\text{equation B2:} \quad v_x + 3v_y + 2P = 7,$$
$$\text{equation B3:} \quad v_x + 3v_y = 5.$$

Eliminating v_x by subtracting equation B1 from B2 and B3 yields

$$\text{equation B1:} \quad v_x + 2v_y + P = 5,$$
$$\text{equation C2:} \quad v_y + P = 2,$$
$$\text{equation C3:} \quad v_y - P = 0.$$

Note that after the elimination operation, P indeed appears in equation C3. Eliminating v_y by subtracting equation C2 from C3 yields

$$\text{equation B1:} \quad v_x + 2v_y + P = 5,$$
$$\text{equation C2:} \quad v_y + P = 2,$$
$$\text{equation D3:} \quad -2P = -2,$$

Obtaining the solution for P from equation D3

$$P = -2/(-2) = 1.$$

Obtaining solution for v_y from equation C2

$$v_y = 2 - P = 2 - 1 = 1.$$

Obtaining the solution for v_x from equation B1

$$v_x = 5 - 2v_y - P = 5 - 2 - 1 = 2.$$

So it works and we obtained all the required solutions, including one for P by the Gaussian elimination (we could also have inferred this from the fact that we have three equations for the three unknowns v_x, v_y and P). Of course, it would be impossible to apply (*without reordering of equations and re-indexing of unknowns*) the same Gaussian elimination approach if the order of global indices (and respectively equations) was inverted, i.e.

> equation 1 (formulated for P): $3v_y + v_x = 5$,
>
> equation 2 (formulated for v_y): $6P + 9v_y + 3v_x = 21$,
>
> equation 3 (formulated for v_x): $2P + 4v_y + 2v_x = 10$.

It should be mentioned, however, that more advanced direct solvers (including '\' command of MATLAB) have internal re-ordering procedures that allow one to obtain solutions in the latter case as well.

An alternative staggered grid structure uses *ghost velocity nodes* to formulate boundary condition equations for v_x and v_y velocity components (Fig. 7.17). These equations are not explicitly added to the global matrix and therefore the respective unknowns for the ghost nodes are not indexed. Instead, these boundary condition equations are used in an implicit manner by taking them into account while discretising the momentum and continuity equations for the internal nodes of the grid located next to the ghost nodes. Values of v_x and v_y velocities in the ghost nodes are recovered from the boundary condition equations after obtaining a global solution for internal nodes. The manner of indexing unknowns is shown in Fig. 7.17 and again is based on a convention that relates staggered nodes to (part of) the basic nodes of the grid, numbered as

$$in_{node} = (j - 1) \times (N_y - 1) + i,$$

where in_{node} is the index for the given node, i and j are indices for the vertical and horizontal gridlines intersecting at the considered node (Fig. 7.17). Note that only $(N_x - 1) \times (N_y - 1)$ basic nodes are numbered. Unknowns attached for each numbered basic node are indexed respectively from 1 to $(N_x - 1) \times (N_y - 1) \times 3$ according to the increasing basic node index

$$in_{v_x} = 3 \times in_{node} - 2,$$
$$in_{v_y} = 3 \times in_{node} - 1,$$
$$in_P = 3 \times in_{node}.$$

This way of indexing again ensures that the index for pressure in a cell is bigger than the indices for all the velocities surrounding the cell. The main advantage of this staggered grid structure is that there is a smaller number of unknowns in the global matrix equal to $(N_x - 1) \times (N_y - 1) \times 3$. In addition, the boundary condition for

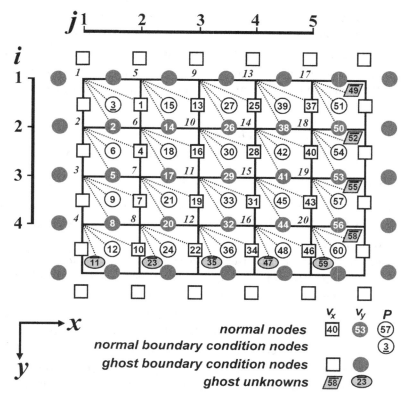

Fig. 7.17 Indexing of unknowns (P, v_x and v_y) for 6×5 2D staggered grid with the use of 'ghost velocity nodes'. Dotted lines show the relationship of the staggered nodes to the basic nodes of the grid (note that only part of the nodes is numbered). Underlined indices denote parameters for which boundary conditions are defined. Over-lined indices correspond to the 'ghost unknowns', which are introduced for the uniformity of numbering but not used in the numerical solution (these unknowns are simply set to zero). Empty symbols correspond to the ghost nodes where velocity boundary condition equations are defined. These boundary condition equations are implicitly used in the numerical solution when formulating momentum and continuity equations for the internal (numbered) nodes of the grid located next to the ghost nodes. Indices i and j correspond to the numbering of gridlines in the vertical and horizontal directions, respectively.

pressure only needs to be defined in a single cell (see first cell in the top left corner in Fig. 7.17). The implementation of ghost nodes, however, requires additional programming as the formulation of the momentum and continuity equations for the 'near-boundary' nodes must be done in a special manner in order to take into account variable velocity boundary conditions.

Compared to the stream function approach (Chapter 5), the use of a *pressure–velocity formulation* (also called *primitive variable formulation*) requires three times more equations for the same grid resolution. The advantages are, however, that

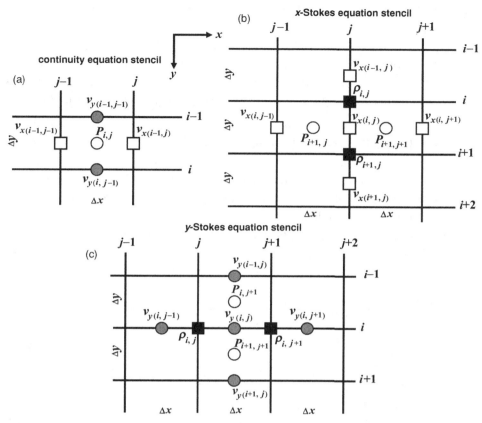

Fig. 7.18 Stencils used for the discretisation of the continuity (a) and Stokes (b),(c) equations on a 2D regular staggered grid for models with constant viscosity based on the pressure–velocity formulation. Indexing of gridlines corresponds to basic (density) nodal points. Indexing of different unknowns is made according to Fig. 7.15.

the solution for pressure is obtained and that lower (second) order derivatives are required compared to the fourth-order derivatives required by the stream function-based equation (5.36).

Programming exercises and homework

Exercise 7.1

Solve the momentum and continuity equations with finite differences for the case of constant viscosity using a pressure–velocity formulation (Eqs. (5.22), (1.28)) on a regular staggered grid (Figs. 7.15, 7.18) of 31×21 points. Program the numerical model for buoyancy-driven flow in a purely vertical gravity field ($g_x = 0$, $g_y = 10$ m/s^2 in Eq. 5.22) for a density structure with two vertical layers (3200 kg/m^3 and 3300 kg/m^3 for the left and right layer, respectively). The model size is 1000×1500 km (i.e. $1\,000\,000 \times 1\,500\,000$ m). Use constant viscosity $\eta = 10^{21}$ Pa s for the entire model.

Compose the matrix of coefficients $\{L\}$, the right-hand side vector $\{R\}$ and obtain the solution vector $\{S\}$ with the direct solver ($S = L\backslash R$) using relations (7.17) and Fig. 7.15 for global indexing of the unknowns. Boundary conditions for the velocity are free slip on all boundaries (Eq. 7.15). Boundary conditions for pressure at the four corners is horizontal symmetry $\frac{\partial P}{\partial x} = 0$, i.e. pressure is the same as in the neighbouring cells in the horizontal direction. Boundary condition for pressure in the additional cell ($i = 2$ and $j = 3$, Fig. 7.15) is $P = 0$. Do not forget that the equations for all ghost unknowns ($P = 0, v_x = 0, v_x = 0$) should also be added to the matrix of coefficients $\{L\}$ and right-hand side $\{R\}$ (see over-lined numbers in Fig. 7.15 for indexing of these unknowns).

The finite-difference representation of the momentum and continuity equations in 2D are formulated by analogy to Eqs. (3.20), (7.5) and (7.6) and follows (Fig. 7.18)

$$\eta \frac{v_{x(i,j-1)} - 2v_{x(i,j)} + v_{x(i,j+1)}}{\Delta x^2} + \eta \frac{v_{x(i-1,j)} - 2v_{x(i,j)} + v_{x(i+1,j)}}{\Delta y^2}$$
$$- K_{cont} \frac{P'_{i+1,j+1} - P'_{i+1,j}}{\Delta x} = 0, \tag{7.18}$$

$$\eta \frac{v_{y(i,j-1)} - 2v_{y(i,j)} + v_{y(i,j+1)}}{\Delta x^2} + \eta \frac{v_{y(i-1,j)} - 2v_{y(i,j)} + v_{y(i+1,j)}}{\Delta y^2}$$
$$- K_{cont} \frac{P'_{i+1,j+1} - P'_{i,j+1}}{\Delta y} = -g_y \frac{\rho_{i,j} + \rho_{i,j+1}}{2}, \tag{7.19}$$

$$K_{cont} \left(\frac{v_{x(i-1,j)} - v_{x(i-1,j-1)}}{\Delta x} + \frac{v_{y(i,j-1)} - v_{y(i-1,j-1)}}{\Delta y} \right) = 0, \tag{7.20}$$

where the scaled pressure $P' = \dfrac{P}{K_{cont}}$ and coefficients $K_{cont} = \dfrac{2\eta}{\Delta x + \Delta y}$ are used to ensure relatively uniform (i.e. not differing by many orders of magnitude) coefficients in all equations. Similarly, both the left and right parts of all boundary condition equations should be multiplied by a coefficient $K_{bond} = \dfrac{4\eta}{(\Delta x + \Delta y)^2}$. This scaling helps obtaining an accurate solution when direct solvers are applied for models with large viscosities. After solving the global matrix for the velocity components and scaled pressure P', the unscaled pressure is computed as $P = P' K_{cont}$.

After obtaining a solution for velocity on staggered nodes, compute the velocity components v_x and v_y for the *internal* basic nodes of the grid (i.e. for 29×19 points) by averaging velocity from surrounding staggered nodes (use the two nearest velocity nodes for each component). Plot the results for pressure (*pcolor*) and

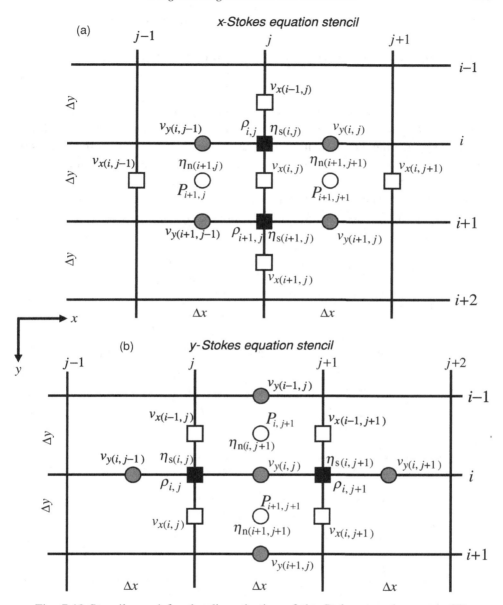

Fig. 7.19 Stencils used for the discretisation of the Stokes equations on a 2D regular staggered grid for models with variable viscosity based on pressure–velocity formulation. Indexing of gridlines corresponds to basic (density) nodal points. Indexing of different unknowns is made according to Fig. 7.15.

velocity (*quiver*) on the same diagram with the vertical axis directed downward (*axis ij*).

An example is in **Stokes_continuity_constant_viscosity.m**.

Exercise 7.2

Modify the previous example for a variable viscosity case. Use different viscosities for the two vertical layers (10^{20} Pa s and 10^{22} Pa s for the left and right layer, respectively). Define the viscosity both at the basic nodes (η_s) and at the centres of cells (η_n) (Fig. 7.19). Indexing for the later one should be the same as for pressure nodes (Fig. 7.15).

The finite-difference representation of the continuity equation is the same as before (Eq. 7.20). Stokes equations are formulated by analogy with Eqs. (7.5) and (7.6) as follows (Fig. 7.19)

$$2\eta_{n(i+1,j+1)}\frac{v_{x(i,j+1)} - v_{x(i,j)}}{\Delta x^2} - 2\eta_{n(i+1,j)}\frac{v_{x(i,j)} - v_{x(i,j-1)}}{\Delta x^2}$$

$$+ \eta_{s(i+1,j)}\left(\frac{v_{x(i+1,j)} - v_{x(i,j)}}{\Delta y^2} + \frac{v_{y(i+1,j)} - v_{y(i+1,j-1)}}{\Delta x \Delta y}\right) - \eta_{s(i,j)} \qquad (7.21)$$

$$\times \left(\frac{v_{x(i,j)} - v_{x(i-1,j)}}{\Delta y^2} + \frac{v_{y(i,j)} - v_{y(i,j-1)}}{\Delta x \Delta y}\right) - K_{cont}\frac{P'_{i+1,j+1} - P'_{i+1,j}}{\Delta x} = 0$$

$$2\eta_{n(i+1,j+1)}\frac{v_{y(i+1,j)} - v_{y(i,j)}}{\Delta y^2} - 2\eta_{n(i,j+1)}\frac{v_{y(i,j)} - v_{y(i-1,j)}}{\Delta y^2}$$

$$+ \eta_{s(i,j+1)}\left(\frac{v_{y(i,j+1)} - v_{y(i,j)}}{\Delta x^2} + \frac{v_{x(i,j+1)} - v_{x(i-1,j+1)}}{\Delta x \Delta y}\right) \qquad (7.22)$$

$$- \eta_{s(i,j)}\left(\frac{v_{y(i,j)} - v_{y(i,j-1)}}{\Delta x^2} + \frac{v_{x(i,j)} - v_{x(i-1,j)}}{\Delta x \Delta y}\right)$$

$$- K_{cont}\frac{P'_{i+1,j+1} - P'_{i,j+1}}{\Delta y} = -g_y\frac{\rho_{i,j} + \rho_{i,j+1}}{2}$$

$$K_{cont}\left(\frac{v_{x(i-1,j)} - v_{x(i-1,j-1)}}{\Delta x} + \frac{v_{y(i,j-1)} - v_{y(i-1,j-1)}}{\Delta y}\right) = 0, \qquad (7.23)$$

where $K_{cont} = \dfrac{2\eta_{min}}{\Delta x + \Delta y}$ and $K_{bond} = \dfrac{4\eta_{min}}{(\Delta x + \Delta y)^2}$ (computed with the value of minimal viscosity in the model, η_{min}) are again used for scaling the coefficients.

An example is in **Stokes_continuity_variable_viscosity.m**.

8

The advection equation
and marker-in-cell method

Theory: Advection equation. Solution methods for continuous and discontinuous variables. Eulerian schemes: upwind differences, higher-
order schemes, flux corrected transport (FCT). Lagrangian schemes:
marker-in-cell method. Runge–Kutta advection schemes. Numerical
interpolation schemes between markers and nodes.
Exercises: Programming of various advection schemes and markers

8.1 Advection equation

As we already know, the deformation of a continuum changes the spatial distribution
of physical properties. These changes can be described by the *advection equation*.
For a scalar function (A) at an Eulerian point, this equation is written as follows:

$$\frac{\partial A}{\partial t} = -\vec{v} \cdot \text{grad}(A) \tag{8.1a}$$

or in 3D,

$$\frac{\partial A}{\partial t} = -v_x \left(\frac{\partial A}{\partial x} \right) - v_y \left(\frac{\partial A}{\partial y} \right) - v_z \left(\frac{\partial A}{\partial z} \right) \tag{8.1b}$$

For a Lagrangian point, the following advection equation relates changes in its
coordinates with material velocities $\vec{v} = (v_x, v_y, v_z)$;

$$\frac{Dx_i}{Dt} = v_i, \tag{8.2a}$$

or in 3D,

$$\frac{Dx}{Dt} = v_x, \tag{8.2b}$$

$$\frac{Dy}{Dt} = v_y, \tag{8.2c}$$

$$\frac{Dz}{Dt} = v_z, \tag{8.2d}$$

where i is a coordinate index and x_i is a spatial coordinate.

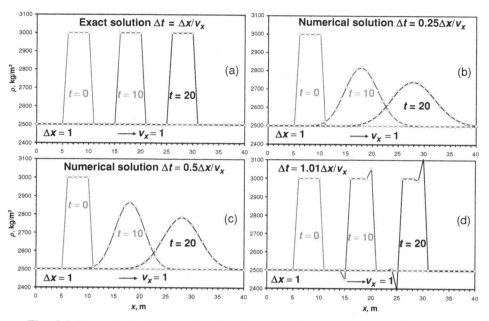

Fig. 8.1 Numerical solutions for the advection of a sharp density wave (i.e. perturbation) obtained using upwind finite differences with different size time steps, Δt.

8.2 Eulerian advection methods

The advection equations appear trivial, but this is an apparent simplicity. Solving them numerically often causes problems (headaches ...). One such problem is the numerical diffusion of sharp gradients during advection. To illustrate the problem, we solve

$$\frac{\partial \rho}{\partial t} = -v_x \left(\frac{\partial \rho}{\partial x} \right), \tag{8.3}$$

on a regular Eulerian 1D grid with constant spacing between the nodes ($\Delta x = 1$), for the case of constant material velocity ($v_x = 1$) by applying upwind differences (Fig. 8.2(a))

$$\rho_i^{t+\Delta t} = \rho_i^t - v_x \Delta t \frac{\rho_i^t - \rho_{i-1}^t}{\Delta x}, \tag{8.4}$$

and using different values of the time step Δt. Figure 8.1(b) demonstrates how a sharp density wave (i.e. perturbation) smoothes out (*diffuses*) during the numerical solution of a 1D advection equation obtained using upwind finite differences.

The intensity of the numerical diffusion depends on the number of numerical steps performed and not on the absolute time of advection (Fig. 8.1(b), (c)).

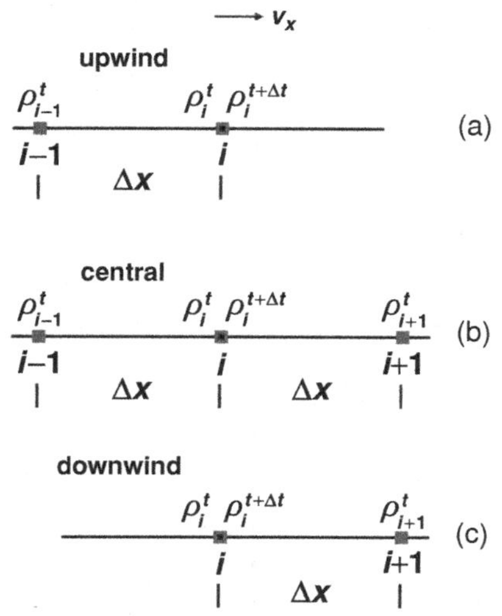

Fig. 8.2 Stencil of a 1D grid used for the discretisation of the Eulerian advection equation with upwind (a), central (b) and downwind (c) differences.

Therefore, smaller time steps give more numerical diffusion for the same total duration of advection (compare Fig. 8.1(b) and (c)). On the other hand, to ensure stability of the numerical solution, the time step should be sufficiently small to satisfy the time limitation condition

$$\Delta t \leq \frac{\Delta x}{v_x},$$ (8.5a)

which states that the material should not move for more than one grid step per one time step (this is also called *the Courant criteria*). Condition (8.5a) should be satisfied *in every Eulerian point* to prevent oscillations (Fig. 8.1(d)). Therefore, one should use the minimal ratio between local grid step (which can be variable) and local flow velocity (which can be variable as well) found in a model. In 2D and 3D, this limitation may be even stricter

$$\Delta t \leq \frac{\Delta x}{2v_x}$$ (8.5b)

to prevent artificial oscillations from appearing in the numerical solutions.

Figure 8.3 shows the progress of numerical diffusion during only four time steps. Upwind differences (Eq. 8.4, Fig. 8.2(a)) are applied along with a constant time step ($\Delta t = 1/2\Delta x/v_x$), which results in the FD formulation:

$$\rho_i^{t+\Delta t} = \rho_i^t - \frac{\rho_i^t - \rho_{i-1}^t}{2},$$ (8.6)

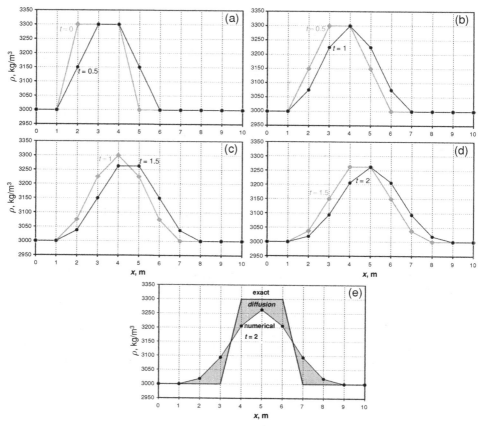

Fig. 8.3 Progress (a)–(d) in the development of the numerical diffusion during four time steps ($\Delta t = 1/2\Delta x/v_x = 0.5\,\mathrm{s}$) in the case of a 1D advection of a square density wave, with constant velocity ($v_x = 1$ m/s) on a regular Eulerian grid ($\Delta x = 1$ m).

which results in strong numerical diffusion (Fig. 8.3(e)). The diffusion term, hidden in Eq. 8.4, can be 'exposed' by analysing this equation on the basis of symmetrical *central differences*

$$\left(\frac{\partial \rho}{\partial x}\right)_i^{central} = \frac{\rho_{i+1}^t - \rho_{i-1}^t}{2\Delta x}.$$

Eq. 8.4 can be reformulated as follows

$$\rho_i^{t+\Delta t} = \rho_i^t - \frac{v_x \Delta t}{2\Delta x}\left(2\rho_i^t - 2\rho_{i-1}^t + \rho_{i+1}^t - \rho_{i+1}^t\right), \tag{8.7a}$$

$$\frac{\rho_i^{t+\Delta t} - \rho_i^t}{\Delta t} = -v_x \frac{\rho_{i+1}^t - \rho_{i-1}^t}{2\Delta x} + \frac{v_x \Delta x}{2}\left(\frac{\rho_{i-1}^t - 2\rho_i^t + \rho_{i+1}^t}{\Delta x^2}\right), \tag{8.7b}$$

taking that $\left(\dfrac{\partial^2 \rho}{\partial x^2}\right)_i = \dfrac{\rho_{i-1}^t - 2\rho_i^t + \rho_{i+1}^t}{\Delta x^2}$ we obtain

$$\left(\frac{\partial \rho}{\partial t}\right)_i = -v_x \left(\frac{\partial \rho}{\partial x}\right)_i^{central} + D\left(\frac{\partial^2 \rho}{\partial x^2}\right)_i, \tag{8.7c}$$

where $D = v_x \Delta x/2$ is the numerical diffusion coefficient. This means that upwind differences inherently contain numerical diffusion compared to central differences. On the other hand, applying central and downwind differences for solving the Eulerian advection equation:

central FD (Fig. 8.2(b)): $\quad \rho_i^{t+\Delta t} = \rho_i^t - v_x \Delta t \dfrac{\rho_{i+1}^t - \rho_{i-1}^t}{2\Delta x}, \tag{8.8}$

downwind FD (Fig. 8.2(c)): $\quad \rho_i^{t+\Delta t} = \rho_i^t - v_x \Delta t \dfrac{\rho_{i+1}^t - \rho_i^t}{\Delta x}, \tag{8.9}$

results in a strong oscillation of the numerical solution compared to the exact one (Fig. 8.4(b),(c)), which is even more dramatic than the numerical diffusion problem which is characteristic for upwind differences (compare Fig. 8.4(a),(b),(c)). The oscillations are caused by the erroneous evaluation of the spatial derivative of density. With both central and downwind differences (in contrast to upwind differences), we take into account the density distribution in the *outgoing material flow* which is useless for predicting density distribution in the *incoming material flow*.

One way to minimise numerical diffusion for the Eulerian advection equation is to use higher-order numerical schemes such as the *Flux-Corrected Transport* (FCT) algorithm (Boris and Book, 1973). The FCT is a conservative *shock-capturing* scheme that can, in particular, be used for solving the advection equation. An FCT algorithm consists of two stages, (I) a transport stage and (II) a flux-corrected anti-diffusion stage. The numerical diffusion errors introduced in the first stage are corrected by the anti-diffusion stage. The implementation of the FCT is more complex than for upwind FD and is based on more nodal points (Fig. 8.5). For the case of constant velocity and constant grid spacing, the algorithm of updating advected parameters (e.g. density) on an Eulerian grid is as follows;

(I) Transport stage – using highly diffusive mass-conservative advection scheme (such as upwind differences, Eqs. (8.4), (8.7)) to obtain preliminary values of density ($\tilde{\rho}_i^{t+\Delta t}$) at the next time instant $t + \Delta t$

$$\tilde{\rho}_i^{t+\Delta t} = \rho_i^t - v_x \Delta t \frac{\rho_{i+1}^t - \rho_{i-1}^t}{2\Delta x} + D\Delta t \left(\frac{\rho_{i-1}^t - 2\rho_i^t + \rho_{i+1}^t}{\Delta x^2}\right), \tag{8.10}$$

where $D = \dfrac{\Delta x^2}{\Delta t}\left[\dfrac{1}{8} + \dfrac{1}{2}\left(\dfrac{v_x \Delta t}{\Delta x}\right)^2\right]$ is a numerical diffusion coefficient.

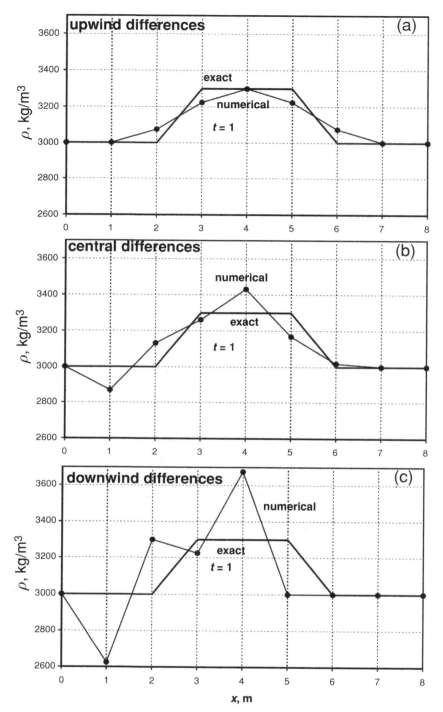

Fig. 8.4 Deviation of the numerical solution from the exact one after only two time steps in the case of upwind (a) central (b) and downwind (c) finite differences for advection of a square wave. The best results are, indeed, obtained by upwind differences, while downwind FD yield strong numerical oscillations. Parameters are as in Fig. 8.3.

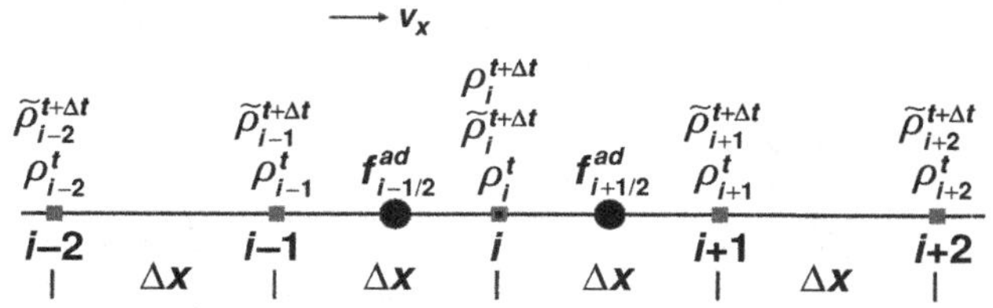

Fig. 8.5 Stencil of a 1D grid used for discretisation of the Eulerian advection equation with FCT algorithm.

(II) Anti-diffusion stage – correcting the numerical diffusion introduced during the transport stage by defining anti-diffusive fluxes to the left ($f_{i-1/2}^{ad}$), and to the right ($f_{i+1/2}^{ad}$) of the nodal point i,

$$\rho_i^{t+\Delta t} = \tilde{\rho}_i^{t+\Delta t} - f_{i+1/2}^{ad} + f_{i-1/2}^{ad}, \tag{8.11}$$

where

$$f_{i-1/2}^{ad} = S_{i-1/2} \max \left[0, \min \left(S_{i-1/2} \Delta \tilde{\rho}_{i-3/2}^{t+\Delta t}, \frac{1}{8} \left| \Delta \tilde{\rho}_{i-1/2}^{t+\Delta t} \right|, S_{i-1/2} \Delta \tilde{\rho}_{i+1/2}^{t+\Delta t} \right) \right],$$

$$f_{i+1/2}^{ad} = S_{i+1/2} \max \left[0, \min \left(S_{i+1/2} \Delta \tilde{\rho}_{i-1/2}^{t+\Delta t}, \frac{1}{8} \left| \Delta \tilde{\rho}_{i+1/2}^{t+\Delta t} \right|, S_{i+1/2} \Delta \tilde{\rho}_{i+3/2}^{t+\Delta t} \right) \right],$$

$$\Delta \tilde{\rho}_{i-3/2}^{t+\Delta t} = \tilde{\rho}_{i-1}^{t+\Delta t} - \tilde{\rho}_{i-2}^{t+\Delta t},$$

$$\Delta \tilde{\rho}_{i-1/2}^{t+\Delta t} = \tilde{\rho}_i^{t+\Delta t} - \tilde{\rho}_{i-1}^{t+\Delta t},$$

$$\Delta \tilde{\rho}_{i+1/2}^{t+\Delta t} = \tilde{\rho}_{i+1}^{t+\Delta t} - \tilde{\rho}_i^{t+\Delta t},$$

$$\Delta \tilde{\rho}_{i+3/2}^{t+\Delta t} = \tilde{\rho}_{i+2}^{t+\Delta t} - \tilde{\rho}_{i+1}^{t+\Delta t},$$

$$S_{i-1/2} = \text{sign} \left(\Delta \tilde{\rho}_{i-1/2}^{t+\Delta t} \right),$$

$$S_{i+1/2} = \text{sign} \left(\Delta \tilde{\rho}_{i+1/2}^{t+\Delta t} \right).$$

Here, sign(A) is a function that gives -1, 0 and 1 if $A < 0$, $A = 0$ and $A > 0$, respectively.

The FCT algorithm stencil (Fig. 8.5) involves five nodal points in a 1D grid, compared to two nodal points in the case of upwind differences (Fig. 8.2(a)). FCT stabilises the numerical diffusion (i.e. advected wave shape stops changing after some amount of time steps) and gives noticeably better results compared to upwind differences (compare Fig. 8.6(a) and (b)). These results, however, are still not perfect and depend on the exact shape of the advected structures (e.g. triangular waves are subjected to a noticeable decrease in amplitude compared to square

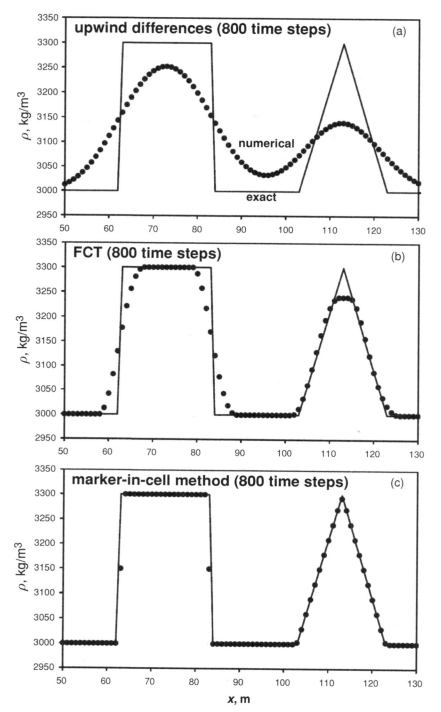

Fig. 8.6 1D advection of a square and triangular density waves with upwind differences (a), FCT (b) and marker-in-cell method (c). The results have been obtained with the programs **Upwind_1D.m**, **FCT_1D.m** and **Markers_1D.m**.

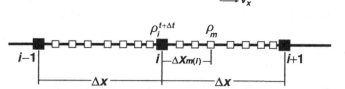

Fig. 8.7 Stencil of 1D Eulerian–Lagrangian grid used for advection with the marker-in-cell technique. Mobile Lagrangian markers (open squares) move with a prescribed velocity v_x and carry information on material density. At every time step, the density at the Eulerian nodes is interpolated from markers found within two grid cells around the nodes.

waves, Fig. 8.6(b)). It should also be mentioned that various existing higher-order Eulerian schemes such as FCT do not eliminate numerical diffusion completely, but rather stabilise it to a certain acceptable level (Fig. 8.6(b)).

8.3 Marker-in-cell techniques

If numerical diffusion needs to be strongly minimised, Lagrangian and Eulerian–Lagrangian advection algorithms can be used. In geomodelling, for example, very accurate advection of non-diffusive properties such as rock type (composition) with strongly discontinuous (e.g., layering) distribution in space is often required. One of the most popular methods in this case is to combine the use of Lagrangian advecting points (*markers, tracers or particles*) with an immobile, Eulerian grid (e.g., Woidt, 1978; Christensen, 1982; Schmeling, 1987; Weinberg and Shmelling, 1992). In this approach, properties are initially distributed on a large amount of Lagrangian points that are advected according to a given/computed velocity field. The advected material properties (e.g. density) are then interpolated from the displaced Lagrangian points to the Eulerian grid (Fig. 8.7) by using a weighted-distance averaging such as the following linear interpolation formula

$$\rho_i^{t+\Delta t} = \frac{\sum\limits_{m} \rho_m w_{m(i)}}{\sum\limits_{m} w_{m(i)}},$$

$$\tag{8.12}$$

$$w_{m(i)} = 1 - \frac{\Delta x_{m(i)}}{\Delta x},$$

where $w_{m(i)}$ is the weight of the m-th marker for the i-th node, $\Delta x_{m(i)}$ denotes the distance from the m-th marker, to the i-th node. In 1D, the density at an Eulerian node is interpolated only with the markers found in the two surrounding cells (i.e.

within one grid space from the node). For a more local interpolation, fewer markers found within the surrounding cell (e.g. half grid space from the node) can be used.

This method is called the *marker-in-cell* (MIC) technique. Obviously, the results of pure advection (e.g. of density, Fig. 8.6(c)) with MIC are not subjected to numerical diffusion (with the exception of non-accumulating interpolation errors between Lagrangian markers and Eulerian nodes) since markers always retain their original density values and only change positions with time.

In order to move a Lagrangian marker A, different advection schemes can be used. The most simple is the first-order accurate advection scheme

$$x_A^{t+\Delta t} = x_A^t + v_{xA}\Delta t, \tag{8.13a}$$
$$y_A^{t+\Delta t} = y_A^t + v_{yA}\Delta t, \tag{8.13b}$$
$$z_A^{t+\Delta t} = z_A^t + v_{zA}\Delta t, \tag{8.13c}$$

where x_A^t, y_A^t and z_A^t are the coordinates of marker A at the current time (t); $x_A^{t+\Delta t}$, $y_A^{t+\Delta t}$ and $z_A^{t+\Delta t}$ are the coordinates of the same marker at the next moment in time ($t + \Delta t$); v_{xA}, v_{yA} and v_{zA} are components of the material velocity vector at the point A, at time t. The velocity of the Lagrangian point A can significantly change during the displacement if there is a strong spatial variation of the velocity field. In this case, a first-order advection scheme will not be very accurate. This problem can be rectified, by either using smaller time steps (Δt), or by using higher-order advection schemes.

One of the most popular in geomodelling is the *Runge–Kutta advection scheme*, given as

$$x_A^{t+\Delta t} = x_A^t + v_x^{eff}\Delta t, \tag{8.14a}$$
$$y_A^{t+\Delta t} = y_A^t + v_y^{eff}\Delta t, \tag{8.14b}$$
$$z_A^{t+\Delta t} = z_A^t + v_z^{eff}\Delta t, \tag{8.14c}$$

where v_x^{eff}, v_y^{eff} and v_z^{eff} are components of the effective material velocity vector for the point A over the period between current (t) and next ($t + \Delta t$) moments of time. Components of the effective material velocity are computed by using the material velocity at several different points in space (varying from 2 to 4, depending on the order of the scheme).

The *second-order Runge–Kutta scheme* uses two points (A and B)

$$v_x^{eff} = v_{xB}, \tag{8.15a}$$
$$v_y^{eff} = v_{yB}, \tag{8.15b}$$
$$v_z^{eff} = v_{zB}, \tag{8.15c}$$

where the coordinates of point B are computed as

$$x_B = x_A^t + v_{xA} \frac{\Delta t}{2}, \; y_B = y_A^t + v_{yA} \frac{\Delta t}{2}, \; z_B = z_A^t + v_{zA} \frac{\Delta t}{2}. \quad (8.15d)$$

The *third-order Runge–Kutta scheme* uses three points (A, B and C)

$$v_x^{eff} = \frac{1}{6}(v_{xA} + 4v_{xB} + v_{xC}), \quad (8.16a)$$

$$v_y^{eff} = \frac{1}{6}(v_{yA} + 4v_{yB} + v_{yC}), \quad (8.16b)$$

$$v_z^{eff} = \frac{1}{6}(v_{zA} + 4v_{zB} + v_{zC}), \quad (8.16c)$$

where the coordinates of points B and C are computed as

$$x_B = x_A^t + v_{xA} \frac{\Delta t}{2}, \; y_B = y_A^t + v_{yA} \frac{\Delta t}{2}, \; z_B = z_A^t + v_{zA} \frac{\Delta t}{2}, \quad (8.16d)$$
$$x_C = x_A^t + (2v_{xB} - v_{xA})\Delta t, \; y_C = y_A^t + (2v_{yB} - v_{yA})\Delta t,$$
$$z_C = z_A^t + (2v_{zB} - v_{zA})\Delta t. \quad (8.16e)$$

And finally, the *classical fourth-order Runge–Kutta scheme* uses four points (A, B, C and D)

$$v_x^{eff} = \frac{1}{6}(v_{xA} + 2v_{xB} + 2v_{xC} + v_{xD}), \quad (8.17a)$$

$$v_y^{eff} = \frac{1}{6}(v_{yA} + 2v_{yB} + 2v_{yC} + v_{yD}), \quad (8.17b)$$

$$v_z^{eff} = \frac{1}{6}(v_{zA} + 2v_{zB} + 2v_{zC} + v_{zD}), \quad (8.17c)$$

where the coordinates of points B, C and D are computed as

$$x_B = x_A^t + v_{xA} \frac{\Delta t}{2}, \; y_B = y_A^t + v_{yA} \frac{\Delta t}{2}, \; z_B = z_A^t + v_{zA} \frac{\Delta t}{2}, \quad (8.17d)$$
$$x_C = x_A^t + v_{xB} \frac{\Delta t}{2}, \; y_C = y_A^t + v_{yB} \frac{\Delta t}{2}, \; z_C = z_A^t + v_{zB} \frac{\Delta t}{2}, \quad (8.17e)$$
$$x_D = x_A^t + v_{xC}\Delta t, \; y_D = y_A^t + v_{yC}\Delta t, \; z_D = z_A^t + v_{zC}\Delta t. \quad (8.17f)$$

The last advection scheme is very accurate in space (fourth-order accuracy) but less accurate in time (first-order accuracy) if the velocity field for the current moment of time (i.e. A-configuration velocity field) is used for B, C and D points. An alternative algorithm is to use the Runge–Kutta scheme in both space and time: all markers are first displaced to their B points and then material properties are re-interpolated to Eulerian nodes and a new B-configuration velocity field is computed by solving the momentum and continuity equations, this B-velocity field is then used for moving markers to their C points for computing a C-configuration velocity

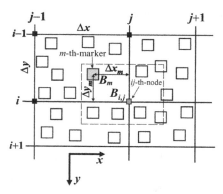

Fig. 8.8 Stencil of 2D grid used for the interpolation of physical properties from the markers to the nodes. The dashed boundary indicates the area from which markers are used for interpolating properties to node (i, j) in the case of a local interpolation scheme.

field etc. Obviously this approach is computationally more expensive. Therefore, in the case of non-steady flows (i.e. with strong variations in velocity field with time), first-order advection schemes with smaller time step are indeed more efficient.

Various interpolation schemes can be used to interpolate physical properties (e.g., density, viscosity, heat capacity) from the Lagrangian markers to the Eulerian nodes. The following standard first-order accurate bilinear scheme is often used to calculate an interpolated value of a parameter $B_{(i, j)}$ for the ij-th-node using values (B_m) assigned to all markers found in the four surrounding cells (Fig. 8.8)

$$B_{i,j} = \frac{\sum\limits_m B_m w_{m(i,j)}}{\sum\limits_m w_{m(i,j)}},$$

$$w_{m(i,j)} = \left(1 - \frac{\Delta x_m}{\Delta x}\right) \times \left(1 - \frac{\Delta y_m}{\Delta y}\right),$$

(8.18)

where $w_{m(i, j)}$ represents a statistical weight of the m-th-marker at the ij-th-node; Δx_m and Δy_m are the distances from the m-th-marker to the ij-th-node. It is worth mentioning that the use of a higher-order interpolation scheme (e.g. Fornberg, 1995) produces undesirable numerical fluctuations in scalar, vector and tensor properties interpolated at the proximity of sharp transitions. This scenario frequently occurs in geodynamic models, hence the first-order interpolation is preferred. A more local interpolation from markers to nodes can again be obtained by using fewer markers located within a limited (e.g. half grid space, see dashed boundary in Fig. 8.8) range of the vertical and horizontal distances around an Eulerian node.

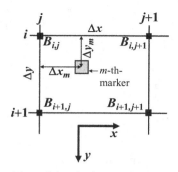

Fig. 8.9 Stencil of a 2D grid used for the interpolation of physical properties from nodes to markers.

Such a local interpolation of properties from markers to nodes, has in many cases, an effect similar to increasing the Eulerian grid resolution.

Interpolation of physical parameters (e.g., velocity, pressure) from the Eulerian nodes to Lagrangian markers and other geometrical points is also commonly required. One of the simplest methods is to use values of the physical parameter B, defined at the four Eulerian nodes surrounding a given marker (or any other geometric point). An effective value of the parameter B for the m-th-marker can then be calculated using the first-order bilinear interpolation scheme as follows

$$B_m = B_{i,j}\left(1 - \frac{\Delta x_m}{\Delta x}\right)\left(1 - \frac{\Delta y_m}{\Delta y}\right) + B_{i,j+1}\frac{\Delta x_m}{\Delta x}\left(1 - \frac{\Delta y_m}{\Delta y}\right)$$

$$+ B_{i+1,j}\left(1 - \frac{\Delta x_m}{\Delta x}\right)\frac{\Delta y_m}{\Delta y} + B_{i+1,j+1}\frac{\Delta x_m \Delta y_m}{\Delta x \Delta y},$$

(8.19)

where B_m denotes the value of the parameter B for the m-th-marker. If one uses an irregularly spaced grid, the indices of the four nodal points that surround a given marker (Fig. 8.9) cannot be calculated directly, as in the case of a regular grid. In this case, a bisection procedure (Fig. 8.10) can be used to define the indices of the two nearest vertical and horizontal grid lines that bound the cell which contain the marker (Fig. 8.9). An alternative, faster approach (which however requires additional memory) is to store the unique cell index for each marker and update it at every time step by checking only the nearest grid lines.

It should also be mentioned that performing interpolation between nodes and markers does introduce numerical diffusion. This problem becomes particularly significant when it is required to interpolate the same time-dependent physical parameter (e.g., temperature or stress), back and forth between the markers and nodes. Such diffusion can indeed be minimised by interpolating *incremental values* before bisection

1 2 3 4   5   6 7 8 9 10 11   12   13 14 15
R | | | | |□| | | | | | | | | | |L
                  M

1 2 3 4   5   6 7 8
R | | | |□| | |L   after check 1
     M

4   5   6 7 8
R |□| | |L   after check 2
     M

4   5   6
R |□| after check 3
  M L

5   6
|□| after check 4

Fig. 8.10 Finding two nearest grid lines around the marker (open square) by a bisection algorithm in the case of an irregularly spaced grid. The search starts by having the first (line 1) and last (line 15) lines of the grid as leftmost ($L = 1$) and rightmost ($R = 15$) limits for the bisection. At each check, the middle line $M = (L + R)/2$ is defined and its horizontal position is compared with that of the marker. Depending on the result of the comparison, line M becomes either L (when M is to the left of the marker) or R (when M is to the right of the marker). The check continues until the difference in the index between L and R become one. The required number of checks (n) is given by the relation $2^n = N$, where N is number of grid lines.

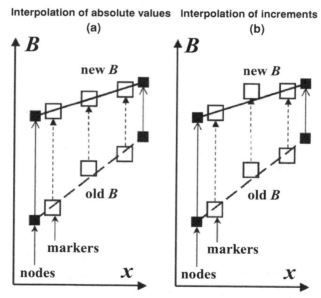

Fig. 8.11 Interpolation of the absolute values of the parameter B (a) and its increments (b) from nodes to markers. Note that the interpolation of increments does not smooth out subgrid variations of B onto the markers.

and not absolute values of the parameter from the nodes to the markers (Figs. 8.9, 8.11).

$$B_m^{t+\Delta t} = B_m^t + \left(B_{i,j}^{t+\Delta t} - B_{i,j}^t\right)\left(1 - \frac{\Delta x_m}{\Delta x}\right)\left(1 - \frac{\Delta y_m}{\Delta y}\right) + \left(B_{i,j+1}^{t+\Delta t} - B_{i,j+1}^t\right)$$

$$\times \frac{\Delta x_m}{\Delta x}\left(1 - \frac{\Delta y_m}{\Delta y}\right) + \left(B_{i+1,j}^{t+\Delta t} - B_{i+1,j}^t\right)\left(1 - \frac{\Delta x_m}{\Delta x}\right)\frac{\Delta y_m}{\Delta y}$$

$$+ \left(B_{i+1,j+1}^{t+\Delta t} - B_{i+1,j+1}^t\right)\frac{\Delta x_m \Delta y_m}{\Delta x \Delta y}, \tag{8.20}$$

where indices t and $t + \Delta t$ correspond to the current and next time instant, respectively.

Programming exercises and homework

Exercise 8.1
Program and compare the simple Eulerian advection schemes (upwind, central and downwind FD, Eqs. (8.4), (8.8), (8.9) in 1D for the case illustrated in Fig. 8.6. The model resolution is 151 nodal points. Other parameters are the same as in Fig. 8.6. An example is in **Upwind_1D.m**.

Exercise 8.2
Program and compare the FCT method and the marker-in-cell schemes in 1D for the same model. Use 5 markers per cell (200 markers for the entire model) and 'recycle' markers which leave the model from one side by adding them to the other side (periodic boundary condition). To recycle markers, update their coordinate as follows: for markers that leave the model through the right boundary and appear at the left boundary, set

$$x_m^{recycled} = x_m - L, \tag{8.21a}$$

For markers that leave the model through the left boundary and enter it at the right boundary, set

$$x_m^{recycled} = x_m + L, \tag{8.21b}$$

where x_m, $x_m^{recycled}$ are marker coordinates before and after recycling, respectively. Use the following formula to define the index j (smallest index is 1) of the nearest node to the left of the marker from its coordinate

$$j = \text{int}\left(\frac{x_m}{\Delta x}\right) + 1, \tag{8.22}$$

where x_m is the marker coordinate, Δx is the nodal (Eulerian) grid space, int() is a function which returns the integer part of a value. Two possible MATLAB implementations of Eq. (8.22) are

$$i = \text{double}(\text{int}16(xm/dx - 0.5)) + 1;$$
$$i = \text{double}(\text{int}16(xm/dx + 0.5));$$

Examples are in **FCT_1D.m** and **Markers_1D.m**.

Exercise 8.3

Modify the previous marker-based code by introducing non-uniform distances between nodal points and markers and by using a bisection algorithm (Fig. 8.10) to define indices of the nearest nodes. Prescribe a slightly variable velocity on the nodal points and interpolate it to markers when displacing them. An example is given in **Markers_1Dirregular.m** associated with this chapter.

Exercise 8.4

Modify the 2D model with the variable viscosity (Exercise 7.2) by including the advection of density and viscosity fields with markers. Create a grid of 200×300 markers with small (up to half of the marker grid distance) random displacements (*rand*) relative to regular positions. Randomisation is introduced to prevent the opening of big gaps between markers during the simulation which often occurs if regular marker grids are used (e.g. due to pure-shear-related stretching that increases the distances between markers). Save the horizontal and vertical coordinate for every marker and assign it with the density and viscosity depending on the position in either the left or the right layer. An alternative approach, which requires less memory, is to assign every marker with a *material type index* depending on the initial position (i.e. 1 and 2 for the left and right layer, respectively). Then the marker density and viscosity (and potentially any other material-dependent property) can be estimated based on the material type index. Interpolate the marker density and viscosity (η_s in Fig. 7.19) to the basic nodes of the grid using Eq. (8.18) (write a loop over the markers and add the density and viscosity of each marker to the four surrounding nodes, Fig. 8.9). For computing i and j indices for the upper left node next to the marker (Fig. 8.9), apply Eq. (8.22) separately for each coordinate. Compute the viscosity for the centres of cells (η_n in Fig. 7.19) by averaging the viscosity from the four surrounding basic nodes (an alternative way is to interpolate this viscosity directly from markers based on Eq. 8.18). After obtaining a velocity field, define a time step in such a manner that the marker displacements do not exceed half the grid spacing. Interpolate the v_x and v_y velocity components for the markers from the staggered nodes (Eq. 8.19) and displace them using a first-order accurate scheme (Eq. (8.13)). Note that for staggered nodes, Eq. (8.22) is modified since these nodes are displaced by half of the grid distance relative to the basic ones

$$i = \text{int}\left(\frac{y_m - \Delta y/2}{\Delta y}\right) + 1 \quad \text{(for } v_x \text{ nodes)}, \tag{8.23a}$$

$$j = \text{int}\left(\frac{x_m - \Delta x/2}{\Delta x}\right) + 1 \quad \text{(for } v_y \text{ nodes)}. \tag{8.23b}$$

Use a uniform velocity interpolation procedure for all markers, including those outside of the grids of staggered v_x- and v_y-nodes (e.g. above the first line of v_x-nodes, within half of vertical grid step from the top of the model, Fig. 7.15). In the latter case, the marker will be located outside the 4-node cell (Fig. 8.9) and normalised distances $\Delta x_m/\Delta x$ and $\Delta y_m/\Delta y$ to the upper left node of the cell in Eq. (8.19) can be either negative or larger than 1. The velocity will indeed be interpolated properly, and will be consistent with the boundary conditions. After displacing all the markers, go to the next time step and interpolate the density and viscosity to the nodes using the new markers positions. An example is in **Stokes_Continuity_Markers.m**.

Exercise 8.5
Update the 2D code developed above to use a fourth-order Runge–Kutta scheme (Eqs. (8.14), (8.17)) for marker displacement. An example is in **Stokes_Continuity_Markers_Runge_Kutta.m**.

9

The heat conservation equation

> **Theory:** Fourier's law of heat conduction. Heat conservation equation and its derivation. Radioactive, viscous and adiabatic heating and their relative importance. Heat conservation equation for the case of a constant thermal conductivity and its relation to the Poisson equation. Analytical examples: steady geotherm and steady temperature profile in case of channel flow.
>
> **Exercises:** Computing shear heating and adiabatic heating distribution for buoyancy driven flow.

9.1 Fourier's law of heat conduction

Heat transport plays a crucial role in geodynamics and is often inherently coupled to material deformation, as for example in mantle convection, granitic cupola growth, subduction etc. Let us first study the equations relevant to heat transport processes. The most basic one is Fourier's law of heat conduction, which relates the heat flux q, (W/m^2) to the temperature gradient $\dfrac{\partial T}{\partial x}$ (K/m) according to

$$q = -k\frac{\partial T}{\partial x},\qquad(9.1)$$

where k (W/m/K) is the thermal conductivity of the material. Thermal conductivity may depend on P, T, composition and structure of the material. The heat flux q is the amount of heat that passes through a unit surface area, per unit time. As we all know, heat is always transferred from a hot body to a colder one. This is reflected by the minus sign in the right part of Equation (9.1), which implies that heat flux is positive in the direction of decreasing temperature, i.e. in the case when the temperature gradient $\dfrac{\partial T}{\partial x}$ is negative. In three dimensions, the heat flux is a vector that can be decomposed into three components

$$\vec{q} = (q_x, q_y, q_z).$$

In this case, Fourier's law relates heat fluxes in different directions to the respective temperature gradients

$$\vec{q} = -k\nabla T \quad \text{or} \quad q_i = -k\frac{\partial T}{\partial x_i}, \tag{9.2}$$

where i is a coordinate index and x_i is a spatial coordinate, or

$$q_x = -k\frac{\partial T}{\partial x},$$

$$q_y = -k\frac{\partial T}{\partial y},$$

$$q_z = -k\frac{\partial T}{\partial z}.$$

9.2 Heat conservation equation

In order to predict changes in temperature due to heat transport, the *heat conservation equation*, also called *temperature equation*, has to be solved. This equation describes the balance of heat in a continuum and relates temperature changes due to internal heat generation, as well as with *advective* and *conductive* heat transport. The Lagrangian temperature equation has the following form

$$\rho C_P \frac{DT}{Dt} = -\frac{\partial q_i}{\partial x_i} + H, \tag{9.3}$$

or spelled out in a complete 3D form

$$\rho C_P \frac{DT}{Dt} = -\frac{\partial q_x}{\partial x} - \frac{\partial q_y}{\partial y} - \frac{\partial q_z}{\partial z} + H,$$

or by using Equation (9.2)

$$\rho C_P \frac{DT}{Dt} = \frac{\partial}{\partial x}\left(k\frac{\partial T}{\partial x}\right) + \frac{\partial}{\partial y}\left(k\frac{\partial T}{\partial y}\right) + \frac{\partial}{\partial z}\left(k\frac{\partial T}{\partial z}\right) + H,$$

where the repeated index i, means a *summation* of derivatives of heat flux components by respective coordinates (x, y, z); ρ is density (kg/m^3); C_P is heat capacity at constant pressure (J/kg/K); H is volumetric heat productions (W/m^3). $\frac{DT}{Dt}$ is the substantive time derivative of temperature corresponding to the standard

Lagrangian–Eulerian relation, which was already discussed in Chapters 1 and 5

$$\frac{DT}{Dt} = \frac{\partial T}{\partial t} + \bar{v} \cdot \text{grad}(T).$$

For example, in 3D

$$\frac{DT}{Dt} = \frac{\partial T}{\partial t} + v_x \frac{\partial T}{\partial x} + v_y \frac{\partial T}{\partial y} + v_z \frac{\partial T}{\partial z}.$$

Accordingly, the temperature equation in an Eulerian form can be written as follows

$$\rho C_P \left(\frac{\partial T}{\partial t} + \bar{v} \cdot \text{grad}(T) \right) = -\frac{\partial q_i}{\partial x_i} + H, \tag{9.4}$$

or in complete 3D form as,

$$\rho C_P \left(\frac{\partial T}{\partial t} + v_x \frac{\partial T}{\partial x} + v_y \frac{\partial T}{\partial y} + v_z \frac{\partial T}{\partial z} \right) = -\frac{\partial q_x}{\partial x} - \frac{\partial q_y}{\partial y} - \frac{\partial q_z}{\partial z} + H,$$

or by using Equation (9.2) as

$$\rho C_P \left(\frac{\partial T}{\partial t} + v_x \frac{\partial T}{\partial x} + v_y \frac{\partial T}{\partial y} + v_z \frac{\partial T}{\partial z} \right)$$
$$= \frac{\partial}{\partial x} \left(k \frac{\partial T}{\partial x} \right) + \frac{\partial}{\partial y} \left(k \frac{\partial T}{\partial y} \right) + \frac{\partial}{\partial z} \left(k \frac{\partial T}{\partial z} \right) + H.$$

The Lagrangian heat conservation equation can be derived by analysing heat fluxes through the small moving rectangular Lagrangian (material) volume of mass m and with dimensions Δx, Δy and Δz (Fig. 9.1). Let us assume that the initial temperature of this volume is T_0.

Heat comes into the volume through the boundaries A, C and E and leaves the volume through the opposed boundaries B, D and F respectively. In addition, there is an internal heat source ΔQ_{int} inside the volume. Heat fluxes affect the amount of heat in the material volume and after a small period of time Δt, the temperature changes to T_1. The amount of heat ΔQ required to change the temperature can be computed from the following thermodynamic relation

$$\Delta Q = m C_P \Delta T = m C_P (T_1 - T_0). \tag{9.4}$$

In accordance with the energy conservation principle, this amount of heat should match the bulk effect of various heat sources and sinks in the volume such that

$$\Delta Q = \Delta Q_{int} + \Delta Q_A - \Delta Q_B + \Delta Q_C - \Delta Q_D + \Delta Q_E - \Delta Q_F, \tag{9.5}$$

where ΔQ_A–ΔQ_F represents the amounts of heat that fluxed through the respective boundaries during the period of time Δt. These can be computed according to the

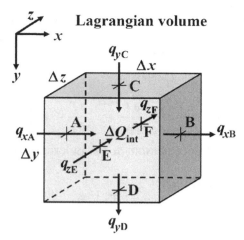

Fig. 9.1 Lagrangian elementary volume considered for derivation of the respective form of the heat conservation equation. Arrows show heat flux components responsible for *heat fluxes through* respective boundaries (A, B, C, D, E and F).

definition of heat fluxes as

$$\Delta Q_A = q_{xA} \Delta y \Delta z \Delta t,$$
$$\Delta Q_B = q_{xB} \Delta y \Delta z \Delta t,$$
$$\Delta Q_C = q_{yC} \Delta x \Delta z \Delta t,$$
$$\Delta Q_D = q_{yD} \Delta x \Delta z \Delta t,$$
$$\Delta Q_E = q_{zE} \Delta x \Delta y \Delta t,$$
$$\Delta Q_F = q_{zF} \Delta x \Delta y \Delta t,$$

(9.6)

where $q_{xA} - q_{zF}$ are the heat flux components responsible for *heat fluxes through* respective boundaries (Fig. 9.1).

Equating the right-hand sides of Equations (9.4), (9.5) and dividing through by Δt and the volume $V = \Delta x \Delta y \Delta z$, we obtain the following equation for the energy conservation in a Lagrangian volume (verify as an exercise)

$$\frac{m}{V} C_P \frac{\Delta T}{\Delta t} = -\frac{(q_{xB} - q_{xA})}{\Delta x} - \frac{(q_{yD} - q_{yC})}{\Delta y} - \frac{(q_{zF} - q_{zE})}{\Delta z} + \frac{\Delta Q_{int}}{V \Delta t},$$

(9.7a)

or in a different notation

$$\frac{m}{V} C_P \frac{\Delta T}{\Delta t} = -\frac{\Delta q_{xBA}}{\Delta x} - \frac{\Delta q_{yDC}}{\Delta y} - \frac{\Delta q_{zFE}}{\Delta z} + \frac{\Delta Q_{int}}{V \Delta t},$$

(9.7b)

where Δq_{xBA}, Δq_{yDC} and Δq_{zFE} are the changes in respective heat fluxes between respective points. Taking into account the relationships $\rho = \dfrac{m}{V}$, and $H = \dfrac{1}{V} \dfrac{DQ_{int}}{Dt}$, and further assuming that Δt, Δx, Δy and Δz all tend towards zero, the differences in Eq. (9.7b) can be replaced by derivatives and we obtain the Lagrangian heat conservation equation

$$\rho C_P \frac{DT}{Dt} = -\frac{\partial q_x}{\partial x} - \frac{\partial q_y}{\partial y} - \frac{\partial q_z}{\partial z} + H.$$

9.3 Heat generation and consumption

There are several types of heat generation/consumption processes that should be taken into account in the temperature equation

$$\rho C_P \frac{DT}{Dt} = -\frac{\partial q_i}{\partial x_i} + H_r + H_s + H_a + H_L, \tag{9.8}$$

where i is a coordinate index, x_i is a spatial coordinate and H_r, H_s, H_a and H_L are the radioactive, shear, adiabatic and latent heat productions (W/m^3), respectively.

The radioactive heat production (H_r) is due to the decay of radioactive elements that are present in rocks. The amount of radioactive heat production depends strongly on the type of rocks and typical, easy-to-remember values are: 2×10^{-6} W/m^3 for granites, 2×10^{-7} W/m^3 for basalts and 2×10^{-8} W/m^3 for mantle rocks (Turcotte and Schubert, 2002).

The shear heat production (H_s) is related to dissipation of the mechanical energy during irreversible non-elastic (e.g., viscous) deformation and can be calculated via the deviatoric stresses and strain rates as follows

$$H_s = \sigma'_{ij} \dot{\varepsilon}'_{ij} \tag{9.9a}$$

where i and j are coordinate indices (x, y, z) and the repeated ij indices denotes summation. In the case of 3D viscous deformation of an incompressible fluid Eq. (9.9a) becomes

$$H_s = \sigma'_{xx}\dot{\varepsilon}_{xx} + \sigma'_{yy}\dot{\varepsilon}_{yy} + \sigma'_{zz}\dot{\varepsilon}_{zz} + 2(\sigma_{xy}\dot{\varepsilon}_{xy} + \sigma_{xz}\dot{\varepsilon}_{xz} + \sigma_{yz}\dot{\varepsilon}_{yz}). \tag{9.9b}$$

The adiabatic heat production/consumption (*adiabatic heating/cooling*) is related to changes in pressure and can be calculated via pressure changes as follows

$$H_a = T\alpha \frac{DP}{Dt}, \tag{9.10}$$

where $\frac{DP}{Dt}$ is the substantive time derivative of pressure. In contrast to shear and radioactive heating, adiabatic effects can be either positive or negative. It is known from thermodynamics that the temperature of a substance under conditions of no thermal exchange increases with increasing pressure and decreases with decreasing pressure, which thus directly reflects the sign of $\frac{DP}{Dt}$. The effects of adiabatic heating can be very significant in cases of strong changes in pressure, a fact which has implications for mantle convection.

The latent heat production/consumption (H_L) is due to the phase transformations in rocks subjected to changes in pressure and temperature. A very common type of latent heat is the latent heat of melting, which is negative (heat sink, $H_L < 0$) for melting and positive (heat production, $H_L > 0$) for crystallisation.

9.4 Simplified temperature equations

In a complete form, the temperature equation looks quite complicated, but at least it does not 'hide' three equations in one in contrast to the momentum equation. In the case of constant thermal conductivity $k = const$, the temperature equation simplifies to

$$\rho C_P \frac{DT}{Dt} = k \frac{\partial^2 T}{\partial x^2} + k \frac{\partial^2 T}{\partial y^2} + k \frac{\partial^2 T}{\partial z^2} + H_r + H_s + H_a + H_L \qquad (9.11a)$$

or,

$$\rho C_P \frac{DT}{Dt} = k \mathbf{\Delta} T + H_r + H_s + H_a + H_L. \qquad (9.11b)$$

When the internal heat production is negligible and there is no advection of material (purely conductive heat transport), the temperature equation takes a form which is similar to the Poisson equation

$$\frac{\partial T}{\partial t} = \kappa \mathbf{\Delta} T, \qquad (9.12)$$

where $\kappa = \dfrac{k}{\rho C_P}$ is thermal diffusivity (m²/sec).

If temperature does not change with time, heat conservation is described by a steady-state temperature equation. The steady-state, Eulerian temperature equation $\frac{\partial T}{\partial t} = 0$, corresponds to the case when temperature remains constant at immobile, Eulerian observation points, while the temperature at Lagrangian points can change. In this case, the temperature equation is as follows,

$$\rho C_P \left(\bar{v} \cdot \mathrm{grad}(T) \right) = -\frac{\partial q_i}{\partial x_i} + H_r + H_s + H_a + H_L, \qquad (9.13)$$

where i is a coordinate index and x_i is a spatial coordinate. This form of the equation is frequently used for computing equilibrium temperature profiles across a deforming medium, for example in the case of steady magma flow in a channel. The steady-state Lagrangian temperature equation $\dfrac{DT}{Dt} = 0$ corresponds to the case when the temperature does not change at Lagrangian points but can vary at Eulerian observation points according to the purely advective heat transport,

$$\frac{\partial T}{\partial t} + \bar{v} \cdot \text{grad}(T) = 0. \tag{9.14}$$

In this case, the temperature equation is as follows

$$-\frac{\partial q_i}{\partial x_i} + H_r + H_s + H_a + H_L = 0. \tag{9.15}$$

The steady-state Eulerian–Lagrangian temperature equation ($\dfrac{\partial T}{\partial t} = 0$ and $\dfrac{DT}{Dt} = 0$) holds for the case when no displacement of the medium occurs, pressure and temperature are constant and therefore $H_s = 0$, $H_a = 0$ and $H_L = 0$. This equation has the simple form

$$-\frac{\partial q_i}{\partial x_i} + H_r = 0, \tag{9.16}$$

and is often used for the calculation of steady-state geotherms that characterise changes of temperature with depth in a layered sequence of rocks with variable radioactive heat production.

Simplified steady-state temperature equations are often used for obtaining analytical solutions which are used for testing accuracy of numerical codes (Chapter 16). Indeed some analytical solutions also exist for more complicated non steady-state (*transient*) cases (e.g. Tikhonov and Samarsky, 1972; Shukla, 2005) which will be further discussed in Chapter 16.

9.5 Heat diffusion timescales

One important aspect that can be analysed analytically concerns the *timescales* of heat diffusion processes. Heat generated within any region is spread by conduction (i.e. diffused) on a characteristic timescale (t_{diff}) that depends on the width L, of the region according to

$$t_{\text{diff}} = \frac{L^2}{\kappa}. \tag{9.17}$$

Therefore, the duration of heat dissipation via conduction grows as the square of the width (L) of the region. For instance, although the shear heat produced within a

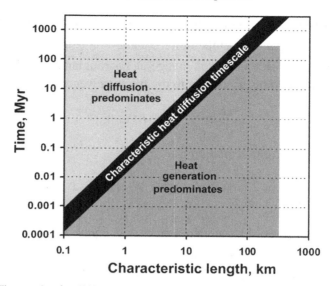

Fig. 9.2 Timescales for different thermal regimes calculated according to the equation $t_{\text{diff}} = L^2/\kappa$ with $\kappa = 10^{-6}$ m²/s. Shaded areas show length and timescales characteristic of collisional orogens (Burg and Gerya, 2005).

100 metre wide shear zone dissipates in only ~1000 yr, the heat generated within a 1 km wide shear zone requires about 100 000 yr for the similar degree of conductive cooling (Fig. 9.2).

Analytical exercises

Exercise 9.1
Integrate Equation (9.16) in order to calculate the steady-state temperature profile across the continental crust with radioactive heat production $H_r = 1 \times 10^{-6}$ W/m³, if the temperature at the surface is 300 K and temperature at the bottom of the crust is 700 K. Take the thermal conductivity of the crust to be $k = 2$ W/(m K).

Exercise 9.2
Compute from Eq. (9.13) the steady-state Eulerian temperature profile across the magmatic channel described in Exercise 5.1 (Fig. 5.2). Assume the temperature at the channel walls, as well as the temperature gradient along the channel to be constant. Use $T_0 = 1300$ K and $\dfrac{\partial T}{\partial y} = 1$ K/m for the analysed horizontal section across the channel. Take the thermal conductivity to be $k = 2$ W/(m K) and the isobaric heat capacity as $C_P = 1000$ J/(kg K).

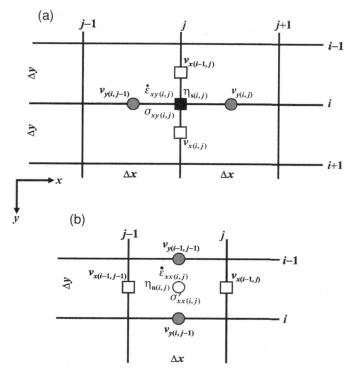

Fig. 9.3 Stencils used for the discretisation of the shear (a) and normal (b) strain rates and the deviatoric stress components. Indexing of gridlines corresponds to the basic (density) nodal points. Indexing of different unknowns is made according to Fig. 7.15.

Programming exercises and homework

Exercise 9.3

Use the variable viscosity model (Exercise 7.2) to compute the strain rate components and deviatoric stress components as follows (Fig. 9.3)

$$\dot{\varepsilon}_{xy(i,j)} = \frac{1}{2}\left(\frac{v_{x(i,j)} - v_{x(i-1,j)}}{\Delta y} + \frac{v_{y(i,j)} - v_{y(i,j-1)}}{\Delta x}\right), \tag{9.18}$$

$$\sigma_{xy(i,j)} = 2\eta_{s(i,j)}\dot{\varepsilon}_{xy(i,j)}, \tag{9.19}$$

$$\dot{\varepsilon}_{xx(i,j)} = \frac{v_{x(i-1,j)} - v_{x(i-1,j-1)}}{\Delta x}, \tag{9.20}$$

$$\sigma'_{xx(i,j)} = 2\eta_{n(i,j)}\dot{\varepsilon}_{xx(i,j)}. \tag{9.21}$$

Compute the $\sigma'_{xx}\dot{\varepsilon}_{xx}$ term for the internal basic nodes (see the solid rectangle in Fig. 9.3(a)) by averaging its values computed at the four surrounding pressure nodes (see open circle in Fig. 9.3(b)). Compute and visualise (*pcolor*) the shear heating distribution (log values) for all internal basic nodes by using the following equation based on Eq. (9.9)

$$H_s = 2\sigma'_{xx}\dot\varepsilon_{xx} + 2\sigma_{xy}\dot\varepsilon_{xy}. \tag{9.22}$$

Eq. 9.22 is valid for an incompressible medium in 2D ($\dot\varepsilon'_{xx} = \dot\varepsilon_{xx} = -\dot\varepsilon'_{yy} = -\dot\varepsilon_{yy}$, $\sigma'_{xx} = -\sigma'_{yy}$).

An example is in **Shear_heating.m**.

Exercise 9.4

Compute the adiabatic heating for the same buoyancy driven flow using the following approximate formula derived from Eq. (9.10) under the assumption that $\dfrac{DP}{Dt} \approx \dfrac{\partial P}{\partial y} v_y \approx \rho g_y v_y$,

$$H_a = T\alpha v_y \rho g_y. \tag{9.23}$$

Use $T = 1300$ K and $\alpha = 3 \times 10^{-5}$ 1/K and compute the adiabatic heating for the internal basic nodes (see solid rectangle in Fig. 9.3(a)). Interpolate the v_y velocity component for these nodes from the two nearest staggered v_y-nodes (see grey circles in Fig. 9.3(a)). Compare the magnitudes of the shear and adiabatic heating. Sum up the shear and adiabatic heating terms and visualise the results.

An example is in **Shear_adiabatic_heating.m**.

10

Numerical solution of the heat conservation equation

> **Theory:** Discretisation of the heat conservation equation with finite differences. Conservative and non-conservative discretisation schemes. Explicit and implicit solution schemes of the heat conservation equation. Advective terms: upwind differences, numerical diffusion. Advection of temperature with markers. Subgrid diffusion. Thermal boundary conditions: constant temperature, constant heat flux, combined boundary conditions. Numerical implementation of thermal boundary conditions. **Exercises:** Programming various thermal boundary conditions. Solving the heat conservation equation in the case of constant and variable thermal conductivity with explicit and implicit solution schemes. Advecting temperature with Eulerian schemes and markers.

10.1 Explicit and implicit formulation of the temperature equation

We now start with the numerical formulation and solution of the temperature equation. Discretisation of this equation with finite differences can be done in an *explicit* and an *implicit* manner. In order to understand the differences, let us consider an example of heat diffusion in a non-deforming medium with constant thermal conductivity (k)

$$\frac{\partial T}{\partial t} = \frac{k}{\rho C_P} \Delta T. \tag{10.1}$$

In 2D, this discretisation is as follows.
  Explicit FD (Fig. 10.1):

$$\frac{T_3^n - T_3^o}{\Delta t} = \frac{k}{\rho C_P} \left(\frac{T_1^o - 2T_3^o + T_5^o}{\Delta x^2} + \frac{T_2^o - 2T_3^o + T_4^o}{\Delta y^2} \right). \tag{10.2}$$

133

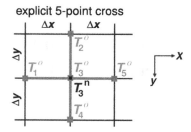

Fig. 10.1 Stencil of the 2D grid used for an explicit discretisation of the temperature equation with constant thermal conductivity.

This form is called explicit because the new temperature (T^n) at the next time instant t_n can be explicitly calculated from the known temperatures (T^o) at the current time instant t_o ($t_n = t_o + \Delta t$, where Δt is time step):

$$T_3^n = \frac{k\Delta t}{\rho C_P}\left(\frac{T_1^o - 2T_3^o + T_5^o}{\Delta x^2} + \frac{T_2^o - 2T_3^o + T_4^o}{\Delta y^2}\right) + T_3^o. \qquad (10.3)$$

The explicit formulation does not require composing and solving of a global system of equations and is therefore very convenient to program. However, this formulation has a strong limitation on the time step that can be used in the calculations. The time step must satisfy,

$$\Delta t < \frac{\Delta x^2}{3\kappa}, \qquad (10.4)$$

where $\kappa = \dfrac{k}{\rho C_P}$ is thermal diffusivity and Δx is the minimum grid spacing. This limitation means that the number of time steps increases as the square of the decrease in the grid spacing, which can be quite inconvenient for high-resolution thermal models. If larger time steps are indeed employed, numerical oscillations occur that increase with the number of time steps (Fig. 10.2, see also program example **Explicit_implicit_1D.m**).

The implicit finite-difference discretisation is given by (Fig. 10.3):

$$\frac{T_3^n - T_3^o}{\Delta t} = \frac{k}{\rho C_P}\left(\frac{T_1^n - 2T_3^n + T_5^n}{\Delta x^2} + \frac{T_2^n - 2T_3^n + T_4^n}{\Delta y^2}\right). \qquad (10.5)$$

This form is called implicit because the new temperature (T^n) for the next time instant t_n cannot be explicitly calculated from the temperatures (T^o) known from the current time instant. In order to obtain new temperatures, the global system of equations written for all points of the model has to be solved:

$$\frac{T_3^n}{\Delta t} - \frac{k}{\rho C_P}\left(\frac{T_1^n - 2T_3^n + T_5^n}{\Delta x^2} + \frac{T_2^n - 2T_3^n + T_4^n}{\Delta y^2}\right) = \frac{T_3^o}{\Delta t}. \qquad (10.6)$$

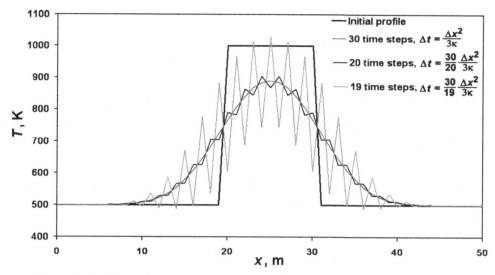

Fig. 10.2 Oscillations of explicit numerical solution of Equation (10.1) due to the use of a too large time step. An example is in **Explicit_implicit_1D.m**.

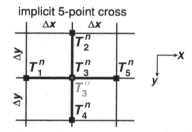

Fig. 10.3 Stencil of the 2D grid used for implicit discretisation of the temperature equation with constant thermal conductivity.

It is important to mention that the implicit formulation places no limitation on the size of the time step (in the absence of internal heat sources and advective terms). Indeed, very large implicit time steps do not necessarily guarantee an accurate solution (since the time derivative in Eq. (10.5) is only first-order accurate in time.

10.2 Conservative finite differences

Conservative finite-difference discretisation should be used in cases when the heat conservation equation contains a variable thermal conductivity. Such finite differences ensure conservation of heat fluxes between nodal points, so allowing a correct numerical solution. In a general sense, this is analogous to the formulation of conservative finite differences for the Stokes equation with variable viscosity, which

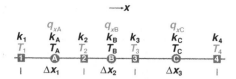

Fig. 10.4 1D staggered grid used for the discretisation of the temperature equation with variable thermal conductivity. 1, 2, 3, 4 are basic nodes (squares) of the grid where the temperature equations are formulated. A, B, C are additional nodes (circles) of the grid where the heat fluxes are defined.

was described in Chapter 7. Below, examples of non-conservative and conservative finite differences are compared for the 1D heat conservation equation (Fig. 10.4)

$$\rho C_P \frac{DT}{Dt} = -\frac{\partial q_x}{\partial x},$$

$$\text{where } q_x = -k\frac{\partial T}{\partial x}.$$

An **erroneous non-conservative** FD formulation of the heat flux terms, in either the explicit or the implicit formulation, for the two basic nodes 2 and 3 can for example be obtained (we can 'arrive' at this equation by assuming erroneously that all we need to do is to use Eq. 10.5 and put different thermal conductivity for different nodes):

$$\text{node 2: } \quad \left(\frac{\partial q}{\partial x}\right)_2 = -k_2 \frac{(T_3 - T_2)/\Delta x_2 - (T_2 - T_1)/\Delta x_1}{(\Delta x_1 + \Delta x_2)/2} \qquad (10.7a)$$

$$\text{node 3: } \quad \left(\frac{\partial q}{\partial x}\right)_3 = -k_3 \frac{(T_4 - T_3)/\Delta x_3 - (T_3 - T_2)/\Delta x_2}{(\Delta x_2 + \Delta x_3)/2}, \qquad (10.7b)$$

which implicitly means that the formulations of the horizontal heat flux q_{xB} for temperature equations at nodes 2 and 3 are different due to the different thermal conductivities k_2 and k_3:

$$\text{node 2: } \quad q_{xB} = -k_2 \frac{T_3 - T_2}{\Delta x_2}$$

$$\text{node 3: } \quad q_{xB} = -k_3 \frac{T_3 - T_2}{\Delta x_2}.$$

This implies that heat flux is *not conserved* and artificially 'jumps' between basic nodes in response to the difference in the thermal conductivity at these nodes.

On the other hand, a **proper conservative** FD formulation of the heat flux term in either the explicit or implicit temperature equations for the two basic nodes 2 and 3 is given by

node 2: $\left(\dfrac{\partial q}{\partial x}\right)_2 = 2\dfrac{q_{xB} - q_{xA}}{(\Delta x_1 + \Delta x_2)}$ or

$$\left(\frac{\partial q}{\partial x}\right)_2 = -\frac{k_B\,(T_3 - T_2)/\Delta x_2 - k_A(T_2 - T_1)/\Delta x_1}{(\Delta x_1 + \Delta x_2)/2},\qquad (10.8a)$$

node 3: $\left(\dfrac{\partial q}{\partial x}\right)_3 = 2\dfrac{q_{xC} - q_{xB}}{(\Delta x_2 + \Delta x_3)}$ or

$$\left(\frac{\partial q}{\partial x}\right)_3 = -\frac{k_C\,(T_4 - T_3)/\Delta x_3 - k_B(T_3 - T_2)/\Delta x_2}{(\Delta x_2 + \Delta x_3)/2},\qquad (10.8b)$$

which imply that the expressions for heat flux q_{xB} at nodes 2 and 3 are identical:

$$q_{xB} = -k_B\frac{T_3 - T_2}{\Delta x_2}.$$

Thus, a conservative FD formulation of the temperature equation (either explicit or implicit) is based on the following three formal rules that are analogous to the rules discussed in Chapter 7 for the Stokes equation with variable viscosity.

(1) The temperature equation is initially discretised in term of heat fluxes at *basic nodes* of the grid (cf. nodes 2, 3, Fig. 10.4),

$$\text{node 2:}\quad \left(\rho C_P \frac{DT}{Dt}\right)_2 = 2\frac{q_{xB} - q_{xA}}{(\Delta x_1 + \Delta x_2)},$$

$$\text{node 3:}\quad \left(\rho C_P \frac{DT}{Dt}\right)_3 = 2\frac{q_{xC} - q_{xB}}{(\Delta x_2 + \Delta x_3)}.$$

(2) These heat fluxes are formulated for *additional (heat flux) nodes* of the grid (cf. nodes A, B, C, Fig. 10.4)

$$\text{node A:}\quad q_{xA} = -k_A\frac{T_2 - T_1}{\Delta x_1},$$

$$\text{node B:}\quad q_{xB} = -k_B\frac{T_3 - T_2}{\Delta x_2},$$

$$\text{node C:}\quad q_{xC} = -k_C\frac{T_4 - T_3}{\Delta x_3}.$$

Note that we have to use thermal conductivity values k_A, k_B and k_C for the additional nodes (A, B, C) at the locations where the heat fluxes are defined. If these values are not known, they can be computed by e.g. arithmetic averaging of known thermal conductivity values from the basic nodes (1, 2, 3, 4)

$$k_A = \frac{k_1 + k_2}{2},$$

$$k_B = \frac{k_2 + k_3}{2},$$

$$k_C = \frac{k_4 + k_3}{2}.$$

Another possibility is to use harmonic averaging

$$k_A = \frac{2k_1 k_2}{k_1 + k_2},$$

$$k_B = \frac{2k_2 k_3}{k_2 + k_3},$$

$$k_C = \frac{2k_3 k_4}{k_3 + k_4}.$$

The harmonic average formula can be derived from the condition that the heat flux to the left of the additional nodes must equal the flux to the right of these nodes. The derivation for node B is done *under the assumption* that the thermal conductivities between nodes 2 and B, and between B and 3 remain constant, and are equal to k_2 and k_3, respectively. Then the following equation can be formulated,

$$q_{xB} = -2k_2 \frac{T_B - T_2}{\Delta x_2},$$

$$q_{xB} = -2k_3 \frac{T_3 - T_B}{\Delta x_2},$$

$$q_{xB} = -k_B \frac{T_3 - T_2}{\Delta x_2}.$$

Solving these equations with respect to T_B and k_B gives (please verify as an exercise)

$$T_B = \frac{T_2 k_2 + T_3 k_3}{k_2 + k_3} \quad \text{and} \quad k_B = \frac{2k_2 k_3}{k_2 + k_3}.$$

Derivations for nodes A and C can be done similarly (please verify as an exercise).

(3) Identical formulations of heat fluxes are used for the temperature equation at different basic nodes.

It is important to mention that conservative finite differences are formulated *in terms of the thermal conductivity (k) and not in terms of the thermal diffusivity* $\kappa = \frac{k}{\rho C_P}$, otherwise one can obtain artificial variations in heat fluxes due to spatial

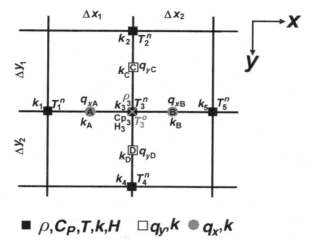

Fig. 10.5 Stencil of a 2D grid used for the implicit discretisation of the Lagrangian temperature equation with variable thermal conductivity. The crossed square corresponds to the node for which the temperature equation is formulated.

variations in density (ρ) and/or heat capacity (C_P). Both density and heat capacity should always be taken from the basic node on which this equation is formulated.

By applying these rules in 2D, the following conservative implicit FD formulation can be derived for the Lagrangian temperature equation (Fig. 10.5)

$$\rho_3 C_{P_3} \left(\frac{DT}{Dt} \right)_3 = -\left(\frac{\partial q_x}{\partial x} \right)_3 - \left(\frac{\partial q_y}{\partial y} \right)_3 + H_3, \qquad (10.9a)$$

$$\rho_3 C_{P_3} \frac{T_3^n - T_3^o}{\Delta t} = -2\frac{q_{xB} - q_{xA}}{\Delta x_1 + \Delta x_2} - 2\frac{q_{yD} - q_{yC}}{\Delta y_1 + \Delta y_2} + H_3, \qquad (10.9b)$$

$$\rho_3 C_{P_3} \frac{T_3^n}{\Delta t} + 2\frac{q_{xB} - q_{xA}}{\Delta x_1 + \Delta x_2} + 2\frac{q_{yD} - q_{yC}}{\Delta y_1 + \Delta y_2} = H_3 + \rho_3 C_{P_3} \frac{T_3^o}{\Delta t}, \qquad (10.9c)$$

where

$$q_{xA} = -k_A \frac{\left(T_3^n - T_1^n \right)}{\Delta x_1},$$

$$q_{xB} = -k_B \frac{\left(T_5^n - T_3^n \right)}{\Delta x_2},$$

$$q_{yC} = -k_C \frac{\left(T_3^n - T_2^n \right)}{\Delta y_1},$$

$$q_{yD} = -k_D \frac{\left(T_4^n - T_3^n \right)}{\Delta y_2}.$$

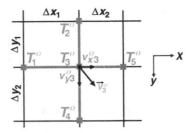

Fig. 10.6 Stencil of a 2D grid used for the explicit discretisation of the advective term for the Eulerian temperature equation.

If values of thermal conductivity for heat flux nodes (A, B, C, D) are not known, they can be computed by averaging known thermal conductivity values from the basic nodes (cf. black squares in Fig. 10.5). For example,

arithmetic average: $k_A = \dfrac{k_1 + k_3}{2}, k_B = \dfrac{k_3 + k_5}{2}, k_C = \dfrac{k_2 + k_3}{2}, k_D = \dfrac{k_3 + k_4}{2};$

harmonic average: $k_A = \dfrac{2k_1 k_3}{k_1 + k_3}, k_B = \dfrac{2k_3 k_5}{k_3 + k_5}, k_C = \dfrac{2k_2 k_3}{k_2 + k_3}, k_D = \dfrac{2k_3 k_4}{k_3 + k_4}.$

Obviously, conservative 2D formulations can also be explicit (derive as an exercise).

10.3 Advection of temperature with Eulerian methods

If the temperature equation is solved in an Eulerian form for a deforming/moving medium, then the advective term $\vec{v} \cdot \mathrm{grad}(T)$ is present in the temperature equation

$$\rho C_P \left(\frac{\partial T}{\partial t} + \vec{v} \cdot \mathrm{grad}(T) \right) = -\frac{\partial q_i}{\partial x_i} + H. \tag{10.10}$$

where i is a coordinate index and x_i is a spatial coordinate, and it should be included in the FD formulation. When discretising this term, explicit finite differences are typically used, based on temperature and velocity at the current time instant. A common approach is to use asymmetric 'upwind' differences (Chapter 8), i.e. to perform the differencing *against the direction* of material flow vector (Fig. 10.6)

$$\vec{v}_3 \cdot \mathrm{grad}(T)_3 = v_{x3} \left(\frac{\partial T}{\partial x} \right)_3 + v_{y3} \left(\frac{\partial T}{\partial y} \right)_3. \tag{10.11}$$

Explicit upwind differences are then as follows

$$\bar{v}_3^o \cdot \operatorname{grad}(T)_3 = v_{x3}^o \left(\frac{\partial T^o}{\partial x}\right)_3 + v_{y3}^o \left(\frac{\partial T^o}{\partial y}\right)_3, \tag{10.11a}$$

$$\left(\frac{\partial T^o}{\partial x}\right)_3 = \frac{T_3^o - T_1^o}{\Delta x_1} \quad \text{when } v_{x3}^o > 0 \quad \text{and} \quad \left(\frac{\partial T^o}{\partial x}\right)_3 = \frac{T_5^o - T_3^o}{\Delta x_2} \quad \text{when } v_{x3}^o < 0, \tag{10.11b}$$

$$\left(\frac{\partial T^o}{\partial y}\right)_3 = \frac{T_3^o - T_2^o}{\Delta y_1} \quad \text{when } v_{y3}^o > 0 \quad \text{and} \quad \left(\frac{\partial T^o}{\partial y}\right)_3 = \frac{T_4^o - T_3^o}{\Delta y_2} \quad \text{when } v_{y3}^o < 0. \tag{10.11c}$$

Such explicit advection terms are simply added to the right-hand side of the temperature equation (10.9).

It should be mentioned that advective terms can also be formulated implicitly. Indeed, irrespective of the formulation these terms always introduce *artificial numerical diffusion* of temperature on the Eulerian grid (Chapter 8). This problem is not relevant for slow flow because *real physical diffusion* is typically faster than numerical diffusion and the latter can be neglected. Numerical diffusion, however, becomes relevant for models with rapid advection (e.g. in subduction models). This problem can be minimised by: (i) using more complicated, higher-order Eulerian advection FD schemes (Chapter 8) and (ii) advecting temperature with Lagrangian points (*method of characteristics, method of markers*) which will be discussed below. Including advective terms in the temperature equations imposes an additional restriction on the time step given by the Courant criteria (Eq. 8.5).

10.4 Advection of temperature with markers

In order to avoid numerical diffusion of temperature one can use the Lagrangian form of heat conservation equation (Chapter 9) and advect temperature with the marker-in-cell technique described in Chapter 8. The temperature equation is then formulated in a Lagrangian form

$$\rho C_P \frac{DT}{Dt} = -\frac{\partial q_i}{\partial x_i} + H, \tag{10.12}$$

and is discretised with Eq. (10.9).

After solving the Lagrangian temperature equation on the Eulerian nodes (Fig. 10.5) the changes in the effective temperature field for the Eulerian nodes are calculated as

$$\Delta T_{i,j} = T_{i,j}^{t+\Delta t} - T_{i,j}^t. \tag{10.13}$$

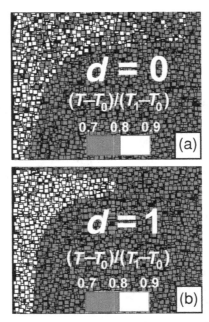

Fig. 10.7 Temperature structure of the markers for a zoomed-in area of the numerical model of convection (Gerya and Yuen, 2003a). Different colours of markers correspond to different values of temperature assigned to the markers. T_1 and T_0 are the maximal and minimal temperatures for the experiment. (a) and (b) show the results calculated without ($d = 0$) and with ($d = 1$) subgrid diffusion (see Eq. 10.16), respectively.

The corresponding temperature increments for the markers ΔT_m, are then interpolated from the nodes using relation 8.20 (Fig. 8.9) in order to calculate new marker temperatures $T_m^{t+\Delta t}$ as

$$T_m^{t+\Delta t} = T_m^t + \Delta T_m. \tag{10.14}$$

The interpolation of the calculated *temperature changes* from the Eulerian nodal points, to the Lagrangian markers reduces numerical diffusion in an efficient manner (Chapter 8). This method does not produce any smoothing of the temperature distribution between adjacent markers (Fig. 8.11), thus resolving the thermal structure of a numerical model in much finer details.

However, a main problem with treating advection-diffusion processes using the simple incremental update scheme of Eq. (10.14) is that small-scale variations of the thermal structures may appear on a *subgrid scale* (i.e. differences in temperature between closely located markers). These variations cannot be damped out by the grid-scale corrections of Eq. (10.14). For example, in the case of strong chaotic mixing of markers (e.g. due to thermal convection), Eq. (10.14) may produce numerical oscillations of thermal field assigned to the adjacent markers (Fig. 10.7a).

These oscillations do not damp out with time based on a characteristic heat diffusion timescale as would be the case if physical diffusion was active. The introduction of a consistent *subgrid diffusion operation* is the way around this. In order to define this operation, we decompose temperature changes computed from Eq. 10.13 into a subgrid part $\Delta T_{i,j}^{subgrid}$ and a remaining part $\Delta T_{i,j}^{remaining}$ such that

$$\Delta T_{i,j} = \Delta T_{i,j}^{subgrid} + \Delta T_{i,j}^{remaining}. \tag{10.15}$$

In order to compute the subgrid part, we apply subgrid diffusion on the markers over a characteristic local heat diffusion timescale Δt_{diff} (Fig. 9.2) and then interpolate the respective temperature changes back to nodes. Subgrid temperature changes on markers are then computed as follows

$$\Delta T_m^{subgrid} = \left(T_{m(nodes)}^t - T_m^t \right) \left[1 - \exp\left(-d \frac{\Delta t}{\Delta t_{diff}} \right) \right]$$

$$\Delta t_{diff} = \frac{C_{P_m} \rho_m}{k_m (2/\Delta x^2 + 2/\Delta y^2)} \tag{10.16}$$

where Δt_{diff} is defined for the corresponding cell of the grid where the marker is located (Fig. 8.9); d is a dimensionless numerical diffusion coefficient (one can use empirical values in the range of $0 \leq d \leq 1$). $T_{m(nodes)}^t$, C_{P_m}, ρ_m and k_m are interpolated for a given marker, respectively, from $T_{i,j}^t$, $C_{P(i,j)}$, $\rho_{(i,j)}$ and $k_{(i,j)}$ values for nodes using the relation (8.19) (Fig. 8.9).

After obtaining $\Delta T_m^{subgrid}$ for all markers $\Delta T_{i,j}^{subgrid}$ are computed by interpolation from markers to nodes using Eq. (8.18) (Fig. 8.8)

$$\Delta T_{i,j}^{subgrid} = \frac{\sum\limits_m \Delta T_m^{subgrid} w_{m(i,j)}}{\sum\limits_m w_{m(i,j)}} \tag{10.17}$$

where

$$w_{m(i,j)} = \left(1 - \frac{\Delta x_m}{\Delta x} \right) \times \left(1 - \frac{\Delta y_m}{\Delta y} \right).$$

Then $\Delta T_{i,j}^{remaining}$ is computed for the nodes from Eq. (10.15)

$$\Delta T_{i,j}^{remaining} = \Delta T_{i,j} - \Delta T_{i,j}^{subgrid}. \tag{10.18}$$

Finally, the new corrected marker temperatures $T_{m(corrected)}^{t+\Delta t}$ are computed according to the modified relation (10.14) that now takes into account Equations (10.15)–(10.18) and thus removes the non-physical subgrid oscillations

$$T_{m(corrected)}^{t+\Delta t} = T_m^t + \Delta T_m^{subgrid} + \Delta T_m^{remaining} \tag{10.19}$$

where $\Delta T_m^{subgrid}$ is given by Eq. (10.16) and $\Delta T_m^{remaining}$ is interpolated from nodal values of $\Delta T_{i,j}^{remaining}$ to markers according to standard bilinear interpolation (Eq. (8.19), Fig. 8.9).

Equation (10.16) requires the decay of differences between marker temperature values T_m^t and interpolated nodal temperature values $T_{m(nodes)}^t$ on the characteristic timescale (Δt_{diff}) of local heat diffusion. It is important to emphasise that the subgrid diffusion does not change the total temperature increments $\Delta T_{i,j}$ computed on nodal points from the heat conservation equation. Instead it splits them into two parts $\Delta T_{i,j}^{subgrid}$ and $\Delta T_{i,j}^{remaining}$. By introducing a subgrid diffusion operation, unrealistic subgrid oscillations are removed (see Fig. 10.7(b)) over the characteristic local heat diffusion timescale without affecting the accuracy of numerical solution of the temperature equation. Realistic subgrid oscillations will, however, be preserved by this scheme if they are related, for example, to the rapid mixing by advection dominating flows.

It is also important to mention that subgrid diffusion is a method for correcting *small* non-physical subgrid oscillations that appear on the markers due to mechanical mixing processes. It is not the way to remove any arbitrary discrepancy between the marker and nodal temperature fields. Such discrepancies can appear, for example, due to prescribing sharp temperature fronts on a fine marker grid which cannot be properly resolved by a coarse nodal grid. Big initial temperature discrepancies between markers and nodes should always be eliminated by re-interpolation of the initial nodal temperatures (with applied boundary conditions) back to markers with the use of Eq. (8.19) before making the first time step.

10.5 Thermal boundary conditions

In order to solve the temperature equation numerically, thermal boundary conditions have to be specified. These conditions depend on the type of numerical problem. The following boundary conditions are frequently used in geo-modelling:

(1) constant temperature
(2) insulating boundary (zero heat flux, symmetry condition)
(3) constant heat flux
(4) infinity-like conditions (external constant temperature)
(5) periodic boundary
(6) combined boundary conditions

Numerical examples of different boundary conditions are shown below (Fig. 10.8).

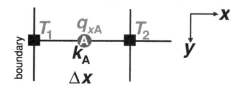

Fig. 10.8 Stencil of a 2D grid used for the discretisation of the thermal boundary conditions.

(1) *A constant temperature condition* implies that the temperature at a boundary is assigned a given value (which may change both along the boundary and in time)

$$T = \text{const}(x, y, z, t) \qquad (10.20a)$$

or in discretised form (Fig. 10.8)

$$T_1 = \text{const}(x, y, z, t) \qquad (10.20b)$$

This condition is typically applied at the lower and upper boundaries of geodynamic models.

(2) *An insulating boundary condition (no heat flux, lateral symmetry condition)* means that heat does not flux through a boundary which implies (from Eq. (9.1)) that no temperature gradient exists across this boundary, i.e.

$$q_x = -k\frac{\partial T}{\partial x} = \frac{\partial T}{\partial x} = 0, \qquad (10.21a)$$

or in discretised form (Fig. 10.8)

$$T_1 - T_2 = 0. \qquad (10.21b)$$

The symmetry condition is used at the lateral boundaries for almost every 2D and 3D Cartesian geodynamic model.

(3) *A constant heat flux condition* does not limit the temperature values at a boundary, but prescribes a heat flux across the boundary (this heat flux can be time and coordinate dependent)

$$q_x = -k\frac{\partial T}{\partial x} = const(x, y, z, t), \qquad (10.22a)$$

or in discretised form (Fig. 10.8)

$$k_A\frac{T_1 - T_2}{\Delta x} = const(x, y, z, t). \qquad (10.22b)$$

(4) *Infinity-like conditions* either mimic the absence of a thermal boundary or imply that this boundary is located very far away. For example, the external constant temperature condition (Burg and Gerya, 2005; Gerya *et al.*, 2008b) implies that condition (10.20a)

and (10.20b) are satisfied at a parallel boundary located at the distance ΔL from the actual boundary of the model, and that the temperature gradient between these two boundaries is constant

$$\frac{\partial T}{\partial x} = \frac{T - T_{external}}{\Delta L}, \tag{10.23a}$$

or in discretised form (Fig. 10.8)

$$\frac{T_2 - T_1}{\Delta x} = \frac{T_1 - T_{external}}{\Delta L}, \tag{10.23b}$$

or

$$T_1 - T_2 \frac{\Delta L}{\Delta L + \Delta x} = \frac{\Delta x}{\Delta L + \Delta x} T_{external}. \tag{10.23c}$$

where $T_{external} = \text{const}\,(x, y, z, t)$ is the prescribed temperature at the parallel external boundary.

(5) *Periodic boundary conditions* are typically established for paired parallel lateral boundaries of a model and imply that temperature fields *at both sides of each boundary* are identical. The physical meaning and usage of this condition is the same as for the respective mechanical boundary condition (Chapter 7).

(6) *Combined conditions* represent a mixture of several types of boundary conditions. Thermal boundary conditions can also be applied *inside the model*.

Programming exercises and homework

Exercise 10.1

Write a program to solve the temperature equation in 2D, in both explicit and implicit form (Figs. 10.1 and 10.3, respectively). Use a regular grid of 51×31 points. The model size is 1000×1500 km (i.e. $1\,000\,000 \times 1\,500\,000$ m). Use constant thermal conductivity $k = 3$ W/m/K, density $\rho = 3200$ kg/m^3 and heat capacity $C_P = 1000$ J/kg/K for the entire model. Test Eq. (10.4) for the time step limitation in the case when explicit FD are used. The initial setup corresponds to background temperature of 1000 K with a rectangular thermal wave (1300 K) in the middle ('wave' means perturbation of the temperature field). Global indexing of the unknowns in the implicit case is the same as for the 2D Poisson equation (Exercise 3.2)

$$k = N_y \times (j - 1) + i, \tag{10.24}$$

where k is the index in the global matrix computed from horizontal (j) and vertical (i) geometrical indices and N_y is the number of nodes in the vertical direction.

Try using constant temperature (1000 K, Eq. (10.20)), and insulating boundary conditions (Eq. (10.21)) at all boundaries. An example is in **Explicit_Implicit2D.m**.

Exercise 10.2

Modify the previous example to take into account variable thermal conductivity using a conservative finite-difference formulation (Eq. (10.9), Fig. 10.5) for a uniform grid (i.e. $\Delta x_1 = \Delta x_2 = \Delta x$ and $\Delta y_1 = \Delta y_2 = \Delta y$ in Eq. (10.9))

$$\rho_3 C_{P3} \frac{T_3^n}{\Delta t} - \frac{(k_3 + k_5)\left(T_5^n - T_3^n\right) - (k_1 + k_3)\left(T_3^n - T_1^n\right)}{2\Delta x^2}$$
$$- \frac{(k_3 + k_4)\left(T_4^n - T_3^n\right) - (k_2 + k_3)\left(T_3^n - T_2^n\right)}{2\Delta y^2} = H_3 + \rho_3 C_{P3} \frac{T_3^o}{\Delta t}.$$
$$(10.25)$$

Use different physical properties for the area of the temperature wave ($k = 10$ W/m/K, $\rho = 3300$ kg/m^3, $C_P = 1100$ J/kg/K) and the surrounding medium ($k = 3$ W/m/K, $\rho = 3200$ kg/m^3, $C_P = 1000$ J/kg/K). An example is in **Variable_ conductivity.m**.

Exercise 10.3

Modify the two previous examples by adding the advective terms into the temperature equation (Eq. (10.11)) and by using a uniform velocity field $v_x = 10^{-9}$ m/s and $v_y = 10^{-9}$ m/s in the entire model. Test Eq. (8.5) for the time step limitation in 1D and refine it for 2D. Experiment with different velocities to see advection- and conduction-dominated regimes. Observe that the numerical diffusion which is clearly visible when the chosen velocity is large (e.g. $v_x = 10$ m/s and $v_y = 10$ m/s). In this case, the timescale for the experiment will be far below the characteristic thermal diffusion timescale (compute it by using a prescribed size of the wave and a total time in your experiment, Eq. (9.17), Fig. 9.2) and thus the moving temperature wave should remain largely unchanged. Examples are in **Conduction_advection2D.m** and **Variable_conductivity_ advection2D.m**.

Exercise 10.4

Modify Exercise 10.2 by adding temperature advection with a marker-in-cell approach combined with subgrid diffusion (Eqs. (10.14)–(10.19)). Use an implicit solution of the Lagrangian temperature equation (10.12). Use marker routines from Exercises 8.4 and 8.5. Interpolate density, heat, capacity and temperature from markers to nodes at every time step using Eq. (8.18). Move the markers with the prescribed constant velocity and recycle them when they leave the model. Use Eqs. (8.21a,b) for both the horizontal and vertical coordinates of the markers. Do not forget to match nodal and marker temperature fields before starting calculations (e.g. interpolate the nodal temperature back to the markers after performing interpolation of nodal temperature from markers for the first time). Also do not forget to apply the boundary conditions for the nodal temperatures interpolated from the markers at every time step. Experiment with different velocities to see advection- and conduction-dominated regimes. Compare these results with Exercise 10.3 for the case when the chosen velocities are large, both without ($d = 0$ in Eq. (10.16)) and with ($d = 1$ in Eq. (10.16)) subgrid diffusion. An example is in **Variable_conductivity_markers2D.m**.

11

2D thermomechanical code structure

> **Theory:** Principal steps of a coupled thermomechanical solution with finite-differences and marker-in-cell techniques. Organisation of a thermomechanical code for the case of viscous, multi-component flows. Adding self-gravity. Handling free planetary surfaces with a weak layer approach.
>
> **Exercises:** Building a 2D thermomechanical code.

11.1 What do we expect from geodynamic codes?

Before describing possible structures for thermomechanical codes, let us discuss what we actually expect from a state-of-the-art, numerical geodynamic modelling tool. Today, as numerical modelling of geodynamic and planetary processes is in the 'new millennium' (although it is only around 40 years old, see Introduction), geoscientists are targeting modelling of realistic situations in lithospheric, mantle and planetary dynamics (e.g. Gerya and Yuen, 2007; Moresi *et al.*, 2008; Zhong *et al.*, 2007; Tackley, 2008). The rheology of crustal and mantle rocks depends strongly on the temperature, strain-rate, volatile content, grain size and the fluid pressure. Physical and dynamical circumstances imposed by the sharply varying viscosity represent a major challenge for solving the momentum equation in geodynamics, unlike those found in the oceanographic or atmospheric sciences. Another complication is due to the variable thermal conductivity in the heat conservation equation. The thermal conductivity of various crustal and mantle rocks is notably different and is also a strong function of temperature, pressure and mineralogy which causes numerical difficulties compared to the constant thermal conductivity situation. Finally, all physical (transport) properties of rocks, including viscosity and conductivity, vary strongly with chemical composition and/or mineralogy. In various geodynamic situations, these result in sharp fronts involving *multi-component*

149

flows (i.e. flows composed of many rock types with contrasting compositions and physical properties). Therefore, from a general geophysical standpoint, we should consider at least three important elements for modelling these kinds of flows:

(1) the ability to conserve stresses under conditions that involve sharply discontinuous viscosity distributions;
(2) the ability to conserve heat fluxes under conditions that involve sharply varying conductivity and temperature gradients at the thermal or chemical layers with temperature-dependent conductivity;
(3) the ability to conserve transport properties, such as temperature field, chemical species, viscosity and density in flows with a strong advection character.

11.2 Thermomechanical code structure

Since we want to address all above requirements in our state-of-the-art '*all-in-one*' thermomechanical code, let us discuss in detail how this can be achieved by using a marker-in-cell algorithm, combined with a conservative finite-differences for primitive variable (pressure–velocity) formulation. The code structure should reflect the physical relations of momentum, continuity, temperature and advection equations. For example, the temperature equation requires values of adiabatic and shear heating that are computed from the velocity, pressure, stress and strain rate fields. Therefore, the temperature equation can only be solved after solving the momentum and continuity equations. On the other hand, the momentum and continuity equations have to be solved simultaneously to obtain values for velocity, that are present in both equations. The advection equation requires a velocity field and should thus also be solved after solving the momentum and continuity equations.

The flow chart in Fig. 11.1 gives an example of a structure for a numerical, thermomechanical viscous 2D code that uses finite-differences and the marker-in-cell technique (FD+MIC) to solve the momentum, continuity and temperature equations. The principal steps of the algorithm are as follows:

(1) Calculating the scalar physical properties (η_m, ρ_m, α_m, C_{Pm}, k_m, etc.) for each marker and interpolating these properties, as well as advected temperature from the markers to Eulerian nodes (Chapters 8, 10). Applying boundary conditions for the nodal temperatures interpolated from markers.
(2) Solving the 2D momentum and continuity equations with a pressure–velocity formulation on a staggered grid by composing and inverting the global matrix with a direct method, which is used because of its stability and high accuracy (Chapter 7).
(3) Defining an optimal displacement time step Δt_m for markers (typically limiting maximal displacement to 0.01–1.0 of minimal grid step) based on the velocity field computed in Step 2 (Chapter 8).

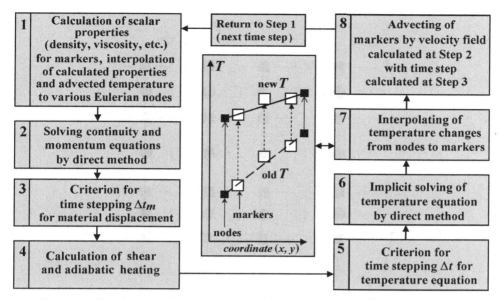

Fig. 11.1 Flow chart that gives an example of a possible structure of a numerical thermomechanical viscous 2D code which employs finite-differences and marker-in-cell technique (FD + MIC) for solving the momentum, continuity and temperature equations. (Gerya and Yuen, 2003a).

(4) Calculating the shear and adiabatic heating terms $H_{s(i,j)}$ and $H_{a(i,j)}$ at the Eulerian nodes (Chapter 9).

(5) Defining an optimal time step Δt for the temperature equation. We take the smallest time step of three time step limiters: given absolute time step limit; given optimal marker displacement time step limit (see Step 3); given absolute nodal temperature change limit (typically 1–20 K) (Chapter 10).

(6) Solving the temperature equation in a Lagrangian formulation, with implicit time stepping and a direct method (Chapter 10).

(7) Interpolating the calculated nodal temperature changes (see Step 7 at Fig. 11.1) from the Eulerian nodes to the markers and calculating new marker temperatures (T_m) taking into account physical diffusion on subgrid (marker) level (Chapter 10).

(8) Using a fourth-order explicit Runge–Kutta scheme (Chapter 8) in space to advect all markers throughout the mesh according to the globally calculated velocity field v (see Step 2). Returning to Step 1 to perform the next time step.

Figure 11.2 shows the geometry of an irregularly spaced, fully staggered numerical grid corresponding to the algorithm outlined above. The irregularly spaced grid is extremely useful in handling geodynamic situations with localisation phenomena and multiple-scale features which will be further discussed in Chapters 16 and 17. Note that by doing the programming exercises for previous chapters *we have*

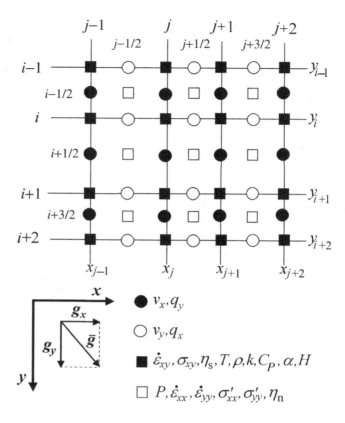

Fig. 11.2 Staggered 2D, irregularly spaced numerical grid corresponding to the algorithm presented in Fig. 11.1.

already implemented (surprise, surprise...) one-by-one all steps of the above computational algorithm and we will now discuss in full detail how to connect these separate steps in order to create a state-of-the-art code out of our 'embryonic' 2D codes. Additional explanations for some of the algorithmic steps are given below.

Step 1: Interpolation of scalar properties from markers to nodes

According to our algorithmic approach, the temperature field and the rock type are represented by values assigned for the multitude of markers initially distributed on a fine regular marker mesh with a small ($\leq 1/2$ marker grid distance) random displacement. Other scalar properties such as density, viscosity, thermal conductivity etc. are computed for each marker at every time step in accordance to the rock type associated with each marker. This approach allows us to minimise the amount of storage associated with markers to only 3 floating point values (two coordinates and temperature) and one integer (rock type). The amount of different rock types in one experiment is typically limited to few tens at most. There are several equations (rheology, density, etc.) associated with each rock type which allow us to compute various properties for a marker of a given type, based on the local temperature and pressure at this marker. The effective values of all these properties at the Eulerian nodal points are computed from the markers at each time step by using a bilinear interpolation (Eq. (8.18)). Note that viscosity for shear (η_s) and normal (η_n) stresses is interpolated separately (Fig. 11.2) by using a local interpolation scheme that takes into account markers found within half a grid spacing distance from the nodes, in both the horizontal and vertical directions (see dashed boundary in Fig. 8.8). The statistical weights of these markers thus vary from 0.25 to 1 (one can also renormalise these weights to vary from 0 to 1). The local interpolation of viscosity allows for a more accurate solution of the momentum equation in the case of a strongly variable viscosity (Gerya and Yuen, 2007; Schmeling *et al.*, 2008). All other properties are interpolated to the basic nodes (cf. solid squares in Fig. 11.2) using a standard scheme that involves markers found in the four grid cells around each node (Fig. 8.8, Eq. (8.18)). In cases when we require a more accurate solution of the temperature equation, for example under the condition of a strongly variable thermal conductivity, localised interpolation schemes from markers should also be used for the thermal conductivity k that will be computed separately for the different heat flux nodes (see solid and open circles in Fig. 11.2). Thermal boundary conditions should be applied to the obtained nodal values after interpolating the temperature from markers. This precludes accumulating an interpolation error which will otherwise grow along the model boundaries as the number of time steps increases.

We use a standard, first-order-accurate bilinear interpolation procedure that involves four nodal points (Fig. 8.9, Eq. (8.19)) for the reverse problem of interpolating scalar properties (including calculated temperature changes), vectors and tensors from the corresponding Eulerian nodal points (see different types of Eulerian nodes in Fig. 11.2), back to the markers and other geometric points (e.g. other nodes). Equation (8.19) is used uniformly, for interpolating velocity, stresses, strain rates, pressure, temperature and other properties from nodal points to markers. Since our staggered grid represents, in fact, the superposition of four simple rectangular grids corresponding to different scalar fields, vectors and tensors (see four different symbols for grid points in Fig. 10.2), these Eulerian grids are used individually for interpolating the respective field variables. Defining the indices of the four nodal points that surround a given marker is done on the basis of the bisection procedure (Fig. 8.10).

Step 2: Solving the momentum and continuity equations

In 2D, we have two Stokes equations of slow, viscous incompressible flow in a uniform gravity field

$$x\text{-Stokes equation}\quad \frac{\partial \sigma'_{xx}}{\partial x} + \frac{\partial \sigma_{xy}}{\partial y} - \frac{\partial P}{\partial x} = -\rho g_x, \tag{11.1}$$

$$y\text{-Stokes equation}\quad \frac{\partial \sigma'_{yy}}{\partial y} + \frac{\partial \sigma_{yx}}{\partial x} - \frac{\partial P}{\partial y} = -\rho g_y, \tag{11.2}$$

$$\sigma'_{xx} = 2\eta \,\dot{\varepsilon}_{xx},$$

$$\sigma'_{yy} = 2\eta \,\dot{\varepsilon}_{yy},$$

$$\sigma_{xy} = \sigma_{yx} = 2\eta \,\dot{\varepsilon}_{xy},$$

$$\dot{\varepsilon}_{xx} = \frac{\partial v_x}{\partial x},$$

$$\dot{\varepsilon}_{yy} = \frac{\partial v_y}{\partial y},$$

$$\dot{\varepsilon}_{xy} = \frac{1}{2}\left(\frac{\partial v_x}{\partial y} + \frac{\partial v_y}{\partial x}\right).$$

Since by definition $\sigma'_{yy} = -\sigma'_{xx}$ and for incompressible fluid $\dot{\varepsilon}_{yy} = -\dot{\varepsilon}_{xx}$ in 2D, we can also avoid any nodal storage for σ'_{yy} and $\dot{\varepsilon}_{yy}$ and re-formulate Eq. (11.2) in a different form

$$-\frac{\partial \sigma'_{xx}}{\partial y} + \frac{\partial \sigma_{xy}}{\partial x} - \frac{\partial P}{\partial y} = -\rho g_y. \tag{11.2a}$$

The conservation of mass is given by the incompressible, 2D continuity equation

$$\frac{\partial v_x}{\partial x} + \frac{\partial v_y}{\partial y} = 0. \tag{11.3}$$

We use the standard procedure described in Chapter 7 (Figs. 7.11, 7.12, Eqs. 7.5, 7.6) for the formulation of FD schemes to represent the momentum equations (11.1) and (11.2) in a conservative form.

The x-Stokes equation is formulated for the horizontal velocity node $v_{x(i+1/2,j)}$

$$\left(\frac{\partial \sigma'_{xx}}{\partial x}\right)_{i+1/2,j} + \left(\frac{\partial \sigma_{xy}}{\partial y}\right)_{i+1/2,j} - \left(\frac{\partial P}{\partial x}\right)_{i+1/2,j} = -(\rho g_x)_{i+1/2,j}, \tag{11.4}$$

or in FD representation

$$2\frac{\sigma'_{xx(i+1/2,j+1/2)} - \sigma'_{xx(i+1/2,j-1/2)}}{x_{j+1} - x_{j-1}} + \frac{\sigma_{xy(i+1,j)} - \sigma_{xy(i,j)}}{y_{i+1} - y_i}$$

$$-2\frac{P_{(i+1/2,j+1/2)} - P_{(i+1/2,j-1/2)}}{x_{j+1} - x_{j-1}} = -\frac{\rho_{(i,j)} + \rho_{(i+1,j)}}{2}g_x$$

where

$$\sigma_{xy(i,j)} = 2\eta_{s(i,j)}\left(\frac{v_{x(i+1/2,j)} - v_{x(i-1/2,j)}}{y_{i+1} - y_{i-1}} + \frac{v_{y(i,j+1/2)} - v_{y(i,j-1/2)}}{x_{j+1} - x_{j-1}}\right),$$

$$\sigma_{xy(i+1,j)} = 2\eta_{s(i+1,j)}\left(\frac{v_{x(i+3/2,j)} - v_{x(i+1/2,j)}}{y_{i+2} - y_i} + \frac{v_{y(i+1,j+1/2)} - v_{y(i+1,j-1/2)}}{x_{j+1} - x_{j-1}}\right),$$

$$\sigma'_{xx(i+1/2,j-1/2)} = 2\eta_{n(i+1/2,j-1/2)}\frac{v_{x(i+1/2,j)} - v_{x(i+1/2,j-1)}}{x_j - x_{j-1}},$$

$$\sigma'_{xx(i+1/2,j+1/2)} = 2\eta_{n(i+1/2,j+1/2)}\frac{v_{x(i+1/2,j+1)} - v_{x(i+1/2,j)}}{x_{j+1} - x_j},$$

where $i, i+1/2$ and $j, j+1/2$ indices denote, respectively, the vertical and horizontal positions of the nodal points corresponding to the different physical parameters (Fig. 11.2) within the staggered grid.

The y-Stokes equation is formulated for the vertical velocity node $v_{y(i,j+1/2)}$

$$\left(\frac{\partial\sigma'_{yy}}{\partial y}\right)_{i,j+1/2} + \left(\frac{\partial\sigma_{xy}}{\partial x}\right)_{i,j+1/2} - \left(\frac{\partial P}{\partial y}\right)_{i,j+1/2} = -(\rho g_y)_{i,j+1/2} \qquad (11.5)$$

or in FD representation

$$2\frac{\sigma'_{yy(i+1/2,j+1/2)} - \sigma'_{yy(i-1/2,j+1/2)}}{y_{i+1} - y_{i-1}} + \frac{\sigma_{xy(i,j+1)} - \sigma_{xy(i,j)}}{x_{j+1} - x_j}$$
$$- 2\frac{P_{(i+1/2,j+1/2)} - P_{(i-1/2,j+1/2)}}{y_{i+1} - y_{i-1}} = -\frac{\rho_{(i,j)} + \rho_{(i,j+1)}}{2}g_y$$

where $\sigma_{xy(i,j)}$ is given above for Eq. (11.4)

$$\sigma_{xy(i,j+1)} = 2\eta_{s(i,j+1)}\left(\frac{v_{x(i+1/2,j+1)} - v_{x(i-1/2,j+1)}}{y_{i+1} - y_{i-1}} + \frac{v_{y(i,j+3/2)} - v_{y(i,j+1/2)}}{x_{j+2} - x_j}\right),$$

$$\sigma'_{yy(i-1/2,j+1/2)} = 2\eta_{n(i-1/2,j+1/2)}\frac{v_{y(i,j+1/2)} - v_{y(i-1,j+1/2)}}{y_i - y_{i-1}},$$

$$\sigma'_{yy(i+1/2,j+1/2)} = 2\eta_{n(i+1/2,j+1/2)}\frac{v_{y(i+1,j+1/2)} - v_{y(i,j+1/2)}}{y_{i+1} - y_i}.$$

The continuity equation is formulated for the pressure node $P_{(i-1/2,j-1/2)}$

$$\left(\frac{\partial v_x}{\partial x}\right)_{i-1/2,j-1/2} + \left(\frac{\partial v_y}{\partial y}\right)_{i-1/2,j-1/2} = 0, \qquad (11.6)$$

or in FD representation

$$\frac{v_{x(i-1/2,j)} - v_{x(i-1/2,j-1)}}{x_j - x_{j-1}} + \frac{v_{y(i,j-1/2)} - v_{y(i-1,j-1/2)}}{y_i - y_{i-1}} = 0.$$

After composing all equations, we *invert* (i.e. solve) the resulting global matrix by an accurate, direct method in order to simultaneously solve the momentum equations (11.4) and (11.5) and the continuity equation (11.6), combined with any boundary conditions for the velocity and pressure (Chapter 7). The momentum equations (11.4) and (11.5) are solved for $v_{x(i+1/2,j)}$ and $v_{y(i,j+1/2)}$, respectively, while the continuity equation (11.6) is solved for $P_{(i-1/2,j-1/2)}$. The incompressible continuity equation (11.6) does not initially contain $P_{(i-1/2,j-1/2)}$ and, as was discussed in Chapter 7, the solution is guaranteed by the order of processing during the inversion of the global matrix based on relating the staggered nodes (open squares and open and solid circles in Fig. 11.2), to the basic nodes of the grid formed by intersections of the horizontal and vertical grid lines (solid squares in Fig. 11.2). This was explained in detail in Chapter 7 (see Figs. 7.15, 7.17).

Steps 4–7: Solving the temperature equation

In order to avoid numerical diffusion of temperature we use the Lagrangian form of heat conservation equation and advect the temperature with markers with the technique described in Chapter 10. The temperature equation is formulated in 2D for the case of variable thermal conductivity and takes into account heat generation H from variable sources including radioactive (H_r), adiabatic (H_a), shear (H_s) and latent (H_L) heat production:

$$\rho C_P \frac{DT}{Dt} = -\frac{\partial q_x}{\partial x} - \frac{\partial q_y}{\partial y} + H_r + H_a + H_s + H_L, \tag{11.7}$$

where

$$q_x = -k\frac{\partial T}{\partial x},$$
$$q_y = -k\frac{\partial T}{\partial y},$$
$$H_r = const,$$
$$H_a = T\alpha\frac{DP}{Dt} = T\alpha\left(\frac{\partial P}{\partial x}v_x + \frac{\partial P}{\partial y}v_y\right),$$
$$H_s = \sigma'_{xx}\dot{\varepsilon}_{xx} + \sigma'_{yy}\dot{\varepsilon}_{yy} + \sigma_{xy}\dot{\varepsilon}_{xy} + \sigma_{yx}\dot{\varepsilon}_{yx} = 2\sigma'_{xx}\dot{\varepsilon}_{xx} + 2\sigma_{xy}\dot{\varepsilon}_{xy}.$$

This equation takes into account the adiabatic heating term, which is in some contradiction with using the incompressible fluid approximation in the momentum and continuity equations (Eqs. (11.1)–(11.3)). However, this is a common simplification (called the *extended Boussinesq approximation*) in numerical geodynamic modelling. It is frequently adopted because of the very small thermal expansion and compressibility of crustal and mantle rocks in the absence of phase transformations (Chapter 2). The calculation of H_a can be simplified by neglecting deviations (which are relatively small in most cases) of the dynamic pressure gradients $\frac{\partial P}{\partial x}$

and $\dfrac{\partial P}{\partial y}$ from ρg_x and ρg_y values

$$H_a \approx T\alpha\rho(g_x v_x + g_y v_y).$$

We use a standard procedure (Fig. 10.5, Eq. (10.9)) for the formulation of a FD scheme in order to represent the temperature equation (11.7) in a conservative form as follows

$$\rho_{i,j}C_{Pi,j}\left(\frac{DT}{Dt}\right)_{i,j} = -\left(\frac{\partial q_x}{\partial x}\right)_{i,j} - \left(\frac{\partial q_y}{\partial y}\right)_{i,j} + H^t_{r(i,j)} + H^t_{a(i,j)} + H^t_{s(i,j)} + H^t_{L(i,j)}$$

$$\tag{11.8}$$

or in FD formulation

$$\rho_{(i,j)}C_{Pi,j}\frac{T^{t+\Delta t}_{i,j} - T^t_{i,j}}{\Delta t} = -2\frac{q_{x(i,j+1/2)} - q_{x(i,j-1/2)}}{x_{j+1} - x_{j-1}} - 2\frac{q_{y(i+1/2,j)} - q_{y(i-1/2,j)}}{y_{i+1} - y_{i-1}}$$

$$+ H^t_{r(i,j)} + H^t_{a(i,j)} + H^t_{s(i,j)} + H^t_{L(i,j)}$$

or, by grouping the known parameters on the right-hand side

$$\frac{\rho_{i,j}C_{Pi,j}}{\Delta t}T^{t+\Delta t}_{i,j} + 2\frac{q_{x(i,j+1/2)} - q_{x(i,j-1/2)}}{x_{j+1} - x_{j-1}} + 2\frac{q_{y(i+1/2,j)} - q_{y(i-1/2,j)}}{y_{i+1} - y_{i-1}}$$

$$= \frac{\rho_{i,j}C_{Pi,j}}{\Delta t}T^t_{i,j} + H^t_{r(i,j)} + H^t_{a(i,j)} + H^t_{s(i,j)} + H^t_{L(i,j)}$$

where

$$q_{x(i,j-1/2)} = -\frac{1}{2}(k_{i,j-1} + k_{i,j})\frac{T^{t+\Delta t}_{i,j} - T^{t+\Delta t}_{i,j-1}}{x_j - x_{j-1}},$$

$$q_{x(i,j+1/2)} = -\frac{1}{2}(k_{i,j} + k_{i,j+1})\frac{T^{t+\Delta t}_{i,j+1} - T^{t+\Delta t}_{i,j}}{x_{j+1} - x_j},$$

$$q_{y(i-1/2,j)} = -\frac{1}{2}(k_{i-1,j} + k_{i,j})\frac{T^{t+\Delta t}_{i,j} - T^{t+\Delta t}_{i-1,j}}{y_i - y_{i-1}},$$

$$q_{y(i+1/2,j)} = -\frac{1}{2}(k_{i,j} + k_{i+1,j})\frac{T^{t+\Delta t}_{i+1,j} - T^{t+\Delta t}_{i,j}}{y_{i+1} - y_i},$$

$$H^t_{a(i,j)} = T^t_{i,j}\alpha_{i,j}\rho_{i,j}\left(\frac{v_{x(i-1/2,j)} + v_{x(i+1/2,j)}}{2}g_x + \frac{v_{y(i,j-1/2)} + v_{y(i,j+1/2)}}{2}g_y\right),$$

$$H^t_{s(i,j)} = \frac{1}{2}\sigma'_{xx(i-1/2,j-1/2)}\dot{\varepsilon}_{xx(i-1/2,j-1/2)} + \frac{1}{2}\sigma'_{xx(i-1/2,j+1/2)}\dot{\varepsilon}_{xx(i-1/2,j+1/2)}$$

$$+ \frac{1}{2}\sigma'_{xx(i+1/2,j-1/2)}\dot{\varepsilon}_{xx(i+1/2,j-1/2)} + \frac{1}{2}\sigma'_{xx(i+1/2,j+1/2)}\dot{\varepsilon}_{xx(i+1/2,j+1/2)}$$

$$+ 2\sigma_{xy(i,j)}\dot{\varepsilon}_{xy(i,j)},$$

where the index $t + \Delta t$ denotes values for the next time instant to be reached by the time-stepping procedure and Δt is the optimal time step (see Step 5 of the algorithm); $H^t_{r(i,j)}$, $H^t_{a(i,j)}$, $H^t_{s(i,j)}$, $\alpha_{i,j}$, $\rho_{i,j}$, $C_{Pi,j}$, $\sigma_{xy(i,j)}$, $\dot{\varepsilon}_{xy(i,j)}$, $v_{x(i-1/2,j)}$ etc. are values of the corresponding parameters for various nodes of the staggered grid ((Fig. 11.2): scalar properties are interpolated from the markers using relation (8.18) (Fig. 8.8), vectors and tensors are taken from the corresponding surrounding nodes.

To solve the temperature equation, we invert directly the global matrix with a direct method. The matrix also contains the linear equations associated with the thermal boundary conditions. The overall numbering of $T^{t+\Delta t}_{i,j}$ for the *global temperature matrix* that defines the order of processing of temperature equations (11.8) is given in Exercise 10.1 (Eq. 10.24).

For advection of temperature, we use the same marker-in-cell technique (Chapter 10) as is commonly used for advection of material field properties, such as the density, viscosity, chemical composition etc. The interpolation of the calculated temperature changes from the Eulerian nodal points to the moving markers reduces numerical diffusion in an efficient manner (Chapter 10), which represents one of the highlights of our computational strategy. This method does not produce any smoothing of the temperature distribution between adjacent markers (Fig. 11.2, diagram for step 7), thus resolving the thermal structure of a numerical model in much finer details. Non-physical small-scale (subgrid) temperature oscillations appearing on markers (Fig. 10.7(a)) in the case of strong chaotic mixing are removed (Fig. 10.7(b)) by using a consistent subgrid diffusion operation (Eqs. (10.15)–(10.19)).

11.3 Adding self-gravity and free surface

For the case when one wants to model the dynamics of self-gravitating planetary bodies, the numerical algorithms presented so far can be easily modified to take a variable gravity field and a free planetary surface into account. This is done by solving the Poisson equation for gravity potential after computing the nodal density field from markers, and before solving the momentum and continuity equations (Fig. 11.3). A staggered grid corresponding to this new algorithm is shown in Fig. 11.4.

In this grid, the gravitational potential Φ (Chapter 2) is defined at the cell centres (i.e. at the pressure nodes) and computed according to the 2D Poisson equation

$$\frac{\partial^2 \Phi}{\partial x^2} + \frac{\partial^2 \Phi}{\partial y^2} = 4K\pi G\rho, \tag{11.16}$$

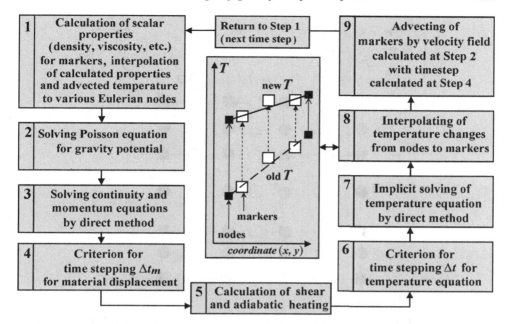

Fig. 11.3 Flow chart representing the modified algorithm of the numerical thermo-mechanical viscous 2D code allowing the modelling of self-gravitating planetary bodies with a fully staggered Cartesian grid (Gerya and Yuen, 2007). Note the new Step 2 compared to Fig. 11.1.

where G is the gravitational constant and K depends on the geometry of self-gravitating body modelled in 2D ($K=1$ and $K=2/3$ stand for cylindrical and spherical geometry, respectively). The factor $K=2/3$ scales the 2D gravity field inside a cylinder of constant density ρ

$$\Phi(r)_{cylindrical} = \pi G\rho r^2, \ g(r)_{cylindrical} = -\frac{\partial \Phi(r)_{cylindrical}}{\partial r} = -2\pi G\rho r,$$

to a 3D gravity field inside a sphere of the same density

$$\Phi(r)_{spherical} = \frac{2}{3}\pi G\rho r^2, \ g(r)_{spherical} = -\frac{\partial \Phi(r)_{spherical}}{\partial r} = -\frac{4}{3}\pi G\rho r,$$

where r is the distance from the centre of the cylinder/sphere. It should be mentioned that this simplified scaling does not allow the exact reproduction of a spherical gravity field in 2D. In particular, the gravitational acceleration is noticeably overestimated *outside the self-gravitating body* since it is proportional to $1/r$ for a cylindrical gravity field and to $1/r^2$ for a spherical one. Our scaling approach allows us to capture changes in an *internal gravity field* that acts on a self-gravitating body with a changing internal mass distribution (Lin *et al.*, 2009). In many cases this is sufficient for the purposes of modelling internal planetary processes.

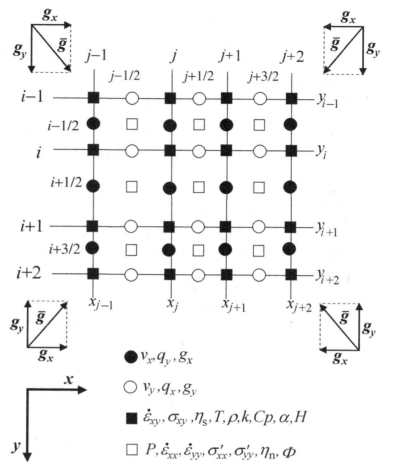

Fig. 11.4 Staggered 2D irregularly spaced numerical grid which corresponds to the algorithm presented in Fig. 11.3.

Discretising Eq. (11.16) in 2D is rather simple and uses a 5-node stencil typical for approximating the Poisson equation with finite differences

$$\left(\frac{\partial^2 \Phi}{\partial x^2}\right)_{i+1/2,j+1/2} + \left(\frac{\partial^2 \Phi}{\partial x^2}\right)_{i+1/2,j+1/2} = 4K\pi G\rho_{i+1/2,j+1/2}, \qquad (11.17)$$

$$\left(\frac{\partial^2 \Phi}{\partial x^2}\right)_{i+1/2,j+1/2} = 2\frac{\Phi_{i+1/2,j+3/2} - \Phi_{i+1/2,j+1/2}}{(x_{j+2} - x_j)(x_{j+1} - x_j)} - 2\frac{\Phi_{i+1/2,j+1/2} - \Phi_{i+1/2,j-1/2}}{(x_{j+1} - x_{j-1})(x_{j+1} - x_j)},$$

$$\left(\frac{\partial^2 \Phi}{\partial y^2}\right)_{i+1/2,j+1/2} = 2\frac{\Phi_{i+3/2,j+1/2} - \Phi_{i+1/2,j+1/2}}{(y_{i+2} - y_i)(y_{i+1} - y_i)} - 2\frac{\Phi_{i+1/2,j+1/2} - \Phi_{i-1/2,j+1/2}}{(y_{i+1} - y_{i-1})(y_{i+1} - y_i)},$$

$$\rho_{i+1/2,j+1/2} = \frac{1}{4}(\rho_{i,j} + \rho_{i,j+1} + \rho_{i+1,j} + \rho_{i+1,j+1}),$$

where the i, $i+1/2$ and j, $j+1/2$ indices denote, respectively, the horizontal and vertical position of the nodal points corresponding to the different physical parameters (Fig. 11.4) within the staggered grid. We invert the global matrix by a direct method for solving Poisson Eqs. (11.17), combined with linear equations for the boundary conditions. The overall numbering of unknowns for the *global gravity potential matrix* is given in Exercise 3.2 (Eq. 3.22).

The gravitational acceleration vector components are then defined at respective Eulerian nodes (see solid and open circles in Fig. 11.4) by numerical differentiation

$$g_{x(i+1/2,j)} = -2\frac{\Phi_{i+1/2,j+1/2} - \Phi_{i+1/2,j-1/2}}{x_{j+1} - x_{j-1}}, \tag{11.18}$$

$$g_{y(i,j+1/2)} = -2\frac{\Phi_{i+1/2,j+1/2} - \Phi_{i-1/2,j+1/2}}{y_{i+1} - y_{i-1}}. \tag{11.19}$$

The new, locally defined gravity acceleration components $g_{x(i+1/2,j)}$ and $g_{y(i,j+1/2)}$ are then included in the right-hand side of the x- and y-Stokes equations (11.4) and (11.5) instead of the globally defined values g_x and g_y, respectively.

Numerical modelling of deformation of a self-gravitating planetary body requires computing the gravity field which changes with time in response to variations in mass distribution inside the planet. Changes in shape of the planet and the related planetary surface deformation should also be considered. In order to tackle these requirements, one can use a 'spherical-Cartesian' approach (Honda *et al.*, 1993; Gerya and Yuen, 2007; Lin *et al.*, 2009) that allows the computation of self-gravitating bodies of arbitrary form on Cartesian grids including the presence of a free surface:

(1) The body is surrounded by the weak medium (e.g. Fig. 11.5) of very low density (≤ 1 kg/m^3) and low viscosity allowing for a high (10^1–10^6) viscosity contrast at the planetary surface.
(2) The gravity field is computed by solving the Poisson equation for the gravitational potential (Eq. (11.17)) associated with the mass (density) distribution portrayed by the markers at each time step.
(3) While solving the momentum equation, the components of the gravitational acceleration vector are computed locally by numerical differentiation of the gravitational potential (Eqs. (11.18), (11.19)) at the corresponding nodal points.

As seen in Fig. 11.5, the spontaneously formed planetary surface is numerically stable under conditions of very strong internal deformation inside the planet. In addition, a spontaneously forming spherical/cylindrical shape of the body is characteristic for a stable density distribution (i.e. when density increases toward the core of the body, see final stages of Fig. 11.5). No evidence for non-spherical Cartesian grid dependence of this stable shape was discerned (Lin *et al.*, 2009).

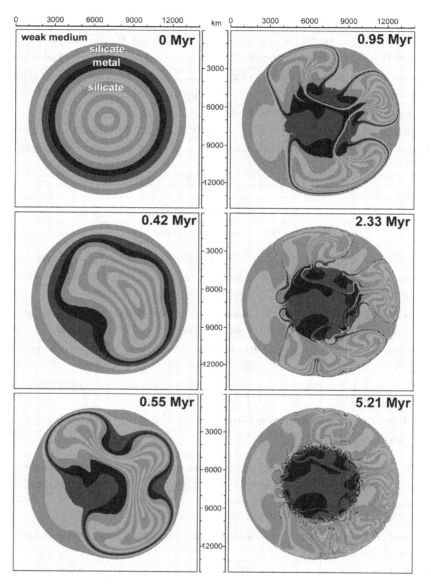

Fig. 11.5 Numerical modelling of a self-gravitating planetary body with the use of spherical-Cartesian method (Gerya and Yuen, 2007).

Using a weak top layer approach for modelling free surfaces is also applicable for normal Cartesian simulations in a prescribed (e.g. purely vertical) gravity field (Schmeling *et al.*, 2008). Combined with various erosion/sedimentation functions, this approach also allows modelling the topography evolution in response to geodynamic processes with the use of the standard thermomechanical algorithm (Fig. 11.1). This point will be argued in Chapter 17.

Programming exercise and homework

Exercise 11.1

Program a thermomechanical code (Fig. 11.1) based on a regularly spaced staggered grid for the case of variable viscosity and thermal conductivity. Combine the thermal and mechanical solutions programmed for Exercise 10.4 and Exercise 8.5, respectively. Use the same vertically layered cross-section and prescribe a different temperature and thermal conductivity for the left ($T = 1000$ K, $k = 1$ W/m/K) and right ($T = 1300$ K, $k = 10$ W/m/K) layer. Use insulating boundary conditions on all boundaries. Include radiogenic (10^{-6} W/m^3 and 10^{-7} W/m^3 for the left and right layer respectively), shear and adiabatic heating terms (Eq. (11.8)) in the temperature equation (they were already programmed for Exercises 9.3 and 9.4). Restrict the time step by limiting the maximal temperature changes to 20 K per time step. If these changes are bigger, then reduce the time step proportionally and repeat the solution for the second time (temperature equation solving is computationally inexpensive compared to the momentum and continuity equations). An example is in **i2vis.m**.

12

Elasticity and plasticity

Theory: Elastic rheology. Maxwell visco-elastic rheology. Rotation of stresses during advection. Plastic rheology. Plastic yielding criterion. Plastic flow potential. Plastic flow rule.
Exercises: Stress build-up/relaxation with a visco-elastic Maxwell rheology.

12.1 Why care about elasticity and plasticity?

As mentioned in the Introduction, rocks behave elastically on a relatively short time scale ($<10^4$ years) and, therefore, modelling of relatively fast processes within the Earth's crust and mantle (e.g. magma intrusion) should take into account the elastic properties of rocks. On the other hand, rocks at cold temperatures can also be subjected to localised *brittle* (at low pressure) and *plastic* (at higher pressure) deformation, which leads to shear zones and fracture zones in natural rock complexes. Therefore, if we want to account for this broad range of geodynamic conditions in our models, we should generally consider the *visco-elasto-plastic rheology* of rocks and be able to model such a complex rheology with our thermomechanical numerical codes. This chapter discusses elastic and plastic rheological behaviours and compares them to the viscous rheology.

12.2 Elastic rheology

The elastic rheology assumes proportionality of stress and strain (Fig. 12.1). This is expressed by the Hooke's law

$$\sigma = E\gamma, \tag{12.1}$$

165

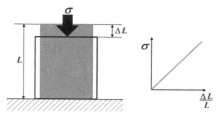

Fig. 12.1 Relationship between applied stress σ and deformation ΔL, of an elastic
body with initial length L.

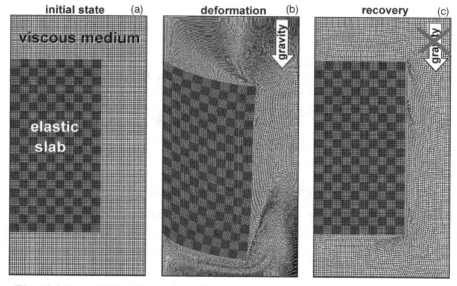

Fig. 12.2 Reversible deformation of initially unstressed (a) elastic slab surrounded
by a weak viscous medium (Gerya and Yuen, 2007). Deformation of the slab in (b)
is caused by a vertical gravity field. When gravity is 'removed', the deformed slab
recovers its original shape (c) while the surrounding medium remains deformed
since viscous deformation is irreversible.

where σ is applied stress, $\gamma = \dfrac{\Delta L}{L}$ is elastic strain (i.e. displacement ΔL normalised to the initial length L of the deforming body) and E is the proportionality
(elasticity) coefficient. In contrast to viscous deformation, the elastic deformation
is reversible: if the load applied on an elastic body is removed, the body instantly
recovers its original state (Fig. 12.2). This shape recovery effect is reproduced with
a spring.

For an isotropic body in 3D, the elastic relationship is written in a tensorial form as

$$\sigma_{ij} = \lambda \delta_{ij} \varepsilon_{kk} + 2\mu \varepsilon_{ij}, \tag{12.2}$$

$$\varepsilon_{ij} = \frac{1}{2} \left(\frac{\partial u_i}{\partial x_j} + \frac{\partial u_j}{\partial x_i} \right), \tag{12.3}$$

$$\varepsilon_{kk} = \varepsilon_{xx} + \varepsilon_{yy} + \varepsilon_{zz}, \tag{12.4}$$

where x_i and x_j are the spatial coordinates (x, y, z), σ_{ij} are components of stress, u_i and u_j are components of *material displacement* vector $\vec{u} = (u_x, u_y, u_z)$ so that $v_i = \dfrac{Du_i}{Dt}$ where v_i is component of velocity vector $\vec{v} = (v_x, v_y, v_z)$, ε_{ij} are components of the strain tensor so that $\dot{\varepsilon}_{ij} = \dfrac{D\varepsilon_{ij}}{Dt}$, where $\dot{\varepsilon}_{ij}$ are components of strain rate tensor (Chapter 4), ε_{kk} is volumetric strain (*cubical dilatation*) and λ and μ are two elastic parameters termed *Lamé's constants* (which depend on pressure temperature and composition). The material displacement characterises the absolute amount of deformation (i.e. is similar to ΔL in Fig. 12.1), while the strain tensor ε_{ij} reflects the relative intensity of this deformation (i.e. is similar to $\dfrac{\Delta L}{L}$ in Fig. 12.1). Formulating pressure through mean normal stress, we can obtain the following relation

$$P = -\frac{\sigma_{kk}}{3} = -\frac{\sigma_{xx} + \sigma_{yy} + \sigma_{zz}}{3}, \tag{12.5}$$

$$P = -\frac{3\lambda + 2\mu}{3}(\varepsilon_{xx} + \varepsilon_{yy} + \varepsilon_{zz}) = -B\varepsilon_{kk}, \tag{12.6}$$

$$B = \lambda + \frac{2}{3}\mu, \tag{12.7}$$

where B is the *bulk modulus* (see Chapter 2) also called *incompressibility*; it establishes the relation between mean stress (pressure) and volumetric strain. Accordingly, the deviatoric stresses σ'_{ij} can be formulated as

$$\sigma'_{ij} = \sigma_{ij} - \frac{\sigma_{kk}}{3}\delta_{ij} = \sigma_{ij} + P\delta_{ij}, \tag{12.8}$$

$$\sigma'_{ij} = \sigma_{ij} - B\varepsilon_{kk}\delta_{ij} = 2\mu\left(\varepsilon_{ij} - \frac{1}{3}\delta_{ij}\varepsilon_{kk}\right), \tag{12.9}$$

$$\sigma'_{ij} = 2\mu\varepsilon'_{ij}, \tag{12.10}$$

$$\varepsilon'_{ij} = \varepsilon_{ij} - \frac{1}{3}\delta_{ij}\varepsilon_{kk}, \tag{12.11}$$

where ε'_{ij} are components of the deviatoric strain tensor and μ is the *shear modulus* or *rigidity*, which is one of the Lamé's constants. The elastic shear modulus

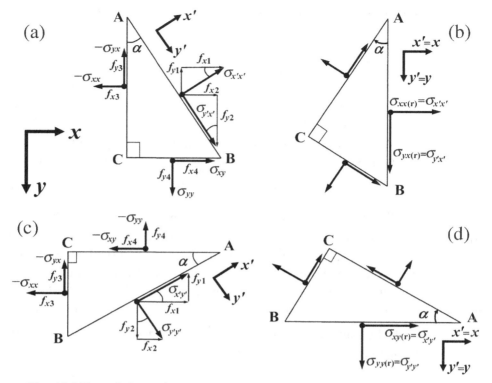

Fig. 12.3 Force balance for a small triangle ABC used for computing σ_{xx}, σ_{yx} (a), (b) and σ_{yy}, σ_{xy} (c), (d) stress components after a clockwise solid body rotation of the triangle by an angle α around point A.

establishes the relationship between the deviatoric stress and deviatoric strain for the elastic rheology (μ is thus somewhat similar to the shear viscosity η which defines the relationship between the deviatoric stress and the deviatoric strain rate for a viscous rheology, see Eq. (5.12)).

12.3 Rotation of elastic stresses

An important peculiarity of stress behaviour in a deforming elastic medium consists in local stress orientation changes due to the rotation of material points. This rotation is caused by a rigid body rotation which is present in the velocity field, which changes the orientation of principal stress axes for moving Lagrangian points. Stress rotation changes stress tensor components but not the stress invariants.

The way of computing of various elastic stress components after rotation of a Lagrangian point by an angle α can be derived on the basis of analysing the force balance for a small triangle ABC, as shown in Fig. 12.3. Let us take a small triangle with two sides parallel to the x and y axes oriented as shown in Fig. 12.3(a). The force equilibrium condition requires that forces acting on the outside of the triangle balance each other and that the resulting force is zero

$$f_x = f_{x1} + f_{x2} + f_{x3} + f_{x4} = 0, \qquad (12.12)$$
$$f_y = f_{y1} + f_{y2} + f_{y3} + f_{y4} = 0, \qquad (12.13)$$

where f are forces acting on different sides of the triangle from the outside. These forces can be computed from shear and normal stresses as discussed in Chapter 4 (note that arrows shown on side AC correspond to the counter-force part of σ_{xx} and σ_{yx} which therefore have a minus sign

$$|AC| = |AB| \cos(\alpha),$$
$$|BC| = |AB| \sin(\alpha),$$
$$f_{x1} = \sigma_{x'x'} |AB| \cos(\alpha),$$
$$f_{x2} = \sigma_{y'x'} |AB| \sin(\alpha),$$
$$f_{x3} = -\sigma_{xx} |AC| = -\sigma_{xx} |AB| \cos(\alpha),$$
$$f_{x4} = \sigma_{xy} |BC| = \sigma_{xy} |AB| \sin(\alpha),$$
$$f_{y1} = -\sigma_{x'x'} |AB| \sin(\alpha),$$
$$f_{y2} = \sigma_{y'x'} |AB| \cos(\alpha),$$
$$f_{y3} = -\sigma_{yx} |AC| = -\sigma_{yx} |AB| \cos(\alpha),$$
$$f_{y4} = \sigma_{yy} |BC| = \sigma_{yy} |AB| \sin(\alpha),$$

where $|AB|$, $|AC|$ and $|BC|$ are the lengths of the respective triangle sides. Then, Equations (12.12) and (12.13) can be converted to yield

$$\frac{f_x}{|AB|} = \sigma_{x'x'} \cos(\alpha) + \sigma_{y'x'} \sin(\alpha) - \sigma_{xx} \cos(\alpha) + \sigma_{xy} \sin(\alpha) = 0, \qquad (12.14)$$

$$\frac{f_y}{|AB|} = -\sigma_{x'x'} \sin(\alpha) + \sigma_{y'x'} \cos(\alpha) - \sigma_{yx} \cos(\alpha) + \sigma_{yy} \sin(\alpha) = 0. \qquad (12.15)$$

By multiplying Eq. (12.14) by $\cos(\alpha)$ and Eq. (12.15) by $\sin(\alpha)$, we obtain

$$\sigma_{x'x'} \cos^2(\alpha) + \sigma_{y'x'} \sin(\alpha) \cos(\alpha) - \sigma_{xx} \cos^2(\alpha) + \sigma_{xy} \sin(\alpha) \cos(\alpha) = 0, \qquad (12.16)$$

$$-\sigma_{x'x'} \sin^2(\alpha) + \sigma_{y'x'} \cos(\alpha) \sin(\alpha) - \sigma_{yx} \cos(\alpha) \sin(\alpha) + \sigma_{yy} \sin^2(\alpha) = 0. \qquad (12.17)$$

By subtracting Eq. (12.17) from Eq. (12.16) we get

$$\sigma_{x'x'} (\cos^2(\alpha) + \sin^2(\alpha)) - \sigma_{xx} \cos^2(\alpha) + \sigma_{xy} \sin(\alpha) \cos(\alpha)$$
$$+ \sigma_{yx} \cos(\alpha) \sin(\alpha) - \sigma_{yy} \sin^2(\alpha) = 0,$$

which can be further be simplified to

$$\sigma_{x'x'} = \sigma_{xx} \cos^2(\alpha) + \sigma_{yy} \sin^2(\alpha) - \sigma_{xy} \sin(2\alpha) \qquad (12.18)$$

by using $\sigma_{yx} = \sigma_{xy}$ and the trigonometric relations $\sin^2(\alpha) + \cos^2(\alpha) = 1$ and $2 \sin(\alpha) \cos(\alpha) = \sin(2\alpha)$.

Similarly, by multiplying Eq. (12.14) by $\sin(\alpha)$ and Eq. (12.15) by $\cos(\alpha)$, and adding them to each other we obtain the following expression for $\sigma_{y'x'}$

$$\sigma_{y'x'} = \frac{1}{2}(\sigma_{xx} - \sigma_{yy}) \sin(2\alpha) + \sigma_{xy} \cos(2\alpha) \qquad (12.19)$$

using the trigonometric relation $\cos^2(\alpha) - \sin^2(\alpha) = \cos(2\alpha)$ (verify as an exercise).

Obviously, after the triangle has been rotated clockwise by an angle α around point A (Fig. 12.3b), the AB side becomes parallel to the y-axis and $\sigma_{x'x'}$, $\sigma_{y'x'}$ will correspond to the respective stress components $\sigma_{xx(r)}$, $\sigma_{yx(r)}$ for the rotated system.

Similarly, by analysing Fig. (12.3c,d), the following expressions for the corrected stress components $\sigma_{y'y'}$ and $\sigma_{x'y'}$ can be obtained (verify as an exercise))

$$\sigma_{y'y'} = \sigma_{xx} \sin^2(\alpha) + \sigma_{yy} \cos^2(\alpha) + \sigma_{xy} \sin(2\alpha), \qquad (12.20)$$

$$\sigma_{x'y'} = \frac{1}{2}(\sigma_{xx} - \sigma_{yy}) \sin(2\alpha) + \sigma_{xy} \cos(2\alpha). \qquad (12.21)$$

Equations (12.19) and (12.21) are obviously equivalent since $\sigma_{yx} = \sigma_{xy}$. Note that rotation does not change the first stress invariant (mean normal stress, pressure) and thus

$$\sigma_{x'x'} + \sigma_{y'y'} = \sigma_{xx} + \sigma_{yy} = -2P.$$

Equations for rotating deviatoric normal stresses are similar to (12.18) and (12.20) since subtracting mean stress does not change the form of these expressions, hence

$$\sigma'_{x'x'} = \sigma'_{xx} \cos^2(\alpha) + \sigma'_{yy} \sin^2(\alpha) - \sigma_{xy} \sin(2\alpha), \qquad (12.22)$$

$$\sigma'_{y'y'} = \sigma'_{xx} \sin^2(\alpha) + \sigma'_{yy} \cos^2(\alpha) + \sigma_{xy} \sin(2\alpha). \qquad (12.23)$$

The condition $\sigma'_{x'x'} + \sigma'_{y'y'} = 0$ is also satisfied due to $\sigma'_{xx} + \sigma'_{yy} = 0$.

If α is very small and tends to 0, then $\cos(\alpha)$ tends to 1, $\sin^2(\alpha)$ tends to 0, and $\sin(\alpha)$ tends to α. In this case, Equations (12.18)–(12.23) can be simplified to the *Jaumann co-rotation formulas*, which are often used in numerical modelling of elastic problems to account for effects of solid body rotation of stresses

$$\sigma_{x'x'} = \sigma_{xx} - \sigma_{xy} 2\alpha, \qquad (12.24)$$

$$\sigma_{y'y'} = \sigma_{yy} + \sigma_{xy} 2\alpha, \qquad (12.25)$$

$$\sigma_{x'y'} = \sigma_{xy} + (\sigma_{xx} - \sigma_{yy})\alpha. \qquad (12.26)$$

In a complex velocity field, the intensity of rotation is defined by the local rotation rate ω which can be computed from the local velocity field as

$$\omega = \frac{\partial \alpha}{\partial t} = \frac{1}{2}\left(\frac{\partial v_y}{\partial x} - \frac{\partial v_x}{\partial y}\right). \tag{12.27}$$

Note that the expressions for stress rotation and formulation of ω depend on

- the convention for normal stresses – if they are taken to be positive (here) or negative (e.g. Turcotte and Schubert, 2002) under extension,
- orientation of x and y axes,
- definition whether clockwise rotation direction is taken to be positive (here) or negative (e.g. Turcotte and Schubert, 2002).

In particular, Equation (12.27) may become invalid, if the definitions of stresses, axes and rotation are different from those used here. In this case, Equations (12.18)–(12.27) should not be used 'mechanically', but should rather be *re-derived* on the basis of a similar analysis.

In 3D, the rotation rate is represented by a rotation rate tensor with components defined as

$$\omega_{ij} = \frac{1}{2}\left(\frac{\partial v_i}{\partial x_j} - \frac{\partial v_j}{\partial x_i}\right), \quad \text{when} \left(\frac{\partial v_i}{\partial x_j} - \frac{\partial v_j}{\partial x_i}\right) > 0 \quad \text{for } clockwise \text{ rotation,} \tag{12.28}$$

$$\omega_{ij} = \frac{1}{2}\left(\frac{\partial v_j}{\partial x_i} - \frac{\partial v_i}{\partial x_j}\right), \quad \text{when} \left(\frac{\partial v_j}{\partial x_i} - \frac{\partial v_i}{\partial x_j}\right) > 0 \quad \text{for } clockwise \text{ rotation,} \tag{12.29}$$

where i and j are coordinate indices; x_i and x_j are spatial coordinates and the view at the ij-plane should be taken in the direction of the third axis. In contrast to symmetric stress and strain rate tensors (i.e. $\sigma_{ij} = \sigma_{ji}$, $\dot{\varepsilon}_{ij} = \dot{\varepsilon}_{ji}$), the rotation rate tensor is anti-symmetric (i.e., $\omega_{ij} = -\omega_{ji}$) and the normal components of this tensor are always equal to zero (i.e., $\omega_{xx} = \omega_{yy} = \omega_{zz} = 0$). One possible way of computing both total and deviatoric stress rotation in 3D is to use the general form of the *Jaumann stress rate*

total stress:

$$\dot{\sigma}_{ij(Jaumann)} = \sigma_{ik}\omega_{kj} - \omega_{ik}\sigma_{kj}, \tag{12.30a}$$

deviatoric stress

$$\dot{\sigma}'_{ij(Jaumann)} = \sigma'_{ik}\omega_{kj} - \omega_{ik}\sigma'_{kj}, \tag{12.30b}$$

where $\dot{\sigma}_{ij(Jaumann)}$ is the rate of change for the σ_{ij} stress component and the repeated index k indicates a summation. Using Eq. (12.30) in 3D for example, the σ'_{xx}

deviatoric stress component yields

$$\dot\sigma'_{xx(Jaumann)} = \sigma'_{xx}\omega_{xx} + \sigma_{xy}\omega_{yx} + \sigma_{xz}\omega_{zx} - \omega_{xx}\sigma'_{xx} - \omega_{xy}\sigma_{yx} - \omega_{xz}\sigma_{zx},$$

$$(12.31a)$$

$$\dot\sigma'_{xx(Jaumann)} = \sigma_{xy}\omega_{yx} + \sigma_{xz}\omega_{zx} - \omega_{xy}\sigma_{yx} - \omega_{xz}\sigma_{zx}, \qquad (12.31b)$$

$$\dot\sigma'_{xx(Jaumann)} = 2\sigma_{xy}\omega_{yx} + 2\sigma_{xz}\omega_{zx}, \qquad (12.31c)$$

where $\omega_{xx} = \dfrac{1}{2}\left(\dfrac{\partial v_x}{\partial x} - \dfrac{\partial v_x}{\partial x}\right) = 0$, $\omega_{yx} = \dfrac{1}{2}\left(\dfrac{\partial v_x}{\partial y} - \dfrac{\partial v_y}{\partial x}\right) = -\omega_{xy}$ and $\omega_{zx} = \dfrac{1}{2}\left(\dfrac{\partial v_x}{\partial z} - \dfrac{\partial v_z}{\partial x}\right) = -\omega_{xz}$, according to Eq. (12.29). Similar derivations can also be done for other deviatoric stress components (verify these as an exercise)

$$\dot\sigma'_{yy(Jaumann)} = 2\sigma_{yx}\omega_{xy} + 2\sigma_{yz}\omega_{zy}, \qquad (12.32)$$

$$\dot\sigma'_{zz(Jaumann)} = 2\sigma_{zx}\omega_{xz} + 2\sigma_{zy}\omega_{yz}, \qquad (12.33)$$

$$\dot\sigma_{xy(Jaumann)} = \dot\sigma_{yx(Jaumann)} = \left(\sigma'_{xx} - \sigma'_{yy}\right)\omega_{xy} + \sigma_{xz}\omega_{zy} - \omega_{xz}\sigma_{zy}, \quad (12.34)$$

$$\dot\sigma_{xz(Jaumann)} = \dot\sigma_{zx(Jaumann)} = \left(\sigma'_{xx} - \sigma'_{zz}\right)\omega_{xz} + \sigma_{xy}\omega_{yz} - \omega_{xy}\sigma_{yz}, \quad (12.35)$$

$$\dot\sigma_{yz(Jaumann)} = \dot\sigma_{zy(Jaumann)} = \left(\sigma'_{yy} - \sigma'_{zz}\right)\omega_{yz} + \sigma_{yx}\omega_{xz} - \omega_{yx}\sigma_{xz}. \quad (12.36)$$

It is worth mentioning that in continuum mechanics, apart from the Jaumann stress rate, typically used in geodynamic modelling, there are a large variety of other *objective stress rate* formulations such as the Truesdell rate, the Green–Naghdi rate, the Oldroyd rate, the convective rate etc. (e.g. Shabana, 2008). However, the other objective derivatives (beside Jaumann) do not preserve the deviatoric property of a tensor. Hence, using them in our case is not straightforward as our formulations assume a splitting of stress into a deviatoric and homogeneous (pressure) part.

12.4 Maxwell visco-elastic rheology

A visco-elastic rheology is obtained by combining viscous (Eq. (5.11)) and elastic (Eq. (12.10)) rheological relations under certain physical assumptions (e.g. Turcotte and Schubert, 2002, Chapter 7, Sections 7–10). In numerical geodynamic modelling, the Maxwell visco-elastic rheology is the most commonly used; it is based on the assumption that both viscous and elastic deformations are happening under the same applied deviatoric stress σ'_{ij} such that bulk strain rate $\dot\varepsilon'_{ij}$ can be represented as a sum of viscous $\dot\varepsilon'_{ij(viscous)}$ and elastic $\dot\varepsilon'_{ij(elastic)}$ strain rates (see Fig. 12.4 for the relationship between the viscous and elastic deformations of a Maxwell body)

$$\dot\varepsilon'_{ij} = \dot\varepsilon'_{ij(viscous)} + \dot\varepsilon'_{ij(elastic)}, \qquad (12.37)$$

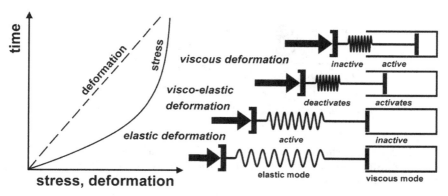

Fig. 12.4 Schematic representation of the Maxwell visco-elastic rheology. The solid line shows a typical pattern of visco-elastic stress buildup under the condition of a linearly growing deformation (dashed line). Deformation initially starts in an elastic mode (see shortening of the spring) but with the growing stress, viscous deformation activates and becomes dominant (see movement of the dashpot) and stress stabilises. The length of the black arrows reflects the magnitude of the applied stress at different moments in time.

which can then be obtained from the rheological relations (5.11) and (12.10) under the assumption that the term $\delta_{ij}\eta_{bulk}\dot{\varepsilon}_{kk(viscous)}$ in Eq. (5.11) is negligible and the shear modulus in Eq. (12.10) is constant;

$$\dot{\varepsilon}'_{ij(viscous)} = \frac{1}{2\eta}\sigma'_{ij},\tag{12.38}$$

$$\dot{\varepsilon}'_{ij(elastic)} = \frac{D\varepsilon'_{ij(elastic)}}{Dt} = \frac{D}{Dt}\left(\frac{\sigma'_{ij}}{2\mu}\right) = \frac{1}{2\mu}\frac{D\sigma'_{ij}}{Dt},\tag{12.39}$$

$$\dot{\varepsilon}'_{ij} = \frac{1}{2\eta}\sigma'_{ij} + \frac{1}{2\mu}\frac{D\sigma'_{ij}}{Dt},\tag{12.40}$$

where $\dfrac{D\sigma'_{ij}}{Dt}$ is the *objective co-rotational time derivative* of the deviatoric stress component σ'_{ij} which includes effects of stress rotation discussed above.

12.5 Plastic rheology

The plastic rheology assumes that an absolute shear stress limit σ_{yield} exists for a body and after reaching this limit *plastic yielding* occurs (Fig. 12.5). Like viscous deformation, plastic yielding is irreversible, but the pattern of deformation is notably different (Fig. 12.6): plastic creep is localised and forms multiple highly deformed shear zones separating relatively undeformed blocks.

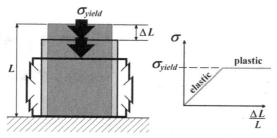

Fig. 12.5 Relationship between the applied stress σ and deformation ΔL, of an elastic–plastic body with initial length L. Elastic deformation changes to plastic yielding after reaching a stress limit σ_{yield}.

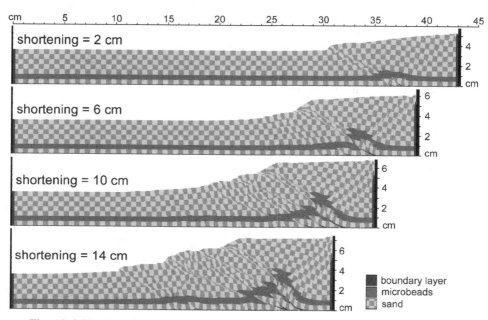

Fig. 12.6 Plastic deformation of sand in a numerical sandbox experiment (Buiter et al., 2006; Gerya and Yuen, 2007). Irreversible localised plastic deformation forms multiple, highly deformed shear zones separating relatively undeformed blocks.

The plastic strength σ_{yield} of a rock generally depends on the mean stress of the solid ($P_{solid} = P$) and on the pore fluid pressure (P_{fluid}) such that:

$$\sigma_{yield} = C + \sin(\varphi)P, \tag{12.41}$$

$$\sin(\varphi) = \sin(\varphi_{dry})(1 - \lambda), \tag{12.42}$$

$$\lambda = \frac{P_{fluid}}{P_{solid}}, \tag{12.43}$$

where C is the cohesion (residual strength at $P = 0$), ϕ is the effective internal friction angle (ϕ_{dry} stands for dry rocks) and λ is the pore fluid pressure factor. For dry fractured crystalline rocks, $\sin(\phi)$ is independent of composition and varies from 0.85 at $P < 200$ MPa to 0.60 at higher pressure (Byerlee law, Byerlee, 1978; Brace and Kohlstedt, 1980). The plastic strength of dry rocks thus strongly increases with pressure to a limit of several GPa. The strength is limited by the *Peierls mechanism* of plastic deformation (Evans and Goetze, 1979; Kameyama *et al.*, 1999; Karato, 2008).

The Peierls mechanism is a temperature-dependent mode of plastic deformation (also called *exponential creep*) which takes over from the dislocation creep mechanism at elevated stresses (typically above 0.1 GPa). Rheological relationships (flow law) for the Peierls creep are commonly represented as (Katayama and Karato, 2008)

$$\dot{\varepsilon}_{\text{II}} = A_{Peierls}\sigma_{\text{II}}^2 \exp\left\{-\frac{E_a + PV_a}{RT}\left[1 - \left(\frac{\sigma_{\text{II}}}{\sigma_{Peierls}}\right)^k\right]^q\right\}, \qquad (12.44)$$

where $\dot{\varepsilon}_{\text{II}}$ and σ_{II} are second invariants of strain rate and stress, respectively and $\sigma_{Peierls}$, $A_{Peierls}$, E_a and V_a are experimentally determined parameters (Chapter 6): $\sigma_{Peierls}$ is the Peierls stress that limits the strength of the material and is similar to σ_{yield} in Eq. (12.41), $A_{Peierls}$ is a material constant for the Peierls creep (Pa^{-2}s^{-1}), E_a is the activation energy (J/mol), V_a is the activation volume (J/Pa/mol). The choice of the exponents k and q in Eq. (12.44) depends on the shape and geometry of obstacles that limit the dislocation motion. Microscopic models show that k and q should have the following ranges $0 < k \leq 1$, $1 \leq q \leq 2$ (Kocks *et al.*, 1975). In contrast to other types of plasticity, Peierls creep is already activated at stresses that are notably lower than the actual strength of material given by $\sigma_{Peierls}$. This deformation mechanism is very important, in particular for deformation of subducting slabs characterised by lowered temperature and elevated stresses compared to the surrounding mantle (e.g. Karato *et al.*, 2001), or for lithospheric-scale shear localisation (Kaus and Podladchikov, 2006).

Further information about various types of plasticity used in geosciences such as Mohr–Coulomb, Von-Mises, Drucker–Prager and Treska models can be found in the books of Turcotte and Schubert (2002) and Ranalli (1995).

12.6 Visco-elasto-plastic rheology

In nature, the general behaviour of rocks is altogether visco-elasto-plastic, which can be formulated by decomposing the total deviatoric strain rate $\dot{\varepsilon}_{ij}'$ into the three respective components:

$$\dot{\varepsilon}'_{ij} = \dot{\varepsilon}'_{ij(viscous)} + \dot{\varepsilon}'_{ij(elastic)} + \dot{\varepsilon}'_{ij(plastic)}, \tag{12.45}$$

where

$$\dot{\varepsilon}'_{ij(viscous)} = \frac{1}{2\eta}\sigma'_{ij} \text{ (under condition of negligible } \delta_{ij}\eta_{bulk}\dot{\varepsilon}_{kk(viscous)} \text{ in Eq. (5.11)),}$$
$$\tag{12.46}$$

$$\dot{\varepsilon}'_{ij(elastic)} = \frac{1}{2\mu}\frac{D\sigma'_{ij}}{Dt}, \tag{12.47}$$

$$\dot{\varepsilon}'_{ij(plastic)} = 0 \text{ for } \sigma_{II} < \sigma_{yield}, \ \dot{\varepsilon}'_{ij(plastic)} = \chi\frac{\partial G_{plastic}}{\partial \sigma'_{ij}} = \chi\frac{\sigma'_{ij}}{2\sigma_{II}} \text{ for } \sigma_{II} = \sigma_{yield},$$
$$\tag{12.48}$$

$$G_{plastic} = \sigma_{II}, \tag{12.49}$$

$$\sigma_{II} = \sqrt{\frac{1}{2}\sigma'^2_{ij}}, \tag{12.50}$$

where $\dfrac{D\sigma'_{ij}}{Dt}$ is the objective co-rotational time derivative of the deviatoric stress component σ'_{ij}, equation (12.48) is the *plastic flow rule*, σ_{yield} is the *plastic yield strength* for a given rock, $G_{plastic}$ is the *plastic flow potential*, which reflects the amount of mechanical energy per unit volume that supports plastic deformation, $Pa = \dfrac{N}{m^2} = \dfrac{J}{m^3}$; σ_{II} is the second deviatoric stress invariant and χ is the plastic multiplier, which satisfies the plastic yielding condition

$$\sigma_{II} = \sigma_{yield}. \tag{12.51}$$

The plastic multiplier is a *variable scaling coefficient* which connects, in a uniform way, components of the plastic strain rate $\dot{\varepsilon}'_{ij(plastic)}$ with the local stress distribution in places where the yielding condition (12.51) is reached. This coefficient is unknown a priori and should be determined locally at each moment of time by solving Equations (12.45)–(12.51), based on local values of stresses σ'_{ij}, strain rates $\dot{\varepsilon}'_{ij}$, viscosity η and shear modulus μ.

This *plastic flow rule* formulation (Eq. (12.48)) includes deviatoric stress and strain rate components only and, consequently, the plastic potential formulation (Eq. (12.49)) is the same for both *dilatant* (i.e. increasing volume during plastic deformation) and *non-dilatant* materials. For plastic deformation of dilatant materials, this formulation is, therefore, combined with the equation describing volumetric changes which are of the form:

$$-\frac{D\ln\rho}{Dt} = div(\bar{v}) = 2\sin(\psi)\dot{\varepsilon}_{II(plastic)}, \tag{12.52}$$

$$\dot{\varepsilon}_{II(plastic)} = \left(\frac{1}{2}\dot{\varepsilon}'_{ij(plastic)}\dot{\varepsilon}'_{ij(plastic)}\right)^{1/2}, \tag{12.53}$$

where ψ is the dilatation angle, which generally depends on total plastic strain, and $\dot{\varepsilon}_{II(plastic)}$ is the second invariant of the deviatoric plastic strain rate tensor.

Analytical exercise

Exercise 12.1

Derive the equation for visco-elastic stress build-up/relaxation with time, using the Maxwell model (Eq. (12.40)) under conditions of constant strain rate $\dot{\varepsilon}'_{ij}$, viscosity η, shear modulus μ, and no stress rotation involved such that $\dfrac{D\sigma'_{ij}}{Dt} = \dfrac{D\sigma'_{ij}}{Dt}$. Take the initial state of stress to be given by σ'_{0ij} and integrate Eq. (12.40) (now reformulated in term of stress σ'_{ij}) to obtain the analytical solution. Reformulate the resulting equation in terms of Maxwell relaxation time

$$\Delta t_{Maxwell} = \frac{\eta}{\mu}, \tag{12.54}$$

which defines the characteristic time scale for visco-elastic stress relaxation.

Programming exercises and homework

Exercise 12.2

Use the analytical formula from the previous example to compute and compare stress–time curves for the following parameters:

(1) $\sigma'_{0ij} = 0$ Pa, $\dot{\varepsilon}'_{ij} = 10^{-14}$ 1/s, $\eta = 10^{21}$ Pa s, $\mu = 10^{10}$ Pa;
(2) $\sigma'_{0ij} = 10^8$ Pa, $\dot{\varepsilon}'_{ij} = 10^{-14}$ 1/s, $\eta = 10^{21}$ Pa s, $\mu = 10^{10}$ Pa;
(3) $\sigma'_{0ij} = 0$ Pa, $\dot{\varepsilon}'_{ij} = 10^{-15}$ 1/s, $\eta = 10^{21}$ Pa s, $\mu = 10^{10}$ Pa;
(4) $\sigma'_{0ij} = 0$ Pa, $\dot{\varepsilon}'_{ij} = 10^{-14}$ 1/s, $\eta = 10^{22}$ Pa s, $\mu = 10^{10}$ Pa;
(5) $\sigma'_{0ij} = 0$ Pa, $\dot{\varepsilon}'_{ij} = 10^{-14}$ 1/s, $\eta = 10^{21}$ Pa s, $\mu = 10^{11}$ Pa;
(6) $\sigma'_{0ij} = 0$ Pa, $\dot{\varepsilon}'_{ij} = 10^{-14}$ 1/s, $\eta = 10^{22}$ Pa s, $\mu = 10^{11}$ Pa.

Try to understand how the different parameters control the stress build-up/relaxation. An example is in **Viscoelastic_stress.m**.

Exercise 12.3

Compute the visco-elasto-plastic stress build-up and observe the changes in the viscous, elastic and plastic strain rates with time when using the parameters from case (4) of the previous example. Assume the condition that the visco-elastic stress σ'_{ij} must not exceed the yield stress limit of 1.5×10^8 Pa. Use Equation (12.46) to compute the viscous strain rate. Use Equations (12.45), (12.47) and (12.48) to compute elastic and plastic strain rates. Consider that after reaching the yielding limit, the stress in the visco-elasto-plastic material should not change any more and therefore $\dfrac{D\sigma'_{ij}}{Dt} = \dfrac{D\sigma'_{ij}}{Dt} = 0$. An example is in **Viscoelastoplastic_strain_rate.m**.

Exercise 12.4

Modify Exercise 6.3 by adding Peierls creep for the high stress region ($>10^8$ Pa). Compute the effective viscosity for this region by analogy to Eqs. (6.16)–(6.18) as follows

$$\frac{1}{\eta_{eff}} = \frac{1}{\eta_{diff}} + \frac{1}{\eta_{disl}} + \frac{1}{\eta_{Peierls}}, \tag{12.55}$$

where $\eta_{Peierls}$ is Peierls creep viscosity defined on the basis of Eqs. (12.44) and (6.4) as

$$\eta_{Peierls} = \frac{1}{2A_{Peierls}\sigma_{II}} \exp\left\{ \frac{E_a + PV_a}{RT} \left[1 - \left(\frac{\sigma_{II}}{\sigma_{Peierls}} \right)^k \right]^q \right\}. \tag{12.56}$$

Use the following Peierls creep parameters (Evans and Goetze, 1979; Katayama and Karato, 2008): dry olivine – $k = 1$, $q = 2$, $A_{Peierls} = 10^{-4.2}$ Pa^{-2}s^{-1}, $E_a = 540\,000$ J/mol, $\sigma_{Peierls} = 9.1 \times 10^9$ Pa; wet olivine – $k = 1$, $q = 2$, $A_{Peierls} = 10^{-4.2}$ Pa^{-2}s^{-1}, $\sigma_{Peierls} = 2.9 \times 10^9$ Pa; $E_a = 430\,000$ J/mol. Note that the activation energy E_a for Peierls creep is the same as for respective dislocation creep (Table 6.1). Note that stress σ_{II} used for computing effective viscosity with Eq. (12.56) should always be limited by $\sigma_{Peierls}$, which corresponds to the upper strength limit. An example is in **Peierls_creep.m**.

13

2D implementation of
visco-elasto-plastic rheology

> **Theory:** Numerical implementation of visco-elasto-plastic rheology. Organisation of a thermomechanical code in case of 2D, visco-elasto-plastic, multi-phase flows.
> **Exercises:** Programming a 2D thermomechanical code with a visco-elasto-plastic rheology.

13.1 Viscous-like reformulation of visco-elasto-plasticity

One way of reformulating the visco-elasto-plastic rheological model (12.45) for easy implementation into a viscous code (which we already have programmed) is based on using finite differences in time. The deviatoric stress σ'_{ij} is expressed as a function of the total deviatoric strain rate $\dot{\varepsilon}'_{ij}$ from the visco-elasto-plastic constitutive relationships (Eq. (12.45)), by using first-order finite differences in time to represent the objective co-rotational time derivatives $\dfrac{D\sigma'_{ij}}{Dt}$ of visco-elastic stresses

$$\frac{D\sigma'_{ij}}{Dt} = \frac{\sigma'_{ij} - \sigma'^{o}_{ij}}{\Delta t}, \tag{13.1}$$

$$\sigma'_{ij} = 2\eta_{vp}\dot{\varepsilon}'_{ij}Z + \sigma'^{o}_{ij}(1 - Z), \tag{13.2}$$

$$Z = \frac{\Delta t \mu}{\Delta t \mu + \eta_{vp}}, \tag{13.3}$$

$$\eta_{vp} = \eta \text{ when } \sigma_{\text{II}} < \sigma_{yield}, \text{ and } \eta_{vp} = \eta \frac{\sigma_{\text{II}}}{\eta \chi + \sigma_{\text{II}}}, \text{ for } \sigma_{\text{II}} = \sigma_{yield}, \tag{13.4}$$

in which Δt is the computational time step, σ'^{o}_{ij} indicates the deviatoric stress from the previous time step corrected for advection and rotation, Z is the visco-elasticity factor and η_{vp} is a viscosity-like Lagrangian parameter that characterises the intensity of the plastic deformation ($\eta_{vp} = \eta$ when no plastic yielding occurs).

179

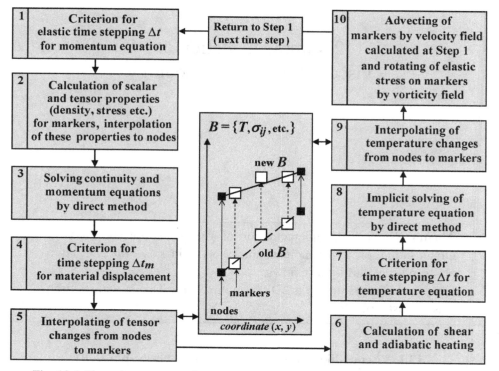

Fig. 13.1 Flow chart representing an example of a possible structure for a numerical thermomechanical visco-elasto-plastic 2D code which uses finite-differences and marker-in-cell technique (FD+MIC) for solving momentum, continuity and temperature equation. (Gerya and Yuen, 2007).

Equation (13.2) can also be applied to a visco-elastic Maxwell rheology by using the condition that $\eta_{vp} = \eta$.

13.2 Structure of visco-elasto-plastic thermomechanical code

When using the *constitutive relationship* (Eq. (13.2)) between the stress and strain rate, we can now formulate the momentum, continuity and temperature equations for the case of visco-elasto-plastic deformation in 2D and implement these equations in a thermomechanical visco-elasto-plastic modelling algorithm (Fig. 13.1). The algorithm is largely based on our viscous thermomechanical code structure discussed in Chapter 11 (Fig. 11.1, 11.2).

The flow chart in Fig. 13.1 gives an example of a possible structure of a numerical thermomechanical visco-elasto-plastic 2D code, which uses finite-differences combined with a marker-in-cell technique (FD+MIC) to solve the momentum, continuity and temperature equations. The principal steps of the algorithm are as follows:

1. Defining an optimal computational time step Δt for the momentum and continuity equations. One can use a minimum time step value, which satisfies the following three conditions: a given absolute time step limit on the order of a minimal characteristic timescale for the processes being modelled; a given relative marker displacement step limit (typically 0.01–1.0 of minimal grid step) that corresponds to the velocity field calculated at the previous time step (see Step 3); a given relative fraction of Lagrangian markers reaching locally the yielding condition (Eq. (12.51)) for the first time (typically 0.0001–0.01 of the total amount of markers representing materials with defined plastic yielding condition).

2. Calculating the physical properties (η_{vp}, μ, ρ, C_P, k, etc.) for the markers and interpolating these newly calculated properties, as well as scalars and tensors defined on the markers (T, σ'_{ij}, etc.) to the Eulerian nodes (Fig. 13.2). The plastic yielding condition (Eq. (12.51)) is locally controlled on markers by using Eq. (13.2) to predict stress changes. This equation is solved in an iterative way for every marker in order to compute η_{vp} based on a local viscous flow law (Eq. 6.5a) and a local plastic flow rule (Eq. 13.4).

3. Solving the 2D Stokes and continuity equations and computing the velocity and pressure by solving the global matrix with a direct method.

4. Defining an optimal displacement time step Δt_m for markers (typically limiting the maximal displacement to 0.01–1.0 of the *minimal grid step*) which can be generally smaller or equal to the computational time step Δt (see Step 1).

5. Calculating the stress changes (Eq. (13.2)) on the Eulerian nodes *for the displacement step Δt_m* (see Step 4) and interpolating these changes to the markers and calculating new tensor values associated with the markers (see central panel in Fig. 13.1).

6. Calculating the shear and adiabatic heating terms $H_{s(i,j)}$ and $H_{a(i,j)}$ at the Eulerian nodes from the computed velocity, pressure, strain rate and stress fields (see Step 3).

7. Defining an optimal time step Δt_T for the temperature equation. One can use a minimum time step value satisfying the following conditions: a given absolute time step limit on the order of a minimal characteristic thermal diffusion timescale for the processes being modelled; a given optimal marker displacement time step limit (see Step 3); a given absolute nodal temperature change limit (typically 1–20 K) (Chapter 10). The temperature equation can be preliminary solved with the displacement time step Δt_m to define possible temperature changes.

8. Solving the temperature equation implicitly in time by a direct method. The temperature equation can be solved in several steps when $\Delta t_T < \Delta t_m$.

9. Interpolating the calculated nodal temperature changes (see central panel in Fig. 13.1) from the Eulerian nodes, to the markers, and calculating new marker temperatures.

10. Advecting all markers throughout the mesh according to the globally calculated velocity field (see Step 3). Components of the stress tensor defined on the markers are recomputed analytically to account for any local stress rotation.

Figure 13.2 shows the geometry of an irregularly spaced, fully staggered numerical grid corresponding to the algorithm. The visco-elasto-plastic code can be developed on the basis of viscous thermomechanical code (Chapter 11). Therefore,

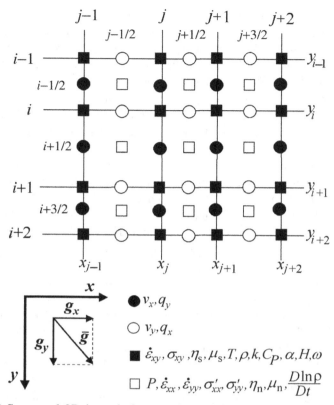

Fig. 13.2 Staggered 2D irregularly spaced numerical grid corresponding to the algorithm presented in Fig. 13.1.

we will concentrate in the following sections on the new modifications that were made to the code described in Chapter 11.

Step 1: Defining an optimal computational time step

Note that the computational time step Δt, for the momentum and continuity equations and the displacement time step Δt_m, for markers are generally independent from each other and can only be related by the condition $\Delta t_m \leq \Delta t$. For an elastic medium, the velocity field numerically computed from the Stokes equation strongly depends on the value of Δt. This time step influences the *numerical viscosity* $\eta_{numerical}$, which can be derived from Eqs. (13.2), (13.3)

$$\eta_{numerical} = \eta_{vp} Z = \frac{\eta_{vp} \mu \Delta t}{\eta_{vp} + \mu \Delta t}.$$

If Δt is much less than the *Maxwell relaxation time*

$$\Delta t_{Maxwell} = \frac{\eta_{vp}}{\mu} \tag{13.5}$$

of a visco-elastic/visco-elasto-plastic medium, then this medium behaves purely elastically, $\eta_{vp} \gg \mu \Delta t$ and the numerical viscosity depends linearly on the time step

$$\eta_{numerical} = \mu \Delta t. \tag{13.6}$$

In this situation, the smaller we take the time step, the smaller the computational viscosity becomes which results in larger velocities. Clearly this causes numerical problems, since if Δt tends to zero, the velocities will tend towards infinity. Physically meaningful solutions can be obtained in two ways. The first approach is to take an extremely small time steps (on the order of seconds or less) and introduce inertial forces into the system by using Navier–Stokes rather than the Stokes equations. Numerical models in this case will only be able to address the seismic modes of deformation, within a very limited period of time (hours, days), which is obviously objectionable if we want to model geodynamic processes that last for millions and even billions (as e.g. mantle convection) of years. The second approach (e.g. Kaus and Becker, 2007) is (when possible) to choose Δt in such a manner that it will be significantly shorter than the Maxwell time of the rheologically strongest materials that are present in the numerical experiment (e.g. a high-viscosity lithosphere), but significantly larger than the Maxwell time of any rheologically weak materials (e.g. low-viscosity asthenosphere). In this case, the weak materials satisfy the condition $\eta_{vp} \ll \mu \Delta t$ and they will behave purely viscously in a broad range of Δt, such that their numerical viscosity is independent of the time step

$$\eta_{numerical} = \eta_{vp}, \tag{13.7}$$

which will tend to stabilise the velocity field computed inside the model. For realistic variations in viscosity $10^{16} < \eta < 10^{26}$ and shear modulus $10^{10} < \mu < 10^{11}$ of solid rocks, this approach allows one to have geologically relevant computational time steps on the order of 10^1–10^5 years.

It should also be pointed out that the actual numerical viscosity contrast in experiments with visco-elastic materials

$$\min\left(\eta_{vp(min)}, \Delta t \mu_{min}\right) \leq \eta_{numerical} \leq \min\left(\eta_{vp(max)}, \Delta t \mu_{max}\right), \tag{13.8}$$

is typically reduced compared to purely viscous experiments

$$\eta_{vp(min)} \leq \eta_{numerical} \leq \eta_{vp(max)}. \tag{13.9}$$

This contrast can be further reduced by decreasing the computational time step Δt. This actually makes the visco-elastic numerical experiments computationally 'easier' than the purely viscous ones. On the other hand, the visco-elastic numerical solutions become equivalent to the viscous one if we choose very large computational time steps, which are bigger than the maximal Maxwell relaxation time (Eq. 13.5) estimated locally (e.g. on markers) within the model.

Step 2: Interpolation of scalar fields, vectors and tensor fields

According to our algorithmic approach, the temperature field as well as components of the σ'_{ij} and $\dot{\varepsilon}'_{ij}$ tensors are represented by values assigned to the markers. The effective values of all these parameters at the Eulerian nodal points are interpolated from the markers at each time step. As in the viscous code, we favour local interpolation schemes (within half grid distance) for some parameters: viscosity, shear modulus and deviatoric stress components. In order to avoid non-physical 'rigid boundary' effects on interfaces between *high-η-low-μ* and *low-η-high-μ* materials, one should use a harmonic average rather than arithmetic mean for the shear modulus (see Eq. (8.18), Fig. 8.8 for notations)

$$\mu_{(i,j)} = \frac{\sum\limits_{m} w_{m(i,j)}}{\sum\limits_{m} \frac{1}{\mu_m} w_{m(i,j)}}. \tag{13.10}$$

As in the viscous code, both the viscosity (η) and shear modulus (μ) are defined at different points (cf. open and solid squares in Fig. 13.2) when used for computing the normal and shear components of the deviatoric stress tensor. The viscosity, shear modulus and the respective stress and strain rate components for these nodal points are interpolated from markers found around the nodes at a distance less than half of the local Eulerian grid step (see dashed boundary in Fig. 8.8).

Step 3: Solving the momentum and continuity equations

In the case of 2D visco-elasto-plastic compressible fluid, the solution of Stokes equations (11.1) and (11.2) does not change. The only alteration concerns the expressions for the deviatoric stresses and strain rates

$$\sigma'_{xx} = 2\eta_{vp}\dot{\varepsilon}'_{xx}Z + \sigma'^{o}_{xx}(1 - Z) = -\sigma'_{yy}, \tag{13.11}$$

$$\sigma_{xy} = 2\eta_{vp}\dot{\varepsilon}_{xy}Z + \sigma^{o}_{xy}(1 - Z) = \sigma_{yx}, \tag{13.12}$$

$$\dot{\varepsilon}'_{xx} = \frac{1}{2}\left(\frac{\partial v_x}{\partial x} - \frac{\partial v_y}{\partial y}\right) = -\dot{\varepsilon}'_{yy}, \tag{13.13}$$

$$\dot{\varepsilon}_{xy} = \frac{1}{2}\left(\frac{\partial v_x}{\partial y} + \frac{\partial v_y}{\partial x}\right). \tag{13.14}$$

The conservation of mass is approximated by a compressible time-dependent 2D continuity equation

$$\frac{D \ln \rho}{Dt} + \frac{\partial v_x}{\partial x} + \frac{\partial v_y}{\partial y} = 0. \tag{13.15}$$

The conservative FD representation of the momentum equations (Eqs. (11.4), (11.5)) modified for a visco-elasto-plastic rheology with Equations (13.11)–(13.14) is as follows (see Fig. 13.2 for the indexing of the grid points):

x-Stokes equation (11.4)

$$\left[\frac{\partial}{\partial x}\left(2\eta\,\dot\varepsilon'_{xx}Z\right)\right]_{i+1/2,j} + \left[\frac{\partial}{\partial y}\left(2\eta\,\dot\varepsilon_{xy}Z\right)\right]_{i+1/2,j} - 2\frac{P_{i+1/2,j+1/2} - P_{i+1/2,j-1/2}}{x_{j+1} - x_{j-1}}$$

$$= -\left[\frac{\partial}{\partial x}\left(\sigma^{'o}_{xx}(1-Z)\right)\right]_{i+1/2,j} - \left[\frac{\partial}{\partial y}\left(\sigma^{o}_{xy}(1-Z)\right)\right]_{i+1/2,j} - \frac{\rho_{i,j} + \rho_{i+1,j}}{2}g_x, \tag{13.16}$$

$$\left[\frac{\partial}{\partial x}\left(2\eta\,\dot\varepsilon'_{xx}Z\right)\right]_{i+1/2,j} = 4\frac{\left[\eta\,\dot\varepsilon'_{xx}Z\right]_{i+1/2,j+1/2} - \left[\eta\,\dot\varepsilon'_{xx}Z\right]_{i+1/2,j-1/2}}{x_{j+1} - x_{j-1}},$$

$$\left[\frac{\partial}{\partial x}\left(\sigma^{'o}_{xx}(1-Z)\right)\right]_{i+1/2,j} = 2\frac{\left[\sigma^{'o}_{xx}(1-Z)\right]_{i+1/2,j+1/2} - \left[\sigma^{'o}_{xx}(1-Z)\right]_{i+1/2,j-1/2}}{x_{j+1} - x_{j-1}},$$

$$\left[\frac{\partial}{\partial y}\left(2\eta\,\dot\varepsilon_{xy}Z\right)\right]_{i+1/2,j} = 2\frac{\left[\eta\,\dot\varepsilon_{xy}Z\right]_{i+1,j} - \left[\eta\,\dot\varepsilon_{xy}Z\right]_{i,j}}{y_{i+1} - y_i},$$

$$\left[\frac{\partial}{\partial y}\left(\sigma^{o}_{xy}(1-Z)\right)\right]_{i+1/2,j} = \frac{\left[\sigma^{o}_{xy}(1-Z)\right]_{i+1,j} - \left[\sigma^{o}_{xy}(1-Z)\right]_{i,j}}{y_{i+1} - y_i},$$

$$\left[\sigma^{'o}_{xx}(1-Z)\right]_{i+1/2,j+1/2} = \frac{\eta_{n(i+1/2,j+1/2)}\sigma^{o}_{xx(i+1/2,j+1/2)}}{\Delta t\mu_{n(i+1/2,j+1/2)} + \eta_{n(i+1/2,j+1/2)}},$$

$$\left[\sigma^{o}_{xy}(1-Z)\right]_{(i,j)} = \frac{\eta_{s(i,j)}\sigma^{o}_{xy(i,j)}}{\Delta t\mu_{s(i,j)} + \eta_{s(i,j)}},$$

$$\left[\eta\,\dot\varepsilon'_{xx}Z\right]_{i+1/2,j+1/2} = \frac{\Delta t\mu_{n(i+1/2,j+1/2)}\eta_{n(i+1/2,j+1/2)}\dot\varepsilon'_{xx(i+1/2,j+1/2)}}{\Delta t\mu_{n(i+1/2,j+1/2)} + \eta_{n(i+1/2,j+1/2)}},$$

$$\left[\eta\,\dot\varepsilon_{xy}Z\right]_{(i,j)} = \frac{\Delta t\mu_{s(i,j)}\eta_{s(i,j)}\dot\varepsilon_{xy(i,j)}}{\Delta t\mu_{s(i,j)} + \eta_{s(i,j)}},$$

$$\dot\varepsilon'_{xx(i+1/2,j+1/2)} = \frac{v_{x(i+1/2,j+1)} - v_{x(i+1/2,j)}}{2(x_{j+1} - x_j)} - \frac{v_{y(i+1,j+1/2)} - v_{y(i,j+1/2)}}{2(y_{i+1} - y_i)},$$

$$\dot\varepsilon_{xy(i,j)} = \frac{v_{x(i+1/2,j)} - v_{x(i-1/2,j)}}{y_{i+1} - y_{i-1}} + \frac{v_{y(i,j+1/2)} - v_{y(i,j-1/2)}}{x_{j+1} - x_{j-1}},$$

where σ_{xy}^o and $\sigma_{xx}^{'o}$ are the deviatoric stress tensor components from the previous time step (corrected for advection and rotation), which are interpolated from markers; Δt is the computational time step. Discretisation of the y-Stokes equation (11.5) is analogous to Eq. (13.16) (derive as an exercise).

The time-dependent compressible continuity Eq. (13.15) is discretised as follows

$$\frac{v_{x(i+1/2,j+1)} - v_{x(i+1/2,j)}}{x_{j+1} - x_j} + \frac{v_{y(i+1,j+1/2)} - v_{y(i,j+1/2)}}{y_{i+1} - y_i} = -\left(\frac{D \ln \rho}{Dt}\right)_{i+1/2,j+1/2},$$

(13.17)

where $\dfrac{D \ln \rho}{Dt}$ are substantive density changes (e.g. Eq. (12.52)) interpolated from markers.

As in the viscous code, the global matrix is solved by a highly accurate, direct method and the numbering of unknowns in this global matrix remains the same.

Step 5: Interpolation of stress changes from nodes to markers

After defining the material displacement time step Δt_m, the changes in the effective stresses and new stress fields for the Eulerian nodes are calculated at the respective nodal points according to Eq. (13.2) as

$$\Delta\sigma'_{xx(i+1/2,j+1/2)} = \left(2\eta_{n(i+1/2,j+1/2)}\dot{\varepsilon}'_{xx(i+1/2,j+1/2)} - \sigma'^o_{xx(i+1/2,j+1/2)}\right)$$

$$\times \frac{\mu_{n(i+1/2,j+1/2)}\Delta t_m}{\eta_{n(i+1/2,j+1/2)} + \mu_{n(i+1/2,j+1/2)}\Delta t_m}, \qquad (13.18)$$

$$\Delta\sigma_{xy(i,j)} = \left(2\eta_{s(i,j)}\dot{\varepsilon}_{xy(i,j)} - \sigma^o_{xy(i,j)}\right)\frac{\mu_{s(i,j)}\Delta t_m}{\eta_{s(i,j)} + \mu_{s(i,j)}\Delta t_m}, \qquad (13.19)$$

$$\sigma'_{xx(i+1/2,j+1/2)} = \sigma'^o_{xx(i+1/2,j+1/2)} + \Delta\sigma'_{xx(i+1/2,j+1/2)}, \qquad (13.20)$$

$$\sigma_{xy(i,j)} = \sigma^o_{xy(i,j)} + \Delta\sigma_{xy(i,j)}. \qquad (13.21)$$

The corresponding stress increments for the markers are then added from the nodes using a standard first-order interpolation scheme (Fig. 8.9, Eq. (8.19)) and the newly updated values of stress components $\sigma'_{xx(m)}$ and $\sigma_{xy(m)}$ are thus obtained for markers. Other stress components are not stored on markers since they are obtained by using the standard relations $\sigma'_{yy} = -\sigma'_{xx}$ and $\sigma_{yx} = \sigma_{xy}$.

The interpolation of the calculated stress component changes from the Eulerian nodal points to the Lagrangian markers is similar to the temperature interpolation strategy described in Chapter 10 and effectively reduces the problem of numerical diffusion and non-physical subgrid oscillations. It uses a subgrid stress relaxation operation occurring over a characteristic Maxwell time. In order to define this operation, stress changes computed from Eqs. (13.18), (13.19) are decomposed into a subgrid part $\Delta\sigma'^{subgrid}_{xx(i+1/2,j+1/2)}$, $\Delta\sigma^{subgrid}_{xy(i,j)}$ and a remaining part $\Delta\sigma'^{remaining}_{xx(i+1/2,j+1/2)}$, $\Delta\sigma^{remaining}_{xy(i,j)}$ so that

$$\Delta\sigma'_{xx(i+1/2,j+1/2)} = \Delta\sigma'^{subgrid}_{xx(i+1/2,j+1/2)} + \Delta\sigma'^{remaining}_{xx(i+1/2,j+1/2)}, \tag{13.22}$$

$$\Delta\sigma'_{xy(i,j)} = \Delta\sigma'^{subgrid}_{xy(i,j)} + \Delta\sigma'^{remaining}_{xy(i,j)}. \tag{13.23}$$

In order to compute the subgrid part, we apply a subgrid stress relaxation on the markers by employing a characteristic, local Maxwell visco-elastic relaxation time scale $\Delta t_{Maxwell(m)}$ and then interpolating the respective stress changes back to the nodes. Subgrid stress changes on the markers are computed as follows

$$\Delta\sigma'^{subgrid}_{xx(m)} = \left(\sigma'^{o}_{xx(nodes)} - \sigma'^{o}_{xx(m)}\right)\left[1 - \exp\left(-d_{ve}\frac{\Delta t_m}{\Delta t_{Maxwell(m)}}\right)\right], \tag{13.24}$$

$$\Delta\sigma^{subgrid}_{xy(m)} = \left(\sigma^{o}_{xy(nodes)} - \sigma^{o}_{xy(m)}\right)\left[1 - \exp\left(-d_{ve}\frac{\Delta t_m}{\Delta t_{Maxwell(m)}}\right)\right], \tag{13.25}$$

where $\Delta t_{Maxwell(m)} = \dfrac{\eta_m}{\mu_m}$ is defined for each marker; d_{ve} is a dimensionless, numerical visco-elastic relaxation coefficient (one can use empirical values in the range of $0 \le d_{ve} \le 1$); $\sigma'^{o}_{xx(nodes)}$ and $\sigma^{o}_{xy(nodes)}$ are interpolated for any given marker from $\sigma'^{o}_{xx(i+1/2,j+1/2)}$ and $\sigma^{o}_{xy(i,j)}$ via the nodal values using relation (8.19), respectively (Fig. 8.9).

After obtaining $\Delta\sigma'^{subgrid}_{xx(m)}$ and $\Delta\sigma^{subgrid}_{xy(m)}$ for all markers, $\Delta\sigma'^{subgrid}_{xx(i+1/2,j+1/2)}$ and $\Delta\sigma^{subgrid}_{xy(i,j)}$ are computed by interpolation from the markers to the nodes using Eq. 8.18 (Fig. 8.8).

Then $\Delta\sigma'^{remaining}_{xx(i+1/2,j+1/2)}$ and $\Delta\sigma^{remaining}_{xy(i,j)}$ are computed for the nodes from Eqs. (13.22), (13.23)

$$\Delta\sigma'^{remaining}_{xx(i+1/2,j+1/2)} = \Delta\sigma'_{xx(i+1/2,j+1/2)} - \Delta\sigma'^{subgrid}_{xx(i+1/2,j+1/2)}, \tag{13.26}$$

$$\Delta\sigma^{remaining}_{xy(i,j)} = \Delta\sigma_{xy(i,j)} - \Delta\sigma^{subgrid}_{xy(i,j)}. \tag{13.27}$$

Finally, new corrected marker stresses $\sigma'^{corrected}_{xx(m)}$ and $\sigma^{corrected}_{xy(m)}$ are computed according to the following relation

$$\sigma'^{corrected}_{xx(m)} = \sigma'^{o}_{xx(m)} + \Delta\sigma'^{subgrid}_{xx(m)} + \Delta\sigma'^{remaining}_{xx(m)}, \tag{13.28}$$

$$\sigma^{corrected}_{xy(m)} = \sigma^{o}_{xy(m)} + \Delta\sigma^{subgrid}_{xy(m)} + \Delta\sigma^{remaining}_{xy(m)}, \tag{13.29}$$

where $\Delta\sigma'^{subgrid}_{xx(m)}$ and $\Delta\sigma^{subgrid}_{xy(m)}$ are given by Eqs. (13.24) and (13.25), respectively, and $\Delta\sigma'^{remaining}_{xx(m)}$ and $\Delta\sigma^{remaining}_{xy(m)}$ are interpolated from nodal values of $\Delta\sigma'^{remaining}_{xx(i+1/2,j+1/2)}$ and $\Delta\sigma^{remaining}_{xy(i,j)}$ to the markers according to standard bilinear interpolation (Eq. 8.19, Fig. 8.9).

Equations (13.24) and (13.25) require the decay of differences between marker stress values $\sigma_{xx(m)}^{\prime o}$, $\sigma_{xy(m)}^{o}$ and the interpolated nodal stress values $\sigma_{xx(nodes)}^{\prime o}$ and $\sigma_{xy(nodes)}^{o}$ on the characteristic local, Maxwell visco-elastic relaxation timescale $\Delta t_{Maxwell(m)}$. It is important to emphasise that the subgrid relaxation does not change the total stress increments $\Delta \sigma_{xx(i+1/2,j+1/2)}^{\prime}$ and $\Delta \sigma_{xy(i,j)}$ computed on nodal points from Eqs. (13.18) and (13.19), respectively, but instead splits them into two parts $\Delta \sigma_{xx(i+1/2,j+1/2)}^{\prime subgrid}$, $\Delta \sigma_{xy(i,j)}^{subgrid}$ and $\Delta \sigma_{xx(i+1/2,j+1/2)}^{\prime remaining}$ and $\Delta \sigma_{xy(i,j)}^{remaining}$. Introducing the subgrid relaxation operation removes unrealistic subgrid stress oscillations over the characteristic local Maxwell visco-elastic relaxation time without affecting the accuracy of the numerical solution for the momentum equations. Similarly to temperature, physically based subgrid variations will, indeed, be preserved by this scheme.

Steps 6 and 7: Solving the temperature equation

Numerical techniques for discretising and solving the temperature equation are identical to those used in the viscous code of Chapter 11. An important difference, however, occurs in computing the shear heating term. Since elastic deformation is reversible and does not contribute to mechanical energy dissipation, this deformation has to be excluded from shear heating calculation and therefore

$$H_s = 2\sigma_{xx}^{\prime}\left(\dot{\varepsilon}_{xx}^{\prime} - \dot{\varepsilon}_{xx(elastic)}^{\prime}\right) + 2\sigma_{xy}(\dot{\varepsilon}_{xy} - \dot{\varepsilon}_{xy(elastic)}). \tag{13.30}$$

With Eq. (12.45), (13.1)–(13.3) this can be further transformed to (derive as an exercise)

$$H_s = 2\sigma_{xx}^{\prime}\left(\dot{\varepsilon}_{xx}^{\prime} - \frac{1}{2\mu}\frac{D\sigma_{xx}^{\prime}}{Dt}\right) + 2\sigma_{xy}\left(\dot{\varepsilon}_{xy} - \frac{1}{2\mu}\frac{D\sigma_{xy}}{Dt}\right), \tag{13.31}$$

$$H_s = 2\sigma_{xx}^{\prime}\left(\dot{\varepsilon}_{xx}^{\prime} - \frac{\sigma_{xx}^{\prime} - \sigma_{xx}^{\prime o}}{2\mu\Delta t}\right) + 2\sigma_{xy}\left(\dot{\varepsilon}_{xy} - \frac{\sigma_{xy} - \sigma_{xy}^{o}}{2\mu\Delta t}\right), \tag{13.32}$$

$$H_s = \sigma_{xx}^{\prime}\frac{\sigma_{xx}^{\prime}}{\eta_{vp}} + \sigma_{xy}\frac{\sigma_{xy}}{\eta_{vp}}. \tag{13.33}$$

Finally, in the FD representation, the shear heating at temperature nodal points can be computed as follows;

$$H_{s(i,j)} = \frac{\left(\sigma_{xx(i-1/2,j-1/2)}^{\prime}\right)^2}{4\eta_{n(i-1/2,j-1/2)}} + \frac{\left(\sigma_{xx(i-1/2,j+1/2)}^{\prime}\right)^2}{4\eta_{n(i-1/2,j+1/2)}} + \frac{\left(\sigma_{xx(i+1/2,j-1/2)}^{\prime}\right)^2}{4\eta_{n(i+1/2,j-1/2)}}$$

$$+ \frac{\left(\sigma_{xx(i+1/2,j+1/2)}^{\prime}\right)^2}{4\eta_{n(i+1/2,j+1/2)}} + \frac{\left(\sigma_{xy(i,j)}\right)^2}{\eta_{s(i,j)}}, \tag{13.34}$$

where $\sigma_{xx(i+1/2,j+1/2)}^{\prime}$ and $\sigma_{xy(i,j)}$ are defined by Eqs. (13.20) and (13.21).

Step 10: Rotation of stresses

Another difference to the viscous algorithm consists in computing local stress changes due to the local rotation of markers. To do this, the rotation rate ω is defined (Eq. (12.27)) at the same grid points as $\dot{\varepsilon}_{xy} = \dfrac{1}{2}\left(\dfrac{\partial v_x}{\partial y} + \dfrac{\partial v_y}{\partial x}\right)$ (see solid squares in Fig. 13.2) and is similarly computed from the velocity field via finite differences

$$\omega_{(i,j)} = \frac{v_{y(i,j+1/2)} - v_{y(i,j-1/2)}}{x_{j+1} - x_{j-1}} - \frac{v_{x(i+1/2,j)} - v_{x(i-1/2,j)}}{y_{i+1} - y_{i-1}}. \tag{13.35}$$

Deviatoric stresses on markers are recomputed before advecting them according to Equations (12.21) and (12.22), or the simplified Jaumann formulas (12.24) and (12.26) (in case of rather small α) by using a first-order accurate scheme to compute rotation angles for the markers α_m

$$\alpha_m = \omega_m \Delta t_m, \tag{13.36}$$

where ω_m is interpolated from the basic nodes using a standard interpolation formula (Eq. 8.19, Fig. 8.9).

13.3 Visco-elasto-plastic iterations

It is important to mention that computation of visco-elasto-plastic solutions may require additional iterations in order to adjust stresses, strain rates and viscosity fields between markers and nodes. These iterations can be both *local* (i.e. done individually on every marker) and *global* (i.e. involving globally solving the momentum and continuity equations and re-interpolating properties between markers and nodes).

Local iterations are commonly needed when the effective viscosity of the medium is non-Newtonian and depends on stresses (e.g. due to the dislocation creep). Note that, in the case of visco-elastic deformation, the effective viscosity should be formulated in *terms of the deviatoric stresses* (Eq. (6.5a)) and *not in terms of the strain rates* (Eq. (6.5b)), since these strain rates are *not purely viscous* and include elastic deformation. In order to avoid numerical oscillations, Eq. (6.5a) should be formulated on markers in terms of the future visco-elastic stresses predicted locally from Eq. (13.2) and be based on the marker's old stresses $\sigma'^o_{xx(m)}$, $\sigma^o_{xy(m)}$, strain rates $\dot{\varepsilon}'_{xx(m)}$, $\dot{\varepsilon}_{xy(m)}$ and shear modulus μ_m. Since Eq. (13.2) also contains the marker viscosity η_m, which in turn depends on the future stresses (Eq. 6.5a), local iterations on every markers are needed to obtain consistent values of $\sigma'_{xx(m)}$, $\sigma_{xy(m)}$ and η_m.

These visco-elastic iterations should be done *before* we check the plastic yielding condition (13.4) on the marker, which requires we use the marker's pressure P_m and consistent values of future visco-elastic stresses $\sigma'_{xx(m)}$, $\sigma_{xy(m)}$ and corresponding $\sigma_{II(m)}$. If $\sigma_{II(m)}$ is larger than $\sigma_{yield(m)}$ for a given marker, a new value of the viscosity-like parameter $\eta_{vp(m)}$, should be computed to satisfy the condition $\sigma_{II(m)} = \sigma_{yield(m)}$ for that marker. A simple way to do this consists of assuming that after reaching $\sigma_{yield(m)}$, the second invariant of stress should stop changing and therefore, the $\dfrac{D\sigma_{II}}{Dt} = 0$ condition will be satisfied locally for each marker. Based on that condition and using Eq. (13.2) and (13.4), we can write the following relations to be applied locally for the marker under the assumption of constant local strain rate invariant $\dot{\varepsilon}_{II(m)}$ and yield stress $\sigma_{yield(m)}$

$$\eta_{vp(m)} = \frac{\sigma_{yield(m)}}{2\dot{\varepsilon}_{II(m)}}, \tag{13.37}$$

$$\sigma'^{o(corrected)}_{xx(m)} = \sigma'^{o}_{xx(m)} \frac{\sigma_{yield(m)}}{\sigma_{II(m)}}, \tag{13.38}$$

$$\sigma^{o(corrected)}_{xy(m)} = \sigma^{o}_{xy(m)} \frac{\sigma_{yield(m)}}{\sigma_{II(m)}}. \tag{13.39}$$

In contrast to computing a non-Newtonian viscosity, Eqs. (13.37)–(13.39) appear simple and can be explicitly applied to any marker without performing any local iteration. However, this procedure strongly affects the marker viscosity ($\eta_{vp(m)}$) and stresses ($\sigma'^{o(corrected)}_{xx(m)}$, $\sigma^{o(corrected)}_{xy(m)}$) local to each marker. This may change the global pressure–velocity solution, which in turn affects the local strain rates $\dot{\varepsilon}'_{xx(m)}$, $\dot{\varepsilon}_{xy(m)}$ and corresponding $\dot{\varepsilon}_{II(m)}$ used in Eqs. (13.2) and (13.37), respectively. Therefore, global iterations that involve solving the momentum and continuity equations and re-interpolating properties between markers and nodes, are often required. One way of performing such iterations is to repeat cycles of global solutions on the nodal points and make local re-adjustments on the markers (without displacing them) using Eqs. (13.37)–(13.39) until convergence is reached. This global iteration method works well when elastic stresses in the model build up gradually, such that they slowly approach the plastic yielding condition. If both the computational (Δt) and the displacement (Δt_m) time steps are small and small marker displacements occur at each time step, global iterations may not be needed since small time steps act as visco-elasto-plastic iterations. If in contrast, stress builds up suddenly (or even instantaneously, which occurs for example in commonly used *inelastic visco-plastic models*), much care should be taken in performing global iterations and more sophisticated iteration procedures such as the Newton–Raphson method (e.g., Belytschko *et al.*, 2000; Souza de Neto *et al.*, 2009) are commonly applied (e.g. Popov and Sobolev, 2008).

Another aspect of treating plasticity with finite-differences and marker-in-cell techniques concerns numerical diffusion. This time it involves diffusion of strain rates. The problem arises from the fact that strain rates are computed on the nodes using the interpolated, viscosity-like parameter η_{vp}, which in turn depends on strain rates. Interpolation of the mutually dependent parameters $\dot{\varepsilon}'_{xx(m)}$, $\dot{\varepsilon}_{xy(m)}$ and η_{vp}, back and forth between markers and nodes introduces systematic numerical diffusion. It has the same origin as in the case discussed in Chapter 8 (Fig. 8.11), when interpolating absolute values of parameters back and forth between markers and nodes. The diffusion defocuses plastic deformation zones, which widen, so deviating notably from strongly localised deformation patterns as reported in rocks. Numerical diffusion can be restricted by using a stress-based approach in which strain rates used in Eqs. (13.37)–(13.39) for each marker are not interpolated from the surrounding nodes directly but are computed from interpolated nodal stresses (σ'_{xx}, σ_{xy}, Eqs. (13.20), (13.21)) and stress changes ($\Delta\sigma'_{xx}$, $\Delta\sigma_{xy}$, Eqs. (13.18), (13.19)) with the use of Eqs. (13.2), (13.3)

$$\dot{\varepsilon}'_{xx(m)} = \frac{\sigma'_{xx}}{2\eta_{vp(m)}} + \frac{\Delta\sigma'_{xx}}{2\Delta t_m \mu_m}, \tag{13.40}$$

$$\dot{\varepsilon}_{xy(m)} = \frac{\sigma_{xy}}{2\eta_{vp(m)}} + \frac{\Delta\sigma_{xy}}{2\Delta t_m \mu_m}, \tag{13.41}$$

where $\eta_{vp(m)}$ and μ_m are values computed for the same marker during the previous iteration/time step; these values are not subjected to numerical diffusion. Other schemes similar to (13.40), (13.41) combining nodal and marker stresses and strain rates can also be proposed (see program example i2elvis.m associated with this chapter). The explicit approach can be easily implemented on markers and allows focusing of shear zones to within 1–2 grid cells. Finally, it is important to mention that strain rates should be computed on markers *before displacing them* since high-strain-rate shear zones are Lagrangian features related to specific material points (and not to immobile Eulerian nodes) which should be advected with the material flow.

Programming exercises and homework

Exercise 13.1
Add elasticity to the viscous thermomechanical code developed in Exercise 11.1. Use an incompressible continuity equation. Implement contrasting shear modulus $\mu = 10^{10}$ Pa and $\mu = 10^{11}$ Pa for the left and right layers, respectively. Program subgrid diffusion of stresses. Note that $\dot{\varepsilon}_{xy}$, σ_{xy} and ω in the staggered grid can be computed from the velocity solution *only for internal basic nodes*, but not for the external nodes. Therefore, use the internal nodes for interpolation of these

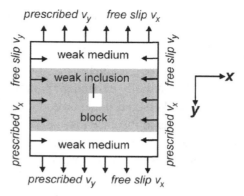

Fig. 13.3 Numerical setup for shortening of a visco-elasto-plastic block in the absence of gravity.

parameters to markers. Also, do not forget that when $\Delta t_T < \Delta t_m$, the temperature equation should be solved in several steps. An example is in **Viscoelastic2D.m**.

Exercise 13.2

Add plasticity to the code. Modify the model setup for shortening in the absence of gravity ($g_x = g_y = 0$) of a visco-elasto-plastic block embedded in a weak medium and containing a weak square inclusion (Fig. 13.3). The model is 1000×1000 km in size with a resolution of 51×51 nodes and 40 000 randomly distributed markers. Use the following material parameters: block (1000×600 km) – $\mu = 10^{10}$ Pa, $\eta = 10^{23}$, $C = 10^7$ Pa, $\sin(\varphi) = 0.6$; weak medium, inclusion (100×100 km) – $\mu = 10^{10}$ Pa, $\eta = 10^{17}$, $C = 10^{10}$ Pa, $\sin(\varphi) = 0$. Program a constant horizontal shortening and vertical extension rates of 5×10^{-9} m/s applied at the vertical and horizontal boundaries respectively (Fig. 13.3, this condition is mass conservative and corresponds to bulk shortening strain rate of 10^{-14} 1/s). Use a small computational time step of 100 years and a marker displacement step of 1% of grid step. Make a global iteration cycle to re-adjust plastic yielding on markers (without displacing them) using Eqs. (13.37)–(13.39). Try different number of iterations for this cycle and see how this affects the numerical solution. Try different values of $\sin(\varphi)$ (from 0 to 0.7) for the block and see how the orientation of the shear bands changes. An example is in **i2elvis.m**.

14

The multigrid method

Theory: Principles of multigrid method. Multigrid method for solving the Poisson equation in 2D. Coupled solving of momentum and continuity equations in 2D with multigrid for the cases with constant and variable viscosity.
Exercises: Programming of multigrid methods for solving the Poisson equation and coupled solving of the momentum and continuity equations in 2D.

14.1 Multigrid – what is it?

The use of direct methods to solve the system of equations places a strong limitation on the maximum possible number of nodal points within a numerical grid due to limitations in computer memory and computational speed. This limitation is particularly critical in 3D where, to reach the same resolution as in 2D (tens and hundreds of grid points in each direction), the amount of linear equations to be solved increases by at least two orders of magnitude and the number of computational operations required to solve the same equations grows by at least three orders of magnitude. Try to increase resolution in your visco-elasto-plastic code (Exercise 13.2) by two to three times in each direction and you will see how much slower it will compute . . .

Therefore, for very high resolution in 2D and for a moderate resolution in 3D, we are forced to use iterative methods which do not have such strong memory limitations. However, many iterative methods also have the problem that they require an increasing number of iterations with an increasing number of grid points (Fig. 14.1). Try to increase resolution in your code which solves the Poisson equation iteratively via Gauss–Seidel (Exercise 3.3) by two to three times in each direction and you will see how many more iterations will be needed with the Gauss–Seidel iterative method to obtain an accurate solution . . .

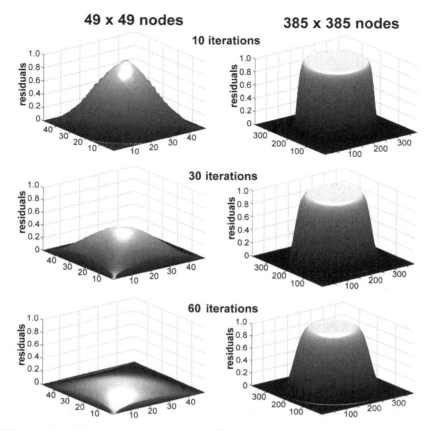

Fig. 14.1 Changes in the distribution of residuals during the iterative solution of 2D Poisson equation for the model with density field corresponding to a circular body (planet) embedded in a mass-less medium (space). Note that for the model with higher resolution (right column) residuals decay much slower than for the model with lower resolution (left column) with the same amount of Gauss–Seidel iterations. The model is computed with the code **Gauss_Seidel_iterations_Poisson.m** associated with this chapter.

What can we do to overcome these problems? One possible way is by using a *multigrid method*. Multigrid is a relatively new type of numerical algorithm explicitly formulated for the first time in 1964 (Fedorenko, 1964) and has been actively developed since the 1980s (a good introduction to multigrid methods can be found in the book of P. Wesseling, 1992). This method greatly speeds up the convergence of iterations and makes the number of iteration cycles independent of the amount of grid points. How does it do it? – By solving the same equations in parallel on several grids (typically having different resolution) and by exchanging information between these grids. This is why it is called MULTI-grid.

Multigrid is based on the simple idea that any linear equation:

$$C_1 x_1 + C_2 x_2 + \cdots + C_n x_n = R, \tag{14.1}$$

where $C_1, C_2, \ldots, C_n$ are coefficients, $x_1, x_2, \ldots, x_n$ are unknowns and R is the right-hand side, can be represented in *additive* form as

$$C_1 \left(x_1' + \Delta x_1 \right) + C_2 \left(x_2' + \Delta x_2 \right) + \cdots + C_n \left(x_n' + \Delta x_n \right) = R, \tag{14.2}$$
$$x_1 = x_1' + \Delta x_1,$$
$$x_2 = x_2' + \Delta x_2,$$
$$x_n = x_n' + \Delta x_n,$$

where $x_1', x_2', \ldots, x_n'$ are the current (known) approximations of $x_1, x_2, \ldots, x_n$ and $\Delta x_1, \Delta x_2, \ldots, \Delta x_n$ are unknown *corrections* needed to satisfy Eq. (14.1). Eq. (14.2) can be further transformed to

$$C_1 \Delta x_1 + C_2 \Delta x_2 + \cdots + C_n \Delta x_n = \Delta R, \tag{14.3}$$
$$\Delta R = R - \left(C_1 x_1' + C_2 x_2' + \cdots + C_n x_n' \right),$$

where ΔR is the current residual of Eq. (14.1) (see Chapter 3).

When some correction approximations are known, Eq. (14.3) can also be represented in the additive form

$$C_1 \left(\Delta x_1' + \Delta \Delta x_1 \right) + C_2 \left(\Delta x_2' + \Delta \Delta x_2 \right) + \cdots + C_n \left(\Delta x_n' + \Delta \Delta x_n \right) = \Delta R, \tag{14.4}$$

$$\Delta x_1 = \Delta x_1' + \Delta \Delta x_1,$$
$$\Delta x_2 = \Delta x_2' + \Delta \Delta x_2,$$
$$\Delta x_n = \Delta x_n' + \Delta \Delta x_n,$$

where $\Delta x_1', \Delta x_2', \ldots, \Delta x_n'$ are current (known) approximations of $\Delta x_1, \Delta x_2, \ldots,$ Δx_n and $\Delta \Delta x_1, \Delta \Delta x_2, \ldots, \Delta \Delta x_n$ are unknown *corrections to corrections* needed to satisfy Eqs. (14.1) and (14.3). Eq. (14.4) can be further transformed to

$$C_1 \Delta \Delta x_1 + C_2 \Delta \Delta x_2 + \cdots + C_n \Delta \Delta x_n = \Delta \Delta R, \tag{14.5}$$
$$\Delta \Delta R = \Delta R - \left(C_1 \Delta x_1' + C_2 \Delta x_2' + \cdots + C_n \Delta x_n' \right),$$

where $\Delta \Delta R$ is the current residual of Eq. (14.3; see Chapter 3). Obviously, the additive representation of Eq. (14.5) can also be done, etc. any desirable amount of times (continue as an exercise).

The multigrid method implies that we use *different numerical grids* to formulate the complementary Equations (14.1), (14.3), (14.5), etc. for the same numerical model. Please note that the coefficients $C_1, C_2, \ldots, C_n$ in Equations (14.1), (14.3) and (14.5) are identical and only the right-hand side of the equations is different.

In the case of numerical solutions, this means that the discretisation scheme for the governing equations will always be the same, independent of whether we formulate (i) equations for unknowns $x_1, x_2, \ldots, x_n$ or (ii) equations for correction to these unknowns $\Delta x_1, \Delta x_2, \ldots, \Delta x_n$ or (iii) equations for correction to these corrections $\Delta \Delta x_1, \Delta \Delta x_2, \ldots, \Delta \Delta x_n$ or (iv) etc. Another important point to note is that for such a hierarchical additive representation, residuals of *approximated* equations go into the right-hand side *of correcting* equations (i.e. residuals of Eq. 14.1 go to the right-hand side of Eq. 14.3, residuals of Eq. 14.3 go to the right-hand side of Eq. 14.5 etc.)

How does multigrid help us? What is required to rapidly obtain a global accurate solution for the steady equations (like Poisson, Stokes and incompressible continuity equations) is that the 'numerical information' propagates quickly across the entire model. During one iteration cycle, the information about updates of unknowns propagates only to (or *is felt by*) neighbouring grid points. Therefore, the finer the grid resolution, the shorter the physical distance over which information propagates during one iteration step. *Residuals with short wavelengths* (in terms of number of grid points) decay relatively fast (within a few iterations), while *residuals with longer wavelength* decay much slower (Fig. 14.2). Therefore, any increase in resolution produces even longer wavelengths in the residual distribution, which will thus require more iterations to decay.

Multigrid resolves this problem by performing additional iterations on several hierarchically arranged coarser grids (levels of resolution) which, therefore, propagate the solution over larger distances and rapidly smooth out longer-wavelength residuals. In this manner, residuals of all wavelengths decay with the same (small) amount of iterations which results in a solution convergence that is independent of the grid resolution. A typical way to program the multigrid method is to use several grids whose resolution increases by a fixed factor (e.g. a factor 2, see Fig. 14.3). The finest grid (Level 1) is the *principal* one, on which an accurate solution is obtained and the coarser grids are used to compute corrections for solutions on finer grids (cf. Eq. (14.1)–(14.5)). The coarser grid that is one level above the finest one will always compute corrections to the real solution, while the other grids will typically compute *corrections to corrections to the real solution, corrections to corrections to corrections to the real solution*, etc. (continue as an exercise).

The equations that are formulated on the various grids (including boundary conditions) are identical with the exception of the right-hand side of the equations; on coarser grids this is substituted by residuals that are interpolated (typically) from the nearest finer grid (cf. Eq. (14.1)–(14.5)). Transport coefficients such as viscosity, shear modulus etc. necessary to formulate the equations are also interpolated from the finer grid. As a result, the solution obtained on the coarser grid (e.g. pressure and velocity values) is in itself a *correction* (small addition) to the solution on the finer grid. It can then be used to update the solution on the finer grid by interpolating

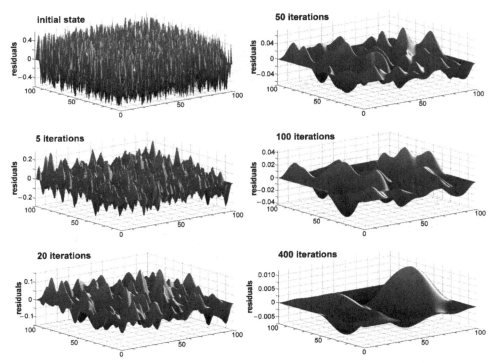

Fig. 14.2 Changes in the distribution of residuals during the iterative solution of 2D Poisson equation for a model with initially random distribution of the density field. Note that long-wavelength residuals decay much slower than short-wavelength ones. The model is computed with the code **Gauss_Seidel_iterations_Poisson.m** associated with this chapter. Model resolution is 100×100 grid points.

the corrections. To sum up: residuals and transport coefficients are interpolated from finer to coarser levels (*restriction operation*) while computed corrections are interpolated back from coarser to finer levels (*prolongation operation*). During one iteration cycle, an accurate solution should only be obtained on the *coarsest* (last) grid where many iterations or a direct matrix inversion can be employed (which can be done at low computational costs since the resolution of this grid is small). On other grids, some *limited number of iterations* (typically increasing by some factor with increasing grid level) should be performed in order to propagate information about the solution update and to compute new residuals. This process is called a *smoothing operation* (the reason for using this term is obvious from Fig. 14.2).

It should be pointed out that an increase in the resolution by an integer factor is not a strict requirement for multigrid. Generally, grid structures on different levels could be made in a fully independent manner (e.g. by using independent irregularly spaced meshes) and the only requirement is that coarser grids should efficiently smooth residuals with larger wavelengths (Fig. 14.2). In this case, special care should be taken when organising the restriction and prolongation operations

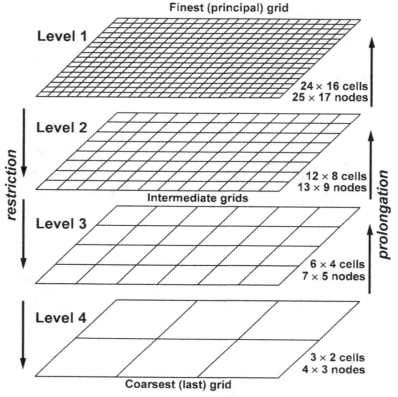

Fig. 14.3 Multigrid structure for a uniformly spaced, 2D rectangular non-staggered grid with four levels of resolution. Resolution of the grid between two nearest levels changes by the factor of 2 (see changes in the number of cells). Coarser levels of resolution are responsible for the decay of larger residual wavelengths (see Fig. 14.2).

between grids since the nodal points of the coarser grid do not overlap those of the finer grid, even in the case of a regular rectangular non-staggered grid. This situation is also very common for staggered grids (even in the case that resolution increases by an even factor) – non-overlap of points between different levels for a staggered grid is unavoidable and, therefore, the resolution at different levels can be chosen quite independently. Indeed, the choice of grid resolution and structure at various levels may notably affect the convergence of the solution. This choice can thus be different for different numerical problems and can be optimised empirically. Examples of programming such 'arbitrary resolution multigrids' for the Poisson equation as well as for the momentum and continuity equations are given respectively in the programs **Poisson_Multigrid_planet_arbitrary.m** and **Variable_viscosity_Multigrid_arbitrary.m** associated with this chapter.

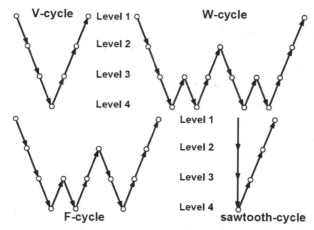

Fig. 14.4 Various *multigrid schedules* (or *cycles*) shown for the case of a four-level multigrid algorithm (Fig. 14.3). Circles denote smoothing operation, arrows – restriction (downward) and prolongation (upward) operations.

The order in which grids are visited is called the *multigrid schedule or cycle*. Several standard schedules exist (Fig. 14.4): the V-cycle, the W-cycle, the F-cycle and the sawtooth-cycle. The V-cycle is the simplest and most commonly used multigrid schedule (Fig. 14.4) – restriction+smoothing go uniformly from the finest level to the coarsest level through all intermediate levels, then prolongation+ smoothing go uniformly from the coarsest level to the finest level, also through all intermediate levels. The order of operations for W- and F-cycles is more complicated and contains several cycles of restriction+smoothing followed by prolongation+smoothing between coarser levels (Fig. 14.4) before returning corrections to the principal (finest) Level 1. The sawtooth-cycle is, in a way, similar to the V-cycle but the smoothing operations are omitted during interpolation of residuals from finer to coarser levels. Thus, this cycle first solves all equations on the coarsest grid and then gradually 'refines' the solution toward the principal level by applying prolongation+smoothing operations. The sawtooth-cycles are typically applied when no good initial approximation of the solution exists on the principal level, which is a common situation for the beginning of a numerical experiment.

Interpolation of residuals and transport coefficients *from finer to coarser grid* (i.e. restriction operation) can be made in the same manner as the interpolation of various parameters *from markers to nodes* (Eq. (8.18), Fig. 8.8). Consequently, the interpolation of corrections *from coarser to finer grid* (i.e. prolongation operation) can be organised by analogy with interpolation *from nodes to markers* (Eq. 8.19, Fig. 8.9). Since interpolation between markers and nodes was already extensively discussed in Chapter 8, programming of restriction and prolongation operations is rather straightforward, which is exemplified by several examples of written MATLAB functions (e.g. **Poisson_restriction_planet**, **Poisson_prolongation_planet**, **Viscosity_restriction**, **Stokes_Continuity_viscous_restriction**, **Stokes_Continuity_prolongation**) used in the codes associated with this chapter. There are also more sophisticated schemes of organising restriction and prolongation operations which give a higher multigrid performance in specific cases (e.g. Wesseling, 1992).

It should also be mentioned that the method described in this chapter is called *geometrical multigrid*, which requires the definition of several grids for the same model, formulation of the same differential equations separately for each grid and storing transport coefficients, solutions and corrections for all grids. Computational and memory costs for the geometric multigrid are relatively small since coarser grids have much less nodal points than the principal one. For example, in case of grid coarsening by a factor of two, all coarser grids will have in 2D and 3D less than 50% and 25% of grid points, respectively, compared to the finest grid.

However, there is also a class of more sophisticated multigrid approaches called *algebraic multigrid* (*AMG*) which do not require the explicit definition of the coarser grids, but rather uses algebraic operations based on multigrid principles to process and solve global matrix constructed for the finest (principal) grid. In an algebraic multigrid scheme, the coarse-level equations are generated from finer-level equations without the use of any geometry or re-discretisation on the coarse levels. This has the advantage that no coarse-level grid has to be generated or stored, and no flux or source term needs be calculated on the coarse levels. This feature makes AMG particularly important for use on unstructured meshes.

How efficient is multigrid? It is extremely efficient for simple cases like solving the Poisson equation on a regular grid (Fig. 14.2) and speeding up convergence by several orders of magnitude (Fig. 14.5). In more complex, thermomechanical modelling cases, it is typically less efficient, particularly when physical phenomena (such as e.g. localisation of deformation) are not properly reproduced on the coarser levels. For many geodynamic applications, the multigrid provides one of the best options to build efficient and robust codes and is therefore widely used in 3D numerical modelling of mantle convection and plate tectonic processes (e.g., Tackley, 2000; 2008).

14.2 Solving the Poisson equation with multigrid

Implementation of a *multigrid solver* for the Poisson equation is generally quite simple: a standard Gauss–Seidel iteration with a relatively high relaxation parameter $\theta_{relaxation}^{Poisson}$ (up to 1.75 uniformly applied for all nodes in all grids) can be used as an efficient smoother and the interpolation of residuals (restriction operation)

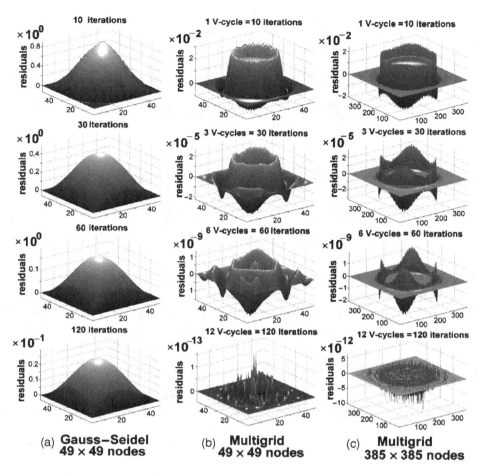

(a) **Gauss–Seidel** 49 × 49 nodes

(b) **Multigrid** 49 × 49 nodes

(c) **Multigrid** 385 × 385 nodes

Fig. 14.5 Changes in the distribution of residuals during the iterative solution of a 2D Poisson equation for a model with a density field that corresponds to a circular body (planet) embedded in a mass-less medium (space). (a) Gauss–Seidel iterations at low grid resolutions. (b), (c) Multigrid iterations at both low (b) and high (c) grid resolution, The multigrid cycle corresponds to a V-cycle with $5+5$ Gauss–Seidel iterations on the finest grid per cycle, four and seven levels of resolution are used for (b) and (c) respectively. Note that the convergence of the numerical solution (decay of residuals) in case of multigrid is several orders of magnitude faster (see bold numbers above vertical axes defining order of magnitude for residuals) than for the same amount of simple Gauss–Seidel iterations performed on the finest level. Also, in contrast to simple Gauss–Seidel iterations, the convergence of the multigrid solutions is independent of grid resolution (compare (b) and (c) with Fig. 14.1). The models are computed with the code **Poisson_ Multigrid.m**.

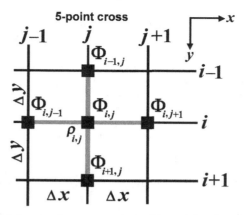

Fig. 14.6 Stencil of the regular rectangular grid used for the discretisation of the Poisson equation for the iterative Gauss–Seidel smoother used with multigrid.

and corrections (prolongation operation) is very straightforward particularly when the resolution of a regular grid between adjacent levels increases by an integer factor (2, 3 etc.) and grid lines of the coarser grid overlap with grid lines of the finer grid (Fig. 14.3). A 5-point stencil in 2D for the discretisation of the Poisson equation on such a regular grid is shown in Fig. 14.6 and the following iterative FD representation is used for updating the solution with a Gauss–Seidel smoother

$$\Delta R_{i,j} = R_{i,j} - \left(\frac{\partial^2 \Phi}{\partial x^2}\right)_{i,j} - \left(\frac{\partial^2 \Phi}{\partial y^2}\right)_{i,j}, \tag{14.6}$$

$$\Phi_{i,j}^{new} = \Phi_{i,j} + \frac{\Delta R_{i,j}}{C_{i,j}} \theta_{relaxation}^{Poisson}, \tag{14.7}$$

$$\left(\frac{\partial^2 \Phi}{\partial x^2}\right)_{i,j} = \frac{\Phi_{i,j-1} - 2\Phi_{i,j} + \Phi_{i,j+1}}{\Delta x^2}, \tag{14.8}$$

$$\left(\frac{\partial^2 \Phi}{\partial y^2}\right)_{i,j} = \frac{\Phi_{i-1,j} - 2\Phi_{i,j} + \Phi_{i+1,j}}{\Delta y^2}, \tag{14.9}$$

$$C_{i,j} = -\frac{2}{\Delta x^2} - \frac{2}{\Delta y^2}, \tag{14.10}$$

where $\Phi_{i,j-1}$, $\Phi_{i-1,j}$, $\Phi_{i,j}$, $\Phi_{i+1,j}$, $\Phi_{i,j+1}$ are the current values of either gravity potential (at finest level) or corrections for this potential (at coarser levels) in respective nodal points, $C_{i,j}$ is the coefficient at $\Phi_{i,j}$ in the discretised Poisson equation, $\Delta R_{i,j}$ is the current residual and $R_{i,j}$ is the right-hand side of the Poisson equation. On the principal level (finest grid), the right-hand side is computed from

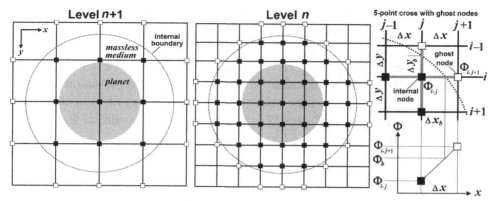

Fig. 14.7 Rectangular grids for two levels of resolution in case of solving Poisson equation for 2D numerical model with an internal boundary based on multigrid. Poisson equation is solved for internal nodes of the grids located inside the boundary (solid squares). Ghost nodes (open squares) located immediately outside the boundary are used to formulate internal boundary conditions when discretising the Poisson equation for the nearest internal nodes.

the standard equation

$$R_{i,j} = 4K\pi G\rho_{i,j}, \tag{14.11}$$

where G is the gravitational constant and K depends on the geometry of self-gravitating body modelled in 2D ($K=1$ and $K=2/3$ stand for cylindrical and spherical geometry, respectively, Eq. (11.16)). For coarser levels, $R_{i,j}$ is composed of residuals interpolated from finer levels. Obviously, grid steps Δx and Δy are also different for different levels of resolution. In a standard case, the simplest possible boundary condition equation $\Phi_{i,j} = 0$ is used for all marginal nodes on all grids, which also poses no difficulty for programming.

A peculiar case, occurs when an internal boundary is present within the model on which a boundary condition to solve the Poisson equation has to be defined (Fig. 14.7). This is, for example the case when we want to compute the gravity potential inside and around a planet, which is a component of a spherical-Cartesian approach for modelling self-gravitating bodies on a rectangular Cartesian grid (Fig. 11.5). In order to force the planet to remain in the centre of the grid and obtain a natural distribution of the gravitational acceleration vector *inside the planet* (Chapter 11), a constant gravity potential boundary condition ($\Phi = \Phi_b$) in 2D can be defined on a circle located at a distance from the planetary surface (Fig. 14.7). In this case, the Poisson equation is solved only for the nodes located in the circle (see solid squares in Fig. 14.7), while the boundary condition $\Phi = \Phi_b$ is applied for all other nodes of the grid. In order to have consistent solutions for all levels of resolution, the FD representation of the Poisson equation should be modified for the nodes located immediately next to the internal boundary with the use of a *ghost node approach*. In this case, the derivative of the gravity potential on the side of a 5-point cross which crosses the internal boundary should be defined in such a manner that it satisfies the boundary condition $\Phi = \Phi_b$. This situation is shown in Fig. 14.7 (right part), where the horizontal derivative of gravity potential to the right of the internal *ij*-th-node should satisfy the boundary condition $\Phi = \Phi_b$ and the following FD equation can be formulated

$$\frac{\partial \Phi}{\partial x} = \frac{\Phi_{i,j+1} - \Phi_{i,j}}{\Delta x} = \frac{\Phi_b - \Phi_{i,j}}{\Delta x_b}, \tag{14.12}$$

where Δx is the horizontal grid step, Δx_b is the distance from *ij*-th-node to the circular boundary and $\Phi_{i,j+1}$ is the gravity potential for an imaginary (ghost) node located at the distance Δx from *ij*-th-node. The value of gravity potential for the ghost node can be then computed as

$$\Phi_{(i,j+1)} = \Phi_b \frac{\Delta x}{\Delta x_b} - \Phi_{(i,j)} \frac{\Delta x - \Delta x_b}{\Delta x_b}. \tag{14.13}$$

Eq. 14.8 for the second *x*-derivative of gravity potential in Eq. 14.6 can be reformulated to the form excluding $\Phi_{i,j+1}$ (verify as an exercise)

$$\left(\frac{\partial^2 \Phi}{\partial x^2}\right)_{i,j} = \Phi_{i,j-1} \frac{1}{\Delta x^2} - \Phi_{i,j} \frac{\Delta x_b + \Delta x}{\Delta x^2 \Delta x_b} + \frac{\Phi_b}{\Delta x \Delta x_b}. \tag{14.14}$$

A similar transformation can be done for Eq. 14.9 for the second *y*-derivative of gravity potential for *ij*-th-node by using $(i\text{-}1, j)$-th-ghost-node (verify as an exercise)

$$\left(\frac{\partial^2 \Phi}{\partial y^2}\right)_{i,j} = \Phi_{i+1,j} \frac{1}{\Delta y^2} - \Phi_{i,j} \frac{\Delta y_b + \Delta y}{\Delta y^2 \Delta y_b} + \frac{\Phi_b}{\Delta y \Delta y_b}. \tag{14.15}$$

Consequently, the coefficient $C_{i,j}$ in Eq. (14.7) will also change to

$$C_{i,j} = -\frac{\Delta x_b + \Delta x}{\Delta x^2 \Delta x_b} - \frac{\Delta y_b + \Delta y}{\Delta y^2 \Delta y_b}. \tag{14.16}$$

Ghost nodes are thus only used for reformulating the Poisson equation in the nearest internal nodes and values of the gravity potential in the ghost nodes are not computed explicitly. Moreover, the 'implied' gravity potential value (Eq. 14.13) in a ghost node is generally different when the same ghost node is used in different Poisson equations formulated for different internal nodes. This is because the gravity potential has a 'kink' on the circular boundary and it is taken to be constant ($\Phi = \Phi_b$) for all nodes outside this boundary including all ghost nodes in the final solution. However, the uniform use of Eqs. (14.14)–(14.16) to reformulate the Poisson equation on different resolution levels is important and ensures the *geometrical compatibility* of solutions between all multigrid levels. This is because the boundary conditions for all grids are formulated on the same internal boundary, irrespective of the resolution and actual positions of the nodal points relative to this boundary. Note that the value of Φ_b should be set to zero for all levels of resolution with the exception of the finest (principal) level since coarser levels are used for computing corrections to the solution and these corrections should tend to zero (and not to Φ_b) with an increasing number of iterations. An example of a multigrid implementation with the circular internal boundary for gravity potential is given in the program **Poisson_Multigrid_planet.m** associated with this chapter.

Solving the Poisson equation on an irregularly spaced grid is not very different from the above procedures. Modifications only concern the manner of computing second derivatives of gravity potential in Eq. 14.6, the $C_{i,j}$ coefficient in Eq. 14.7, and the way of finding a correspondence between nodal points of coarser and finer grids when programming restriction and prolongation operations. The necessary modifications of the Poisson equation can be made on the basis of respective FD equations given in Chapter 11 (Fig. 11.4 Eq. (11.17)) while finding a correspondence between nodal points is analogous to that between Lagrangian markers (= nodes of finer grid) and Eulerian grid points (= nodes of coarser grid) and can be based on the same bisection procedure presented in Chapter 8 (Fig. 8.8–8.10, Eqs. 8.18, 8.19).

14.3 Solving Stokes and continuity equations with multigrid

The main challenge for solving coupled momentum and continuity equations with a multigrid method consists of programming a robust smoother that uses a primitive variable (pressure–velocity) formulation. In the case of constant viscosity, this challenge can be avoided by using a stream function formulation that requires double solving of the Poisson equation (Chapter 5). However, we are more interested in creating a *primitive variable smoother* that allows explicitly computation of pressure distribution in the model and is further applicable (with some modifications) to the variable viscosity case, which is much more relevant for geodynamic applications. The main obstacle to building such a primitive variable smoother comes from solving the incompressible continuity equation $\text{div}(\vec{v}) = 0$ which does not contain pressure and therefore (without modifications) cannot be converted into a pressure solution update procedure as e.g. the Stokes equation (for velocity) or Poisson equation (for gravity potential, see Eqs. 14.6, 14.7). This problem can be overcome with the *computational compressibility approach* according to which an iterative pressure update in a specific location is made proportional to the current residual of the continuity equation computed for the same location

$$\Delta R_{i,j}^{continuity} = R_{i,j}^{continuity} - \operatorname{div}(\bar{v})_{i,j}, \tag{14.17}$$

$$P_{i,j}^{new} = P_{i,j} + \frac{\Delta R_{i,j}^{continuity}}{\beta_{i,j}^{computational}} \theta_{relaxation}^{continuity}, \tag{14.18}$$

where $R_{i,j}^{continuity}$ is the right-hand side of the continuity equation (it is zero on the finest level and is made of residuals for coarser levels), $\beta_{i,j}^{computational}$ is the computational compressibility and $\theta_{relaxation}^{continuity}$ is the relaxation coefficient used for the continuity equation. Despite the artificial origin of this scheme for the incompressible viscous medium, a surprisingly efficient choice of $\beta_{i,j}^{computational}$ can be made for all levels of resolution on the basis of the simple relation:

$$\beta_{i,j}^{computational} = \frac{1}{\eta_{i,j}}, \tag{14.19}$$

where $\eta_{i,j}$ is the local viscosity. Equations (14.17)–(14.19) are applicable for both variable and constant viscosity cases (in the constant viscosity case $\eta_{i,j}$ is obviously equal to the global viscosity η for the entire model) and give stable convergence of solutions for coupled momentum and continuity equations if the relaxation coefficient $\theta_{relaxation}^{continuity}$ is chosen to be 0.1–0.3. An explanation for this surprising efficiency is that the computational compressibility provides a natural way of coupling between pressure and velocity equations: in places where for the current iteration step fluid converges and $\operatorname{div}(\bar{v}) < 0$ Eqs. (14.17)–(14.19) produce an increase in pressure, thus creating an outward directed pressure gradient that forces (through the Stokes equation) divergence of velocity and improves the solution of the continuity equation; in places where fluid diverges and $\operatorname{div}(\bar{v}) > 0$, the reaction is opposite.

An efficient numerical representation of the momentum and continuity equations for the multigrid can be based on a staggered grid with external velocity points (Fig. 14.8), as discussed in Chapter 7 (Fig. 7.17). Since the global numbering of unknowns is not needed in the case of iterative methods, the indexing of arrays for different parameters can be done separately and these arrays will also have different dimensions (Fig. 7.17): $N_x \times N_y$ for ρ and η_s (viscosity used for shear stress formulation) located in basic nodes, $N_x \times (N_y + 1)$ for v_x, $(N_x + 1) \times N_y$ for v_y and $(N_x - 1) \times (N_y - 1)$ for P and η_n (the viscosity used for normal deviatoric stress components formulation), where N_x and N_y is respectively horizontal and vertical resolution of the basic grid at a given multigrid level (see black rectangles in Fig. 14.8).

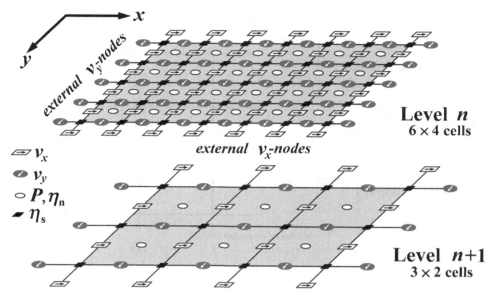

Fig. 14.8 Geometry of staggered grids with external velocity points used for two adjacent levels of multigrid in case of coupled solving of momentum and continuity equations. Internal (working) part of the grid is shown in grey. Note that distances from the grid boundaries to the external velocity nodes are different for different levels of resolution.

Constant viscosity case

A simple pressure–velocity update scheme based on the Gauss–Seidel iteration and a computational compressibility approach (Eqs. (14.17)–(14.19)) can be constructed on a regular grid for the case of constant viscosity (Fig. 14.9)

$$\Delta R_{i,j}^{x\text{-}Stokes} = R_{i,j}^{x\text{-}Stokes} - \eta \left(\frac{\partial^2 v_x}{\partial x^2}\right)_{i,j} - \eta \left(\frac{\partial^2 v_x}{\partial y^2}\right)_{i,j} + \left(\frac{\partial P}{\partial x}\right)_{i,j}, \quad (14.20)$$

$$v_{x(i,j)}^{new} = v_{x(i,j)} + \frac{\Delta R_{i,j}^{x\text{-}Stokes}}{C_{v_x(i,j)}} \theta_{relaxation}^{Stokes}, \quad (14.21)$$

$$\Delta R_{i,j}^{y\text{-}Stokes} = R_{i,j}^{y\text{-}Stokes} - \eta \left(\frac{\partial^2 v_y}{\partial x^2}\right)_{i,j} - \eta \left(\frac{\partial^2 v_y}{\partial y^2}\right)_{i,j} + \left(\frac{\partial P}{\partial y}\right)_{i,j}, \quad (14.22)$$

$$v_{y(i,j)}^{new} = v_{y(i,j)} + \frac{\Delta R_{i,j}^{y\text{-}Stokes}}{C_{v_y(i,j)}} \theta_{relaxation}^{Stokes}, \quad (14.23)$$

$$\Delta R_{i,j}^{continuity} = R_{i,j}^{continuity} - \left(\frac{\partial v_x}{\partial x}\right)_{i,j} - \left(\frac{\partial v_y}{\partial y}\right)_{i,j}, \quad (14.24)$$

$$P_{i,j}^{new} = P_{i,j} + \eta \Delta R_{i,j}^{continuity} \theta_{relaxation}^{continuity}, \quad (14.25)$$

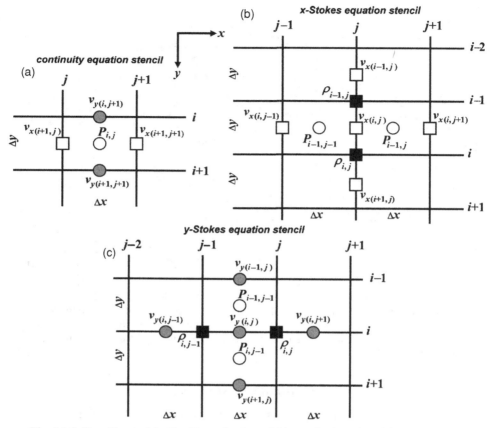

Fig. 14.9 Stencils used for the discretisation of the continuity (a) and Stokes (b), (c) equations on a 2D regular staggered grid (Fig. 14.8) for the models with constant viscosity. Indexing of grid lines corresponds to a basic (density) nodal points. Indexing of different unknowns is made separately depending on the amount of respective nodal points in the staggered grid (Fig. 14.8).

$$\left(\frac{\partial^2 v_x}{\partial x^2}\right)_{i,j} = \frac{v_{x(i,j-1)} - 2v_{x(i,j)} + v_{x(i,j+1)}}{\Delta x^2}, \quad \text{(Fig. 14.9(b))} \quad (14.26)$$

$$\left(\frac{\partial^2 v_x}{\partial y^2}\right)_{i,j} = \frac{v_{x(i-1,j)} - 2v_{x(i,j)} + v_{x(i+1,j)}}{\Delta y^2}, \quad \text{(Fig. 14.9(b))} \quad (14.27)$$

$$\left(\frac{\partial^2 v_y}{\partial x^2}\right)_{i,j} = \frac{v_{y(i,j-1)} - 2v_{y(i,j)} + v_{y(i,j+1)}}{\Delta x^2}, \quad \text{(Fig. 14.9(c))} \quad (14.28)$$

$$\left(\frac{\partial^2 v_y}{\partial y^2}\right)_{i,j} = \frac{v_{y(i-1,j)} - 2v_{y(i,j)} + v_{y(i+1,j)}}{\Delta y^2}, \quad \text{(Fig. 14.9(c))} \quad (14.29)$$

$$\left(\frac{\partial P}{\partial x}\right)_{i,j} = \frac{P_{i-1,j} - P_{i-1,j-1}}{\Delta x}, \quad \text{(Fig. 14.9(b))} \tag{14.30}$$

$$\left(\frac{\partial P}{\partial y}\right)_{i,j} = \frac{P_{i,j-1} - P_{i-1,j-1}}{\Delta y}, \quad \text{(Fig. 14.9(c))} \tag{14.31}$$

$$\left(\frac{\partial v_x}{\partial x}\right)_{i,j} = \frac{v_{x(i+1,j+1)} - v_{x(i+1,j)}}{\Delta x}, \quad \text{(Fig. 14.9(a))} \tag{14.32}$$

$$\left(\frac{\partial v_y}{\partial y}\right)_{i,j} = \frac{v_{y(i+1,j+1)} - v_{y(i,j+1)}}{\Delta y}, \quad \text{(Fig. 14.9(a))} \tag{14.33}$$

$$C_{v_x(i,j)} = -\frac{2\eta}{\Delta x^2} - \frac{2\eta}{\Delta y^2}, \tag{14.34}$$

$$C_{v_y(i,j)} = -\frac{2\eta}{\Delta x^2} - \frac{2\eta}{\Delta y^2}, \tag{14.35}$$

where $\theta_{relaxation}^{Stokes}$ is a relaxation parameter for the Stokes equations. $v_{x(i,j)}$, $v_{y(i,j)}$, $P_{i,j}$ etc. are current values of either velocity components and pressure (at finest level) or corrections for these values (at coarser levels) at respective nodal points. $C_{v_x(i,j)}$ and $C_{v_y(i,j)}$ are coefficients at respectively $v_{x(i,j)}$ and $v_{y(i,j)}$ in the discretised x- and y-Stokes equations, respectively. $\Delta R_{i,j}^{x\text{-}Stokes}$, $\Delta R_{i,j}^{y\text{-}Stokes}$, $\Delta R_{i,j}^{continuity}$ and $R_{i,j}^{x\text{-}Stokes}$, $R_{i,j}^{y\text{-}Stokes}$, $R_{i,j}^{continuity}$ are current residuals, and right-hand side for the momentum and continuity equations, respectively. On the finest level of resolution, these right-hand side contributions are computed from the standard equations

$$R_{i,j}^{x\text{-}Stokes} = -g_x \frac{\rho_{i,j} + \rho_{i-1,j}}{2}, \quad \text{(Fig. 14.9(b))} \tag{14.36}$$

$$R_{i,j}^{y\text{-}Stokes} = -g_y \frac{\rho_{i,j-1} + \rho_{i,j}}{2}, \quad \text{(Fig. 14.9(c))} \tag{14.37}$$

$$R_{i,j}^{continuity} = 0, \tag{14.38}$$

where g_x and g_y are respective components of the gravitational acceleration vector. At coarser levels, $R_{i,j}^{x\text{-}Stokes}$, $R_{i,j}^{y\text{-}Stokes}$ and $R_{i,j}^{continuity}$ are composed of respective residuals interpolated from finer levels. Obviously, grid steps Δx and Δy are different for each multigrid level. Standard boundary condition equations for no slip and free slip conditions are always applied to the same external boundaries of the basic grid (see grey areas in Fig. 14.8) and can be formulated uniformly for all levels as follows:

upper boundary

$$v_{y(1,j)} = 0,$$

$$v_{x(i=1,j)} = v_{x(i=2,j)} \text{ for free slip, i.e. } \frac{\partial v_x}{\partial y} = 0 \text{ across the boundary,}$$

$$v_{x(i=1,j)} = -v_{x(i=2,j)} \text{ for no slip, i.e. } v_x = 0 \text{ on the boundary;}$$

left boundary

$$v_{x(i,1)} = 0,$$

$$v_{y(i,j=1)} = v_{y(i,j=2)} \text{ for free slip, i.e. } \frac{\partial v_y}{\partial x} = 0 \text{ across the boundary,}$$

$$v_{y(i,j=1)} = -v_{y(i,j=2)} \text{ for no slip, i.e. } v_y = 0 \text{ on the boundary;}$$

and conditions for lower and right boundaries are done similarly. These boundary-condition equations can either be called directly in the Gauss–Seidel iteration cycle or (which often gives better convergence of the solution) implemented within the x- and y-Stokes equations (Eqs. (14.20)–(14.35)) discretised for the nearest internal velocity nodes. For example, a free slip condition at the upper boundary can be implemented as

$$\left(\frac{\partial^2 v_x}{\partial y^2}\right)_{i=2,j} = \frac{-v_{x(i=2,j)} + v_{x(i=3,j)}}{\Delta y^2}, \qquad \text{(Fig. 14.9(b) compare with Eq. (14.27)),}$$

$$(14.39)$$

and respectively $C_{v_x(i=2,j)} = -\frac{2\eta}{\Delta x^2} - \frac{\eta}{\Delta y^2}$ (compare with Eq. (14.34)),

$$(14.40)$$

$$\left(\frac{\partial^2 v_y}{\partial y^2}\right)_{i=2,j} = \frac{-2v_{y(i=2,j)} + v_{y(i=3,j)}}{\Delta y^2}, \qquad \text{(Fig. 14.9(c), compare with Eq. (14.29)),}$$

$$(14.41)$$

$$\left(\frac{\partial v_y}{\partial y}\right)_{i=1,j} = \frac{v_{y(i=2,j+1)}}{\Delta y}, \qquad \text{(Fig. 14.9(a), compare with Eq. (14.33)).}$$

$$(14.42)$$

Velocity conditions for other boundaries can be implemented in a similar way. An example of such boundary conditions implementation is given in the MATLAB function **Stokes_Continuity_smoother_ghost.m**.

In order to be able to compute pressure fields, we also need to prescribe pressure value *in one selected cell*. This can be done on the finest level at the end of each smoothing cycle by subtracting uniformly from all pressure nodes an estimated current difference between required and actual pressure values in the selected cell. This operation does not change pressure gradients in Stokes equations and thus does not affect the accuracy of the solution. During the smoothing procedure, an iterative pressure update is done uniformly with Eq. (14.25) for all pressure nodes, including the selected one, as such uniformity typically gives better convergence of multigrid. Also, faster convergence is often obtained when the hydrostatic pressure

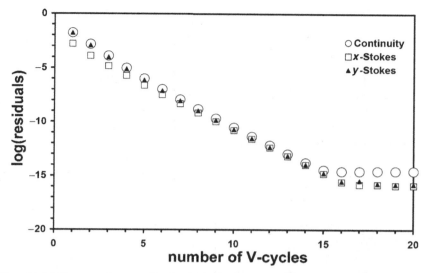

Fig. 14.10 Decay of normalised residuals for Stokes and continuity equations versus the number of multigrid V-cycles for a constant viscosity model. Residuals stabilise at computer accuracy level. Four-level multigrid with resolution 49×49 points on the finest level are used with relaxation parameters $\theta_{relaxation}^{continuity} = 0.3$ and $\theta_{relaxation}^{Stokes} = 0.9$. Numerical setup: rectangular block having higher-density sinks in lower-density fluid. Iterations start from a hydrostatic pressure field and zero velocities. Results are obtained with the program **Constant_Viscosity_Multigrid_ghost.m**.

distribution is initially defined in the computational domain. This distribution can be computed from the density field and gravity vector as follows:

- a pressure value is first defined in a first cell $P_{i=1,j=1}$ and then computed in the first row of cells based on density and horizontal (g_x) component of gravity vector

$$P_{i=1,j} = P_{i=1,j-1} + g_x \frac{\rho_{i=1,j} + \rho_{i=2,j}}{2} \Delta x \quad \text{(Fig. 14.9b),} \qquad (14.43)$$

- pressure in the remaining cells is computed by columns based on density and vertical (g_y) component of gravity vector

$$P_{(i,j)} = P_{i-1,j} + g_y \frac{\rho_{i,j} + \rho_{i,j+1}}{2} \Delta y \quad \text{(Fig. 14.9c).} \qquad (14.44)$$

A multigrid solver based on the described procedures is very efficient for the case of constant viscosity and residuals of both Stokes and continuity equations decay very rapidly (up to an order of magnitude per one V-cycle, Fig. 14.10) to computer accuracy in 15–20 cycles. Examples of the

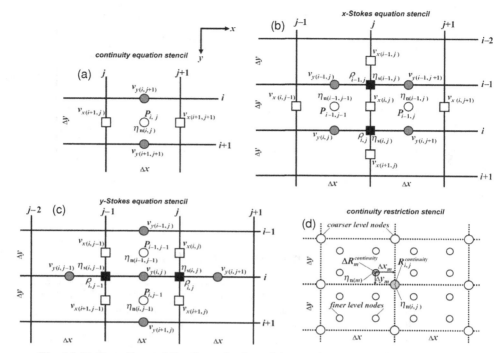

Fig. 14.11 Stencils used for discretisation of the continuity (a) and Stokes (b), (c) equations and for restriction of continuity residuals (d) on a 2D regular staggered grid (Fig. 14.8) for the models with variable viscosity. Indexing of solid grid lines corresponds to basic (density) nodal points. Indexing of different unknowns is done separately depending on the amount of respective nodal points in the staggered grid (Fig. 14.8).

described multigrid implementation for the case of constant viscosity are given in the programs **Stokes_Continuity_Multigrid.m** (with boundary condition equations called directly in the Gauss–Seidel iteration cycle) and **Constant_Viscosity_Multigrid_ghost.m** (with boundary condition equations implemented to momentum and continuity equations).

Adding variable viscosity

A variable viscosity multigrid solver is based essentially on the same principles as for constant viscosity but conservative finite differences discussed in Chapter 7 should be used to re-formulate the Stokes equations. Pressure–velocity update schemes based on Gauss–Seidel iterations in the case of variable viscosity and regular grids (Fig. 14.11) can be written as follows (only equations which are different from Eqs. (14.20)–(14.33) are shown)

$$\Delta R_{i,j}^{x\text{-}Stokes} = R_{i,j}^{x\text{-}Stokes} - \left(\frac{\partial \sigma'_{xx}}{\partial x}\right)_{i,j} - \left(\frac{\partial \sigma_{xy}}{\partial y}\right)_{i,j} + \left(\frac{\partial P}{\partial x}\right)_{i,j}, \quad (14.45)$$

$$\Delta R_{i,j}^{y\text{-}Stokes} = R_{i,j}^{y\text{-}Stokes} - \left(\frac{\partial \sigma'_{yy}}{\partial y}\right)_{i,j} - \left(\frac{\partial \sigma_{yx}}{\partial x}\right)_{i,j} + \left(\frac{\partial P}{\partial y}\right)_{i,j}, \quad (14.46)$$

$$P_{i,j}^{new} = P_{i,j} + \eta_{n(i,j)} \Delta R_{i,j}^{continuity} \theta_{relaxation}^{continuity}, \quad (\text{Fig. 14.11(a)}) \quad (14.47)$$

$$\left(\frac{\partial \sigma'_{xx}}{\partial x}\right)_{i,j} = 2\eta_{n(i-1,j)}\frac{v_{x(i,j+1)} - v_{x(i,j)}}{\Delta x^2}$$

$$- 2\eta_{n(i-1,j-1)}\frac{v_{x(i,j)} - v_{x(i,j-1)}}{\Delta x^2}, \quad (\text{Fig. 14.11(b)}) \quad (14.48)$$

$$\left(\frac{\partial \sigma'_{xy}}{\partial y}\right)_{i,j} = \eta_{s(i,j)} \left(\frac{v_{x(i+1,j)} - v_{x(i,j)}}{\Delta y^2} + \frac{v_{y(i,j+1)} - v_{y(i,j)}}{\Delta x \Delta y}\right)$$

$$- \eta_{s(i-1,j)} \left(\frac{v_{x(i,j)} - v_{x(i-1,j)}}{\Delta y^2} + \frac{v_{y(i-1,j+1)} - v_{y(i-1,j)}}{\Delta x \Delta y}\right),$$
$$\quad (\text{Fig. 14.11(b)}) \quad (14.49)$$

$$\left(\frac{\partial \sigma'_{yy}}{\partial y}\right)_{i,j} = 2\eta_{n(i,j-1)}\frac{v_{y(i+1,j)} - v_{y(i,j)}}{\Delta y^2}$$

$$- 2\eta_{n(i-1,j-1)}\frac{v_{y(i,j)} - v_{y(i-1,j)}}{\Delta y^2}, \quad (\text{Fig. 14.11(c)}) \quad (14.50)$$

$$\left(\frac{\partial \sigma'_{yx}}{\partial x}\right)_{i,j} = \eta_{s(i,j)} \left(\frac{v_{y(i,j+1)} - v_{y(i,j)}}{\Delta x^2} + \frac{v_{x(i+1,j)} - v_{x(i,j)}}{\Delta x \Delta y}\right)$$

$$- \eta_{s(i,j-1)} \left(\frac{v_{y(i,j)} - v_{y(i,j-1)}}{\Delta x^2} + \frac{v_{x(i+1,j-1)} - v_{x(i,j-1)}}{\Delta x \Delta y}\right),$$
$$\quad (\text{Fig. 14.11(c)}) \quad (14.51)$$

$$C_{v_x(i,j)} = -2\frac{\eta_{n(i-1,j)} + \eta_{n(i-1,j-1)}}{\Delta x^2} - \frac{\eta_{s(i,j)} + \eta_{s(i-1,j)}}{\Delta y^2}, \quad (14.52)$$

$$C_{v_y(i,j)} = -2\frac{\eta_{n(i,j-1)} + \eta_{n(i-1,j-1)}}{\Delta y^2} - \frac{\eta_{s(i,j)} + \eta_{s(i,j-1)}}{\Delta x^2}. \quad (14.53)$$

Note that the viscosity is defined (Fig. 14.11) both in the cell-centres (η_n) and in the basic nodes (η_s) which are separately used to formulate normal ($\sigma'_{xx}, \sigma'_{yy}$) and shear ($\sigma_{xy} = \sigma_{yx}$) deviatoric stress components, respectively. These two types of viscosity should be interpolated from finer to coarser levels (i.e. *viscosity restriction*) before starting any multigrid iterations. Note that continuity equation residuals are multiplied to a local viscosity η_n defined in the centre of the respective cell when computing pressure updates for this cell (Eq. (14.47)). Using a uniform (e.g. average) viscosity to compute pressure updates, instead, typically gives worse convergence. Moreover, in the case of a large viscosity contrast, convergence is notably improved when continuity residuals are rescaled based on local viscosities at both finer and coarser levels, during restriction operations. The following first order of accuracy bilinear scheme is used to calculate the right-hand side of the continuity equation $R_{i,j}^{continuity}$ for the ij-th-pressure node at a coarser level based on the continuity equation residuals $\Delta R_m^{continuity}$ computed for finer-level nodes located within one grid step distance around the coarser-level node (Fig. 14.11d)

$$R_{i,j}^{continuity} = \frac{\sum_m \eta_{n(m)} \Delta R_m^{continuity} w_{m(i,j)}}{\eta_{n(i,j)} \sum_m w_{m(i,j)}}, \tag{14.54}$$

$$w_{m(i,j)} = \left(1 - \frac{\Delta x_m}{\Delta x}\right) \times \left(1 - \frac{\Delta y_m}{\Delta y}\right), \tag{14.55}$$

where $w_{m(i,j)}$ represents a statistical weight of m-th-finer-level node at the ij-th-coarser-level node; Δx_m and Δy_m are distances from m-th-node to ij-th-node. Like interpolation from markers to nodes, Equation (14.55) only accounts for finer-level nodes located within a limited (one coarser grid step) distance around the coarser-level node. Equation (14.54) guarantees that pressure corrections computed at coarser levels will always be proportional to the product of continuity residuals and local viscosities at the finest (principal) level as required by Eq. (14.47). Obviously, in the constant viscosity case, Eq. (14.54) turns into a standard bilinear interpolation scheme (Chapter 8, Eq. (8.18), Fig. 8.8). It should also be mentioned that more sophisticated pressure update and restriction/prolongation schemes (Tackley, 2008) are based on computational compressibility (Eq. (14.18)) defined as local pressure derivative of velocity divergence

$$\beta_{i,j}^{computational} = \left(\frac{\partial \mathrm{div}(\vec{v})}{\partial P}\right)_{i,j}. \tag{14.56}$$

This derivative can be computed numerically in the centre of a cell by using the discretised Stokes equations for four surrounding velocity nodes (Fig. 14.11(a)) that contain the pressure value for this specific cell (Eqs. (14.45), (14.46)). Indeed, Eq. (14.56) always predicts an inverse proportionality between $\beta_{i,j}^{computational}$ and local viscosity $\eta_{n(i,j)}$ (Tackley, 2008) which explains why the simplified update scheme of Eq. (14.47) is sufficiently robust.

One more modification to the multigrid solution algorithm which helps to obtain a solution at the first time step is a gradual increase in the viscosity contrast. When we initialise a computation that has a large viscosity contrast ($>10^3$), we typically do not have any initial approximation of the velocity field (since we cannot use the velocity field from the previous time step). If we start from a zero velocity and a hydrostatic pressure field, the convergence of the solution can be very slow and the velocity field may remain unrealistic (too slow) for many iterations. This is particularly the case when the velocity field is defined *by the weakest, rather than by the strongest* medium. This happens, for example, in case of a hard Stokes sphere/cylinder that passes through a low-viscosity fluid (cf. Stokes cylinder test, Popov and Sobolev, 2008; Schmeling *et al.*, 2008) or in the case of a rigid *isolated* dense slab/block sinking in a weak medium (cf. falling block test, Gerya and Yuen, 2003a). Stokes-sphere-like setups with isolated rigid objects are in strong contrast with Rayleigh–Taylor-like models where a strong layer is attached to the model boundaries and the velocity field is therefore given by the rate of its internal deformation. In the latter case, the multigrid solution converges rapidly even for large viscosity contrasts. In the former case, a gradual increase in the *computational viscosity contrast* may indeed notably improve and speed up the solution (Fig. 14.12). Initially (in the beginning of multigrid cycles) the viscosity field is rescaled to a low/no viscosity contrast for which accurate velocity and pressure fields can be rapidly computed, starting from a hydrostatic pressure and zero velocity fields (Fig. 14.10). Then, after either a limited number of iterations or after reaching some level of accuracy, the computational viscosity contrast is gradually increased by a certain factor (1.5 to 10) and the original viscosity field is rescaled to this new contrast. The operations are repeated until the *original viscosity contrast* of the model is recovered. Rescaling of viscosity for the model can be made on the basis of the following formula

$$\eta_{i,j} = \eta_{min}^{computational} \exp\left[\frac{\ln\left(\eta_{max}^{computational}/\eta_{min}^{computational}\right)}{\ln\left(\eta_{max}^{original}/\eta_{min}^{original}\right)} \ln\left(\frac{\eta_{i,j}}{\eta_{min}^{original}}\right)\right],$$

(14.57)

where $\eta_{min}^{original}$, $\eta_{max}^{original}$ and $\eta_{min}^{computational}$, $\eta_{max}^{computational}$ are respectively the original and computational minimal and maximal viscosity for the model. An example of using such an algorithm (Fig. 14.12) is given in the program **Variable_viscosity_Multigrid_arbitrary.m**. It should however be mentioned that at large ($\gg 10^3$) and sharp (on one/few nodal points) viscosity contrasts, the accuracy of the multigrid solution is typically lowered compared to cases with lower viscosity contrast (see decreasing depth of residual minimisation 'spikes' with increasing viscosity contrast in Fig. 14.12). A reasonably high level of accuracy (10^{-4}–10^{-7}) for such sharply inhomogeneous models can indeed be reached and the use of more complex multigrid schedules such as F- and W-cycles can also improve convergence. Time steps following the first time step, typically do not require viscosity rescaling as

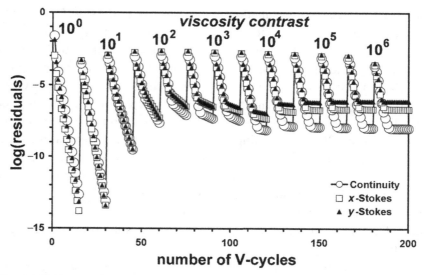

Fig. 14.12 Decay of normalised residuals for Stokes and continuity equations with the number of multigrid V-cycles for a model with variable viscosity. Residuals stabilise above the computer accuracy level. Four-level multigrid with resolution 49×49 points on the finest level is used with relaxation parameters $\theta_{relaxation}^{continuity} = 0.3$ and $\theta_{relaxation}^{Stokes} = 0.9$. Numerical setup: rectangular block having higher density and viscosity (by factor 10^6) sinks in lower density and viscosity fluid. Iterations start from a hydrostatic pressure field, zero velocities and no viscosity contrast. Spikes in the solutions are caused by an increase in viscosity contrast by the factor of 3.333 every 15 multigrid cycles. Results are obtained with program **Variable_viscosity_Multigrid_arbitrary.m**.

they have a much better initial guess for pressure and velocity. In addition, as discussed in Chapter 13, the numerical viscosity contrast can be efficiently decreased by using visco-elastic rheological models in which the upper limit of the *numerical viscosity* decreases proportionally with a decreasing computational time step (see Eqs. (13.6)–(13.9)).

Another efficient possibility to improve convergence in case of large viscosity contrasts is to use *repetitive cycles of gradual increase in a computational viscosity contrast*. In the beginning of each cycle (the first one excepted) residuals obtained for the finest grid level in the end of the previous cycle are assigned to the right-hand side of respective equations *at the same finest level*. At the end of the cycle, corrections computed at the finest level are added to the global solution and new residuals are then computed at the finest level to be used in the next cycle.

This method can be defined as a 'multi-multigrid' approach which uses a hierarchical representation of governing equations on the same numerical grid, which is analogous to the derivation of Eqs. (14.1)–(14.5). In many cases, this approach allows us to reach computer accuracy solution within a finite amount of iterations, even for very large viscosity contrasts (Fig. 14.13). An example of using such algorithm (Fig.14.13) is given in the program **Variable_viscosity_MultiMultigrid_arbitrary.m**.

Finally, it is also important to mention that besides large viscosity contrasts, further convergence problems can be caused by strongly irregular grid spacing, by significant differences in grid spacing used for different dimensions, by strong (e.g. plastic) localisation of deformation characterising a mechanical solution at the finest grid which is not captured on coarser levels etc. Therefore, be prepared that some of your thermomechanical models which utilise multigrid will be 'demanding', and will require special efforts in tuning and adjusting the iteration procedures.

Programming exercises and homework

Exercise 14.1

Program the multigrid solution based on a V-cycle for solving the Poisson equation in 2D for the case of a circular planetary body embedded in a mass less-like medium (Eqs. (14.6)–(14.16), Figs. 14.6–14.7). Use a ghost-node approach to define the boundary conditions $\Phi = 0$, along a circular boundary located at a distance from the planet. Program a Poisson equation smoother based on Gauss–Seidel iteration (Exercise 3.3 for Chapter 3) as an external MATLAB function and call it for different levels of resolution. Program external functions for restriction (Eq. (8.18), Fig. 8.8) and prolongation (Eq. (8.19), Fig. 8.9) operations to be called for different multigrid levels. Model parameters: radius of the planet = 6000 km, density of the planet = 6000 kg/m^3, model size = 18000 × 18000 km, radius for gravity potential boundary = 8999 km, number of resolution levels = 4, resolution on the coarsest (last) grid = 7 × 7 nodal points, factor of increase in resolution between the levels = 2, relaxation coefficient for Gauss–Seidel iterations, $\theta_{relaxation}^{Poisson} = 1.5$, number of smoothing iterations on the finest level = 5, factor of increase in the number of iterations with the level coarsening = 2. An example is in **Poisson_Multigrid_planet.m**.

Exercise 14.2

Program a multigrid solution for solving the Stokes and continuity equations in 2D for a constant viscosity case using a pressure–velocity formulation and a ghost-node approach (Eqs. (14.20)–(14.44), Figs. 14.8–14.9). The setup corresponds to a dense rectangular block sinking in the lower density medium. Model parameters: model size = 100 × 100 km, block size = 20 × 20 km (located in the middle of the model), block density = 3100 kg/m^3, medium density = 3000 kg/m^3, acceleration of gravity (vertical) = 9.81 m/s^2, model viscosity = 10^{20} Pa s, boundary conditions

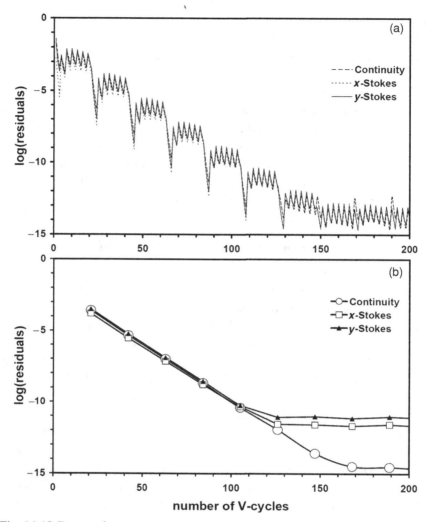

Fig. 14.13 Decay of normalised residuals for Stokes and continuity equations with the number of multigrid V-cycles for a model with variable viscosity in the case of a 'multi-multigrid' approach, which uses repetitive cycles of gradual increase in computational viscosity contrast. Residuals stabilise at the computer accuracy level. Five-level multigrid with resolution 49×49 points on the finest level are used with relaxation parameters $\theta^{continuity}_{relaxation} = 0.3$ and $\theta^{Stokes}_{relaxation} = 0.9$. Numerical setup: rectangular block having higher density and viscosity (by factor 10^6) sinks into a lower density and viscosity fluid. Iterations start from a hydrostatic pressure field and zero velocities. (a) Decay of local residuals computed with current corrections and right-hand side *within* each cycle (steps) of gradual viscosity contrast increase (spikes). (b) Decay of global residuals computed *after* each cycle of gradual viscosity contrast increase. Spikes in (a) are caused by an increase in viscosity contrast by the factor of 10 every 3 multigrid cycles. Results are obtained with the program **Variable_viscosity_MultiMultigrid_arbitrary.m**.

= free slip at all boundaries, number of resolution levels = 4, resolution of the basic grid on the coarsest (last) level = 7×7 nodal points, factor of increase in resolution between the levels = 2, relaxation coefficients for Gauss–Seidel iterations, $\theta_{relaxation}^{Stokes}$ = 1.2 and $\theta_{relaxation}^{continuity}$ = 0.3, number of smoothing iterations on the finest (basic) level = 5, factor of increase in the number of iterations with the level coarsening = 2. An example is in **Constant_Viscosity_Multigrid_ghost.m**.

Exercise 14.3

Modify the previous example to include a variable viscosity (Eqs. (14.45)–(14.55), Fig. 14.11). Use a high viscosity for the block (10^{23} Pa s) in comparison to the surrounding medium. Use $\theta_{relaxation}^{Stokes}$ = 1.0 and $\theta_{relaxation}^{continuity}$ = 0.3 and program a gradual increase in the viscosity contrast by the factor of $10^{1/2}$ (Eq. (14.57)) to reach an accurate solution (Fig. 14.12). Example is in **Variable_viscosity_Multigrid_arbitrary.m**.

15

Programming of 3D problems

Theory: Formulation of thermomechanical problems in 3D and its numerical implementation. Numerical methods for solving temperature, Poisson, momentum and continuity equations in 3D.
Exercises: Programming of numerical methods for temperature and Poisson equations and coupled solving of momentum and continuity equations in 3D.

15.1 Why simply not always 3D?

We know very well that the Earth is a 3D, nearly spherical object and, therefore, all the dynamic processes inside our planet are inherently three dimensional. Therefore, it is very logical to assume that realistic geodynamic modelling should always be done in 3D. Also, if you talk to geoscientists studying various natural geological objects, you are frequently told that such objects can only be modelled in 3D. This is a normal expectation since they are perfectly aware of the spatial 2D variability of geological structures on the Earth's surface and they thus know that a similar variability also exists in depth. Therefore, 3D modelling appears to be the natural choice for 'observers'. What about 'modellers'? Why don't they always use 3D modelling? What's wrong with it? The 'uncensored' truth about 3D modelling is the following:

- 3D thermomechanical modelling is quite easy from a methodological point of view – it is fairly straightforward to formulate and discretise the governing equations in a 3D Cartesian geometry for both simple viscous and more realistic visco-elasto-plastic rheologies (especially by using the same relatively simple finite-differences and marker-in-cell techniques that we extensively discussed in this book).

221

- 3D modelling is much more difficult from a technical point of view. This mainly concerns the coupled solving of momentum and continuity equations. Highly accurate direct solvers that are applicable in 2D are too slow and consume too much memory to yield the same spatial resolution of numerical grids in 3D. Iterative solvers, on the other hand, are very efficient for simple rheologies (such as constant viscosity problems) but do not always converge well for realistic geodynamic problems, which involve large viscosity contrasts on sharp interfaces.

It is obvious that when changing from 2D to 3D models, *we do not want to reduce the numerical resolution*. This requirement, however, immediately implies several orders of magnitude increase in the amount of grid points and markers and, thus, in the amount of equations that have to be solved:

- A 100×100 2D grid with 5×5 markers per cell implies around $3 \times 100 \times 100 = 30\,000$ momentum and continuity equations to be solved and $5 \times 5 \times 100 \times 100 = 250\,000$ markers to be followed at each time step;
- A $100 \times 100 \times 100$ 3D grid with $5 \times 5 \times 5$ markers per cell involves about $4 \times 100 \times 100 \times 100 = 4\,000\,000$ momentum and continuity equations to be solved (i.e. 130 times more than in 2D) and $5 \times 5 \times 5 \times 100 \times 100 \times 100 = 125\,000\,000$ markers to be followed at each time step (i.e. 500 times more than in 2D).

This is the reason why modellers prefer to apply 2D, rather than 3D approaches, where geodynamic problems can be justifiably simplified to lower dimensions. 3D geodynamic modelling is now developing very actively and significant progress is already achieved in the field of mantle convection in both Cartesian and spherical geometry, in large-scale modelling of plate tectonics processes (especially in modelling subduction) and in some other directions. Several groups are currently working on the development of more efficient and universal all-in-one 3D numerical geodynamic codes and 3D thermomechanical modelling is likely to become a standard tool in all fields of computational geodynamics (see Kaus *et al.*, 2008a for an overview of the current state of the art).

This chapter gives a practical summary that allows a relatively simple implementation of 3D thermomechanical modelling, based on conservative finite-differences and marker-in-cell techniques combined with iterative *multigrid* solvers similar to those discussed in Chapter 14 for 2D problems.

15.2 3D staggered grid and discretisation of momentum, continuity, temperature and Poisson equations

Let us first discuss the discretisation of various equations in 3D. Figure 15.1 shows an elementary volume (cell) of a 3D staggered grid that can be used for discretisation of momentum, continuity, Poisson and temperature equations in the case of viscous flow with variable viscosity and variable thermal conductivity. The

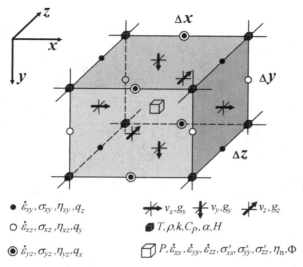

$\bullet\ \dot{\varepsilon}_{xy},\sigma_{xy},\eta_{xy},q_z$

$\circ\ \dot{\varepsilon}_{xz},\sigma_{xz},\eta_{xz},q_y$

$\circledcirc\ \dot{\varepsilon}_{yz},\sigma_{yz},\eta_{yz},q_x$

$\nearrow v_x,g_x \quad \nearrow v_y,g_y \quad \nearrow v_z,g_z$

$\blacksquare\ T,\rho,k,C_P,\alpha,H$

$P,\dot{\varepsilon}_{xx},\dot{\varepsilon}_{yy},\dot{\varepsilon}_{zz},\sigma'_{xx},\sigma'_{yy},\sigma'_{zz},\eta_n,\Phi$

Fig. 15.1 Elementary volume (cell) of 3D staggered grid used for discretisation of momentum, continuity, Poisson and temperature equations in the case of incompressible viscous flow with variable viscosity and thermal conductivity.

grid is constructed in a specific way that allows a natural representation of all governing equations with conservative finite differences:

- pressure, deviatoric normal stresses and strain rates and gravity potential (when needed) are located at the centre of the cell,
- components of velocity vector v_x, v_y and v_z and variable gravitational acceleration vector g_x, g_y and g_z (when needed) are located in the middle of the faces orthogonal to x, y and z axes, respectively,
- shear stresses and strain rates are located in the middle of the edges formed by the intersection of faces containing respective velocity components, i.e. by intersection of v_x- and v_y-faces in case of σ_{xy} and $\dot{\varepsilon}_{xy}$ etc.,
- viscosity is defined in four different places corresponding to positions of normal (η_n, in the centre of the cell) and shear (η_{xy}, η_{xz}, η_{yz}, in the middle of respective edges) stress components,
- heat fluxes q_x, q_y and q_z are located in the middle of edges parallel to x, y and z axes, respectively,
- other material properties and temperature are located at the cell corners, which are the basic nodes of the grid.

Before discretising the governing equations on a 3D staggered grid, an important step is (although it might sound really boring...) *to properly understand the indexing* of different field variables located around a grid cell (Fig. 15.1):

- Indexing of variables located in the basic nodes (T, ρ, α, C_P, etc.) of the grid is simple since the respective arrays have dimension of $N_x \times N_y \times N_z$, where N_x, N_y and N_z are the number of nodes of the basic grid in the respective directions.

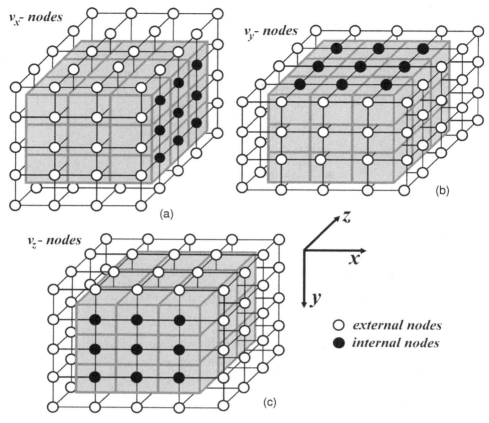

Fig. 15.2 Distribution of various velocity nodal points in 3D in the case when external nodes (open circles) are used to formulate boundary conditions for v_x (a), v_y (b) and v_z (c) velocity components. The basic grid of the model (see solid lines in Fig. 15.1) is shown in grey.

- Arrays for the variables located in cell centres (P, σ'_{xx}, η_n, Φ, etc.) will be $(N_x - 1) \times (N_y - 1) \times (N_z - 1)$.

- Arrays for various shear stresses, strain rates and respective viscosity values located on cell edges will be $N_x \times N_y \times (N_z - 1)$ for σ_{xy}, $\dot{\varepsilon}_{xy}$ and η_{xy}, $N_x \times (N_y - 1) \times N_z$ for σ_{xz}, $\dot{\varepsilon}_{xz}$ and η_{xz}, $(N_x - 1) \times N_y \times N_z$ for σ_{yz}, $\dot{\varepsilon}_{yz}$ and η_{yz}.

- Finally, the indexing of the velocity nodes should take into account nodes located outside the basic grid, which are used for formulating boundary conditions and interpolation of velocity components to markers (similarly to ones that we discussed in Chapter 14 for 2D grids, Fig. 14.8). Consequently, velocity arrays will be larger in two directions compared to the basic grid resolution (Fig. 15.2): $N_x \times (N_y + 1) \times (N_z + 1)$ for v_x, $(N_x + 1) \times N_y \times (N_z + 1)$ for v_y and $(N_x + 1) \times (N_y + 1) \times N_z$ for v_z.

Based on these considerations, we can now understand the logic of indexing for various grid points (Fig. 15.3) which will be then used to construct conservative 3D

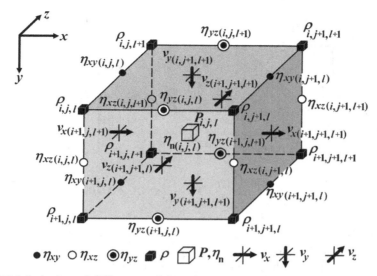

Fig. 15.3 Indexing of different variables for a 3D staggered grid (Fig. 15.1) with external velocity nodes (Fig. 15.2).

finite-difference schemes for the momentum, continuity, Poisson and temperature equations.

After extensive discussions on composing conservative FD schemes in 1D and 2D, the discretisation of various equations on a 3D staggered grid shown in Figs. 15.1–15.3 is quite straightforward and therefore we only discuss it briefly.

The representation of the incompressible 3D continuity equation on a stencil with six velocity nodes around a cell (Fig. 15.4) is

$$\frac{v_{x(i+1,j+1,l+1)} - v_{x(i+1,j,l+1)}}{\Delta x_{j+1/2}} + \frac{v_{y(i+1,j+1,l+1)} - v_{y(i,j+1,l+1)}}{\Delta y_{i+1/2}}$$

$$+ \frac{v_{z(i+1,j+1,l+1)} - v_{z(i+1,j+1,l)}}{\Delta z_{l+1/2}} = 0, \tag{15.1}$$

where i, j and l are indices in respectively y, x and z directions.

Discretisation of the Stokes equation for an incompressible fluid with variable viscosity uses a stencil containing 15 velocity nodes and 2 pressure nodes. An example of this stencil is shown in Fig. 15.5 for the case of the x-Stokes equation

$$\frac{\partial \sigma'_{xx}}{\partial x} + \frac{\partial \sigma_{xy}}{\partial y} + \frac{\partial \sigma_{xz}}{\partial z} - 2\frac{P_{i-1,j,l-1} - P_{i-1,j-1,l-1}}{\Delta x_{j-1/2} + \Delta x_{j+1/2}}$$

$$= \frac{1}{4}(\rho_{i-1,j,l-1} + \rho_{i,j,l-1} + \rho_{i-1,j,l} + \rho_{i,j,l})g_x, \tag{15.2}$$

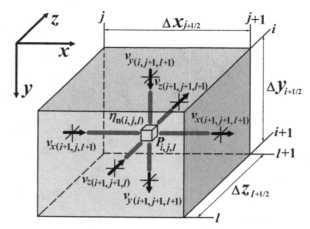

Fig. 15.4 Stencil of a 3D staggered grid used for the discretisation of the continuity equation for iterative solution. The small open cube in the centre corresponds to the pressure node at which the continuity equation is formulated. Notation of different nodal points is as in Fig. 15.1. Indexing of different variables corresponds to 3D staggered grid with external velocity nodes (Fig. 15.3).

$$\frac{\partial \sigma'_{xx}}{\partial x} = 4\eta_{n(i-1,j,l-1)}\frac{v_{x(i,j+1,l)} - v_{x(i,j,l)}}{\Delta x_{j+1/2}(\Delta x_{j-1/2} + \Delta x_{j+1/2})}$$

$$- 4\eta_{n(i-1,j-1,l-1)}\frac{v_{x(i,j,l)} - v_{x(i,j-1,l)}}{\Delta x_{j-1/2}(\Delta x_{j-1/2} + \Delta x_{j+1/2})}, \tag{15.3}$$

$$\frac{\partial \sigma_{xy}}{\partial y} = 2\eta_{xy(i,j,l-1)}\left(\frac{v_{x(i+1,j,l)} - v_{x(i,j,l)}}{\Delta y_{i-1/2}(\Delta y_{i-1/2} + \Delta y_{i+1/2})} + \frac{v_{y(i,j+1,l)} - v_{y(i,j,l)}}{\Delta y_{i-1/2}(\Delta x_{j-1/2} + \Delta x_{j+1/2})}\right)$$

$$- 2\eta_{xy(i-1,j,l-1)}\left(\frac{v_{x(i,j,l)} - v_{x(i-1,j,l)}}{\Delta y_{i-1/2}(\Delta y_{i-3/2} + \Delta y_{i-1/2})} + \frac{v_{y(i-1,j+1,l)} - v_{y(i-1,j,l)}}{\Delta y_{i-1/2}(\Delta x_{j-1/2} + \Delta x_{j+1/2})}\right),$$
$$\tag{15.4}$$

$$\frac{\partial \sigma_{xz}}{\partial z} = 2\eta_{xz(i-1,j,l)}\left(\frac{v_{x(i,j,l+1)} - v_{x(i,j,l)}}{\Delta z_{l-1/2}(\Delta z_{l-1/2} + \Delta z_{l+1/2})} + \frac{v_{z(i,j+1,l)} - v_{z(i,j,l)}}{\Delta z_{l-1/2}(\Delta x_{j-1/2} + \Delta x_{j+1/2})}\right)$$

$$- 2\eta_{xz(i-1,j,l-1)}\left(\frac{v_{x(i,j,l)} - v_{x(i,j,l-1)}}{\Delta z_{l-1/2}(\Delta z_{l-3/2} + \Delta z_{l-1/2})} + \frac{v_{z(i,j+1,l-1)} - v_{z(i,j,l-1)}}{\Delta z_{l-1/2}(\Delta x_{j-1/2} + \Delta x_{j+1/2})}\right).$$
$$\tag{15.5}$$

Discretisation of the y-Stokes and z-Stokes is rather obvious and the respective conservative FD schemes can be constructed and indexed by analogy to Fig. 15.5 and Eqns. (15.2)–(15.5) (derive as an exercise).

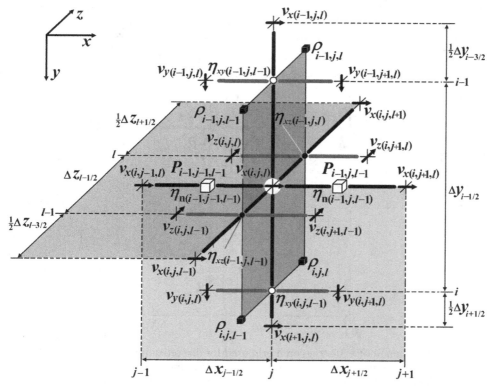

Fig. 15.5 Stencil of a 3D staggered grid used for the discretisation of the x-Stokes equations with variable viscosity. The white circle in the centre corresponds to a horizontal velocity node for which the x-Stokes equation is formulated. The notation of different nodal points is as in Fig. 15.1. Indexing of different variables corresponds to a 3D staggered grid with external velocity nodes (Fig. 15.3).

Given that temperature advection is solved with markers (see Chapter 10), discretisation of the 3D temperature equation with a variable thermal conductivity can be done in a simple Lagrangian form, which does not include advective terms and uses a stencil with 7 temperature nodes (7-point cross, Fig. 15.6). In implicit form, the conservative FD can be written as follows

$$\rho_{i,j,l} C_{P_{i,j,l}} \frac{T_{i,j,l} - T_{i,j,l}^{o}}{\Delta t} + \frac{\partial q_x}{\partial x} + \frac{\partial q_y}{\partial y} + \frac{\partial q_z}{\partial z} = H_{i,j,l}, \qquad (15.6)$$

$$\frac{\partial q_x}{\partial x} = \frac{(k_{i,j-1,l} + k_{i,j,l})(T_{i,j,l} - T_{(i,j-1,l)})}{\Delta x_{j-1/2}(\Delta x_{j-1/2} + \Delta x_{j+1/2})} - \frac{(k_{i,j,l} + k_{i,j+1,l})(T_{i,j+1,l} - T_{i,j,l})}{\Delta x_{j+1/2}(\Delta x_{j-1/2} + \Delta x_{j+1/2})}, \qquad (15.7)$$

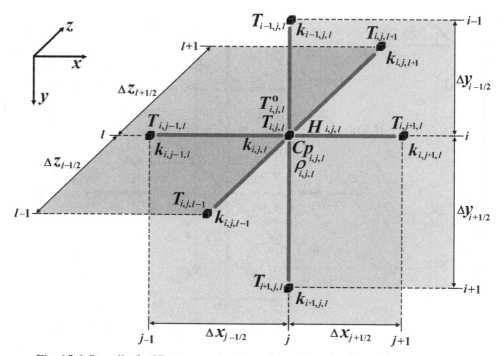

Fig. 15.6 Stencil of a 3D staggered grid used for discretisation of the temperature equation with variable thermal conductivity. The temperature equation is formulated for the central node $T_{i,j,l}$ which is one of the basic nodes of the 3D staggered grid (Fig. 15.1). The notation of different nodal points is the same as in Fig. 15.1. Indexing of different variables corresponds to a 3D staggered grid with external velocity nodes (Fig. 15.3).

$$\frac{\partial q_y}{\partial y} = \frac{(k_{i-1,j,l} + k_{i,j,l})(T_{i,j,l} - T_{i-1,j,l})}{\Delta y_{i-1/2}(\Delta y_{i-1/2} + \Delta y_{i+1/2})} - \frac{(k_{i,j,l} + k_{i+1,j,l})(T_{i+1,j,l} - T_{i,j,l})}{\Delta y_{i+1/2}(\Delta y_{i-1/2} + \Delta y_{i+1/2})},$$
$$(15.8)$$

$$\frac{\partial q_z}{\partial z} = \frac{(k_{i,j,l-1} + k_{i,j,l})(T_{i,j,l} - T_{i,j,l-1})}{\Delta z_{l-1/2}(\Delta z_{l-1/2} + \Delta z_{l+1/2})} - \frac{(k_{i,j,l} + k_{i,j,l+1})(T_{i,j,l+1} - T_{i,j,l})}{\Delta z_{l+1/2}(\Delta z_{l-1/2} + \Delta z_{l+1/2})}.$$
$$(15.9)$$

where $T_{i,j,l}^o$ is temperature for the current moment of time in the central *ijl*-th node of the cross and $T_{i,j,l-1}$, $T_{i,j,l+1}$, etc. are temperatures in 7 nodal points for the next moment of time, $k_{i,j,l-1}$, $k_{i,j,l+1}$, etc. stand for thermal conductivity that can be different at different nodes, $\rho_{i,j,l}$, $C_{Pi,j,l}$, and $H_{i,j,l}$ is density, isobaric heat capacity and heat production values for the central node of the cross, respectively.

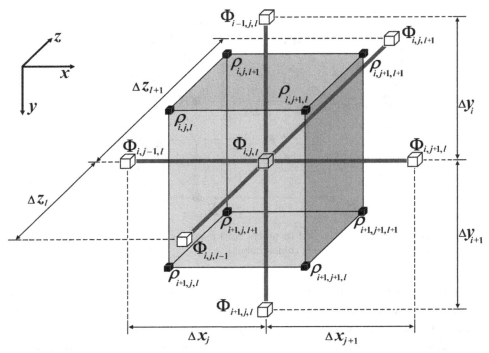

Fig. 15.7 Stencil of a 3D staggered grid used for the discretisation of the Poisson equation. The Poisson equation is formulated for the central node $\Phi_{i,j,l}$ located in one of the cell centres (pressure nodes) of the 3D staggered grid (Fig. 15.1). The notation of different nodal points is the same as in Fig. 15.1. Indexing of different variables corresponds to 3D staggered grid with external velocity nodes (Fig. 15.3).

Like for the temperature equation, the discretisation of the Poisson equation in 3D is also based on a 7-point cross (Fig. 15.7)

$$\frac{\partial^2\Phi}{\partial x^2} + \frac{\partial^2\Phi}{\partial y^2} + \frac{\partial^2\Phi}{\partial z^2} = \pi G(\rho_{i,j,l} + \rho_{i+1,j,l} + \rho_{i,j+1,l} + \rho_{i+1,j+1,l} + \rho_{i,j,l+1}$$

$$+ \rho_{i+1,j,l+1} + \rho_{i,j+1,l+1} + \rho_{i+1,j+1,l+1})/2, \qquad (15.10)$$

$$\frac{\partial^2\Phi}{\partial x^2} = 2\frac{\Phi_{i,j+1,l} - \Phi_{i,j,l}}{\Delta x_{j+1}(\Delta x_j + \Delta x_{j+1})} - 2\frac{\Phi_{i,j,l} - \Phi_{i,j-1,l}}{\Delta x_j(\Delta x_j + \Delta x_{j+1})}, \qquad (15.11)$$

$$\frac{\partial^2\Phi}{\partial y^2} = 2\frac{\Phi_{i+1,j,l)} - \Phi_{i,j,l}}{\Delta y_{i+1}(\Delta y_i + \Delta y_{i+1})} - 2\frac{\Phi_{i,j,l} - \Phi_{i-1,j,l}}{\Delta y_i(\Delta y_i + \Delta y_{i+1})}, \qquad (15.12)$$

$$\frac{\partial^2\Phi}{\partial z^2} = 2\frac{\Phi_{i,j,l+1} - \Phi_{i,j,l}}{\Delta z_{l+1}(\Delta z_l + \Delta z_{l+1})} - 2\frac{\Phi_{i,j,l} - \Phi_{i,j,l-1}}{\Delta z_l(\Delta z_l + \Delta z_{l+1})}. \qquad (15.13)$$

The density at the central node of the cross in Eq. (15.10) is computed as an arithmetic average from 8 surrounding basic nodes. Alternatively, additional density

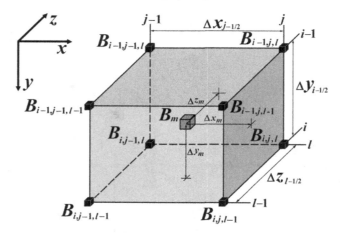

Fig. 15.8 Stencil of a 3D grid (8-nodes cell) used for the interpolation between marker (grey cube) and nodes (black cubes).

points can be defined in the same gravity potential nodes (i.e. in the centre of cells, Fig. 15.1) and respective density values can be separately interpolated from markers when deformation of self-gravitating body is modelled in 3D.

Finally, interpolation between markers and nodes in 3D is also based on the same principles as those discussed for 2D interpolation in Chapter 8. It can be done with the following standard first-order of accuracy trilinear interpolation schemes (Fig. 15.8): interpolation to ijl-th node from markers found in 8 cells surrounding this node

$$B_{(i,j,l)} = \frac{\sum\limits_m B_m w_{m(i,j,l)}}{\sum\limits_m w_{m(i,j,l)}}, \qquad (15.14)$$

interpolation to a marker in a cell from 8 nodes surrounding the cell

$$\begin{aligned}
B_m = \; & B_{i,j,l} w_{m(i,j,l)} + B_{i-1,j,l} w_{m(i-1,j,l)} + B_{i,j-1,l} w_{m(i,j-1,l)} \\
& + B_{i-1,j-1,l} w_{m(i-1,j-1,l)} + B_{i,j,l-1} w_{mi,j,l-1} + B_{i-1,j,l-1} w_{mi-1,j,l-1} \\
& + B_{i,j-1,l-1} w_{mi,j-1,l-1} + B_{i-1,j-1,l-1} w_{mi-1,j-1,l-1}, \qquad (15.15)
\end{aligned}$$

where the statistical weight of m-th-marker for ijl-th-node depends on Δx_m, Δy_m, Δz_m distances to this nodes as

$$w_{m(i,j,l)} = \left(1 - \frac{\Delta x_m}{\Delta x_{j-1/2}}\right) \times \left(1 - \frac{\Delta y_m}{\Delta y_{i-1/2}}\right) \times \left(1 - \frac{\Delta z_m}{\Delta z_{l-1/2}}\right). \qquad (15.16)$$

15.3 Solving discretised 3D equations

After discussing discretisation of variable equations required for thermomechanical geodynamic modelling in 3D, let us consider methods for solving these equations. All approaches considered are iterative.

Temperature equation. Efficient solving of the non-steady 3D temperature equation (15.6) does not require a multigrid since, in contrast to steady Stokes and Poisson equations, the time-dependent solutions for temperature are controlled locally by local heat fluxes rather than globally and can be implemented on the basis of a Gauss–Seidel iteration with a relatively high relaxation parameter $\theta_{relaxation}^{temperature}$ that ranges from 0.5 to 1.5. The respective iterative temperature update scheme *for a regularly spaced grid* based on Eqs. (15.6)–(15.9) is as follows

$$T_{i,j,l}^{new} = T_{i,j,l} + \frac{\Delta R_{i,j,l}^{temperature}}{C_{T(i,j,l)}}\theta_{relaxation}^{temperature}, \tag{15.17}$$

$$\Delta R_{i,j,l}^{temperature} = H_{i,j,l} - \rho_{i,j,l}C_{Pi,j,l}\frac{T_{i,j,l} - T_{i,j,l}^{o}}{\Delta t} - \frac{\partial q_x}{\partial x} - \frac{\partial q_y}{\partial y} - \frac{\partial q_z}{\partial z}, \tag{15.18}$$

$$\frac{\partial q_x}{\partial x} = \frac{(k_{i,j,l} + k_{i,j-1,l})(T_{i,j,l} - T_{i,j-1,l}) - (k_{i,j+1,l} + k_{i,j,l})(T_{i,j+1,l} - T_{i,j,l})}{2\Delta x^2}, \tag{15.19}$$

$$\frac{\partial q_y}{\partial y} = \frac{(k_{i,j,l} + k_{i-1,j,l})(T_{i,j,l} - T_{i-1,j,l}) - (k_{i+1,j,l} + k_{i,j,l})(T_{i+1,j,l} - T_{i,j,l})}{2\Delta y^2}, \tag{15.20}$$

$$\frac{\partial q_z}{\partial z} = \frac{(k_{i,j,l} + k_{i,j,l-1})(T_{i,j,l} - T_{i,j,l-1}) - (k_{i,j,l+1} + k_{i,j,l})(T_{i,j,l+1} - T_{i,j,l})}{2\Delta z^2}, \tag{15.21}$$

$$C_{T(i,j,l)} = \frac{\rho_{i,j,l}C_{Pi,j,l}}{\Delta t} + \frac{k_{i,j-1,l} + 2k_{i,j,l} + k_{i,j+1,l}}{2\Delta x^2}$$
$$+ \frac{k_{i-1,j,l} + 2k_{i,j,l} + k_{i+1,j,l}}{2\Delta y^2} + \frac{k_{i,j,l-1} + 2k_{i,j,l} + k_{i,j,l+1}}{2\Delta z^2}, \tag{15.22}$$

where Δx, Δy, and Δz are regular grid steps in respective directions, $T_{i,j,l}^{o}$ is the temperature at the *ijl*-th node for the current moment of time, $T_{i,j,l}$ and $T_{i,j,l}^{new}$ are old and new (updated) values of temperature at the *ijl*-th node for the next moment of time. To satisfy the boundary condition equations, these equations are called for marginal temperature nodes in the same Gauss–Seidel iteration cycle

upper boundary ($i = 1$)

$$\text{no vertical heat flux } T_{i=1,j,l} = T_{i=2,j,l}$$
$$\text{constant temperature } T_{i=1,j,l} = T_{top}$$

left boundary ($j = 1$)

$$\text{no horizontal heat flux } T_{i,j=1,l} = T_{i,j=2,l}$$

$$\text{constant temperature } T_{i,j=1,l} = T_{left}$$

front boundary ($l = 1$)

$$\text{no lateral heat flux } T_{i,j,l=1} = T_{i,j,l=2}$$

$$\text{constant temperature } T_{i,j,l=1} = T_{front}$$

etc. for other boundaries. An example implementation of above algorithm is given in the program **Temperature3D_Gauss_Seidel.m**. As can be seen from Fig. 15.9, Gauss–Seidel iterations allow us to obtain high-accuracy solutions with 10–20 iterations per time step while the computer accuracy solution is obtained with a few tens of iterations.

Poisson equation. Like in 2D (Chapter 14), the efficient solving of the 3D Poisson equation (15.10) can be based on a multigrid approach, that can again be implemented on the basis of Gauss–Seidel iterations with a relatively high relaxation parameter $\theta_{relaxation}^{Poisson}$ ranging from 0.5 to 1.5. The respective iterative gravity potential update scheme *for regularly spaced grid* based on Eqs. (15.10)–(15.13) is as follows

$$\Phi_{i,j,l}^{new} = \Phi_{i,j,l} + \frac{\Delta R_{i,j,l}}{C_{\Phi(i,j,l)}} \theta_{relaxation}^{Poisson}, \tag{15.23}$$

$$\Delta R_{i,j,l}^{Poisson} = R_{i,j,l}^{Poisson} - \frac{\partial^2 \Phi}{\partial x^2} - \frac{\partial^2 \Phi}{\partial y^2} - \frac{\partial^2 \Phi}{\partial z^2}, \tag{15.24}$$

$$\frac{\partial^2 \Phi}{\partial x^2} = \frac{\Phi_{i,j-1,l} - 2\Phi_{i,j,l} + \Phi_{i,j+1,l}}{\Delta x^2}, \tag{15.25}$$

$$\frac{\partial^2 \Phi}{\partial y^2} = \frac{\Phi_{i-1,j,l} - 2\Phi_{i,j,l} + \Phi_{i+1,j,l}}{\Delta y^2}, \tag{15.26}$$

$$\frac{\partial^2 \Phi}{\partial z^2} = \frac{\Phi_{i,j,l-1} - 2\Phi_{i,j,l} + \Phi_{i,j,l+1}}{\Delta z^2}. \tag{15.27}$$

$$C_{\Phi(i,j,l)} = -\frac{2}{\Delta x^2} - \frac{2}{\Delta y^2} - \frac{2}{\Delta z^2}, \tag{15.28}$$

$\Phi_{i,j-1,l}$, $\Phi_{i-1,j,l}$, etc. are current values of either gravity potential (at finest level) or corrections for this potential (at coarser levels) at respective nodal points, $C_{\Phi(i,j,l)}$ is the coefficient at $\Phi_{i,j,l}$ in the discretised Poisson equation, $\Delta R_{i,j,l}$ is the current residual and $R_{i,j,l}$ is the right-hand side of the Poisson equation. On the finest principal level of resolution, the right-hand side is computed from the standard equation

$$R_{i,j,l} = 4\pi G \rho_{i,j,l}, \tag{15.29}$$

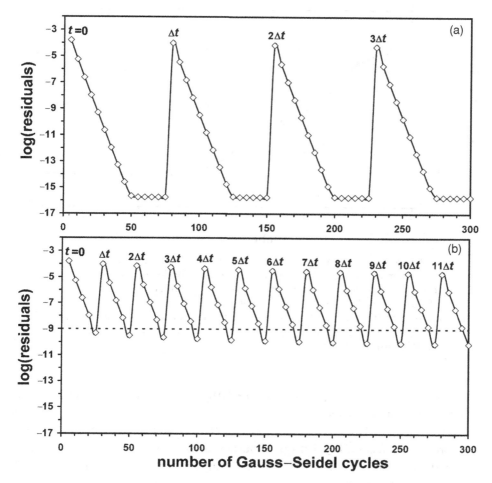

Fig. 15.9 Decay of normalised residuals for the temperature equations with the number of Gauss–Seidel cycles for a 3D model with variable thermal conductivity. Residuals in (a) stabilise at computer accuracy level (10^{-16}) after around 50 iterations (per time step). Iterations in (b) are terminated after reaching given level of tolerance (10^{-9}) that takes around 25 iterations per time step. Model resolution is $51 \times 51 \times 51$ points with regularly spaced grid. Relaxation parameter $\theta^{temperature}_{relaxation} = 1.25$ is used (in Eq. 15.17). Numerical setup: rectangular block having higher temperature is placed in lower temperature surrounding. Results are obtained with program **Temperature3D_Gauss_Seidel.m**.

where G is the gravitational constant and $\rho_{i,j,l}$ the density defined at the same location as $\Phi_{i,j,l}$ (alternatively use Eq. (15.10) when the gravity potential and density are defined at different points). For coarser levels, $R_{i,j,l}$ is composed of residuals interpolated from finer levels. Obviously, grid steps Δx, Δy and Δz are different at different levels of resolution. Various boundary conditions are defined

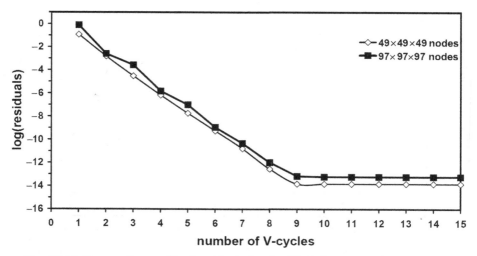

Fig. 15.10 Decay of normalised residuals for the 3D Poisson equation at various resolutions versus number of multigrid V-cycles for a model with a spherical planet embedded in a mass less medium (space). Residuals stabilise at computer accuracy level. 5- and 6-level multigrid with resolutions respectively $49 \times 49 \times 49$ (open diamonds) and $97 \times 97 \times 97$ (solid squares) nodes on the finest level are used with a relaxation parameter $\theta_{relaxation}^{Poisson} = 1.5$. Results are obtained with program **Poisson3D_Multigrid_planet_arbitrary.m**.

as in 2D (see Fig. 14.7, Eqs. (14.12)–(14.16)). An example implementation of the above algorithm, for the case of a self-gravitating planet and gravity potential boundary condition defined on the internal spherical surface inside the grid is given in the program **Poisson3D_Multigrid_planet_arbitrary.m**. As can be seen from Fig. 15.10, multigrid allows us to obtain computer accuracy within around 10 V-cycles independent of the model resolution.

Momentum and continuity equations. As in 2D (Chapter 14), the efficient solution of the 3D Stokes and continuity equations (15.1)−(15.5) can be based on a multigrid approach that can be implemented on the basis of Gauss–Seidel iterations with pressure updates computed from local divergence scaled to local viscosity. The respective iterative pressure and velocity update schemes *for a regularly spaced grid* can be derived on the basis of Eqs. (15.1)–(15.5)

$$P_{i,j,l}^{new} = P_{i,j,l} + \eta_{n(i,j,l)} \Delta R_{i,j,l}^{continuity} \theta_{relaxation}^{continuity}, \tag{15.30}$$

$$\Delta R_{i,j,l}^{continuity} = R_{i,j,l}^{continuity} - \frac{\partial v_x}{\partial x} - \frac{\partial v_y}{\partial y} - \frac{\partial v_z}{\partial z}, \tag{15.31}$$

$$v_{x(i,j,l)}^{new} = v_{x(i,j,l)} + \frac{\Delta R_{i,j,l}^{x\text{-}Stokes}}{C_{v_x(i,j,l)}} \theta_{relaxation}^{Stokes}, \tag{15.32}$$

$$v_{y(i,j,l)}^{new} = v_{y(i,j,l)} + \frac{\Delta R_{i,j,l}^{y\text{-}Stokes}}{C_{v_y(i,j,l)}} \theta_{relaxation}^{Stokes}, \tag{15.33}$$

$$v_{z(i,j,l)}^{new} = v_{z(i,j,l)} + \frac{\Delta R_{i,j,l}^{z\text{-}Stokes}}{C_{v_z(i,j,l)}} \theta_{relaxation}^{Stokes}. \tag{15.34}$$

For models with constant viscosity η, the respective residuals and coefficients in Eqs. (15.30)–(15.34) become

$$\Delta R_{i,j,l}^{x\text{-}Stokes} = R_{i,j,l}^{x\text{-}Stokes} - \eta \left(\frac{\partial^2 v_x}{\partial x^2} + \frac{\partial^2 v_x}{\partial y^2} + \frac{\partial^2 v_x}{\partial z^2} \right) + \frac{\partial P}{\partial x}, \tag{15.35}$$

$$\Delta R_{i,j,l}^{y\text{-}Stokes} = R_{i,j,l}^{y\text{-}Stokes} - \eta \left(\frac{\partial^2 v_y}{\partial x^2} + \frac{\partial^2 v_y}{\partial y^2} + \frac{\partial^2 v_y}{\partial z^2} \right) + \frac{\partial P}{\partial y}, \tag{15.36}$$

$$\Delta R_{i,j,l}^{z\text{-}Stokes} = R_{i,j,l}^{z\text{-}Stokes} - \eta \left(\frac{\partial^2 v_z}{\partial x^2} + \frac{\partial^2 v_z}{\partial y^2} + \frac{\partial^2 v_z}{\partial z^2} \right) + \frac{\partial P}{\partial z}, \tag{15.37}$$

$$\eta_{n(i,j,l)} = \eta, \tag{15.38}$$

$$\frac{\partial^2 v_x}{\partial x^2} = \frac{v_{x(i,j-1,l)} - 2v_{x(i,j,l)} + v_{x(i,j+1,l)}}{\Delta x^2}, \quad \text{(Fig. 15.5)} \tag{15.39}$$

$$\frac{\partial^2 v_x}{\partial y^2} = \frac{v_{x(i-1,j,l)} - 2v_{x(i,j,l)} + v_{x(i+1,j,l)}}{\Delta y^2}, \quad \text{(Fig. 15.5)} \tag{15.40}$$

$$\frac{\partial^2 v_x}{\partial z^2} = \frac{v_{x(i,j,l-1)} - 2v_{x(i,j,l)} + v_{x(i,j,l+1)}}{\Delta z^2}, \quad \text{(Fig. 15.5)} \tag{15.41}$$

$$\frac{\partial P}{\partial x} = \frac{P_{i-1,j,l-1} - P_{i-1,j-1,l-1}}{\Delta x}, \quad \text{(Fig. 15.5)} \tag{15.42}$$

$$C_{v_x(i,j,l)} = -\frac{2\eta}{\Delta x^2} - \frac{2\eta}{\Delta y^2} - \frac{2\eta}{\Delta z^2}, \tag{15.43}$$

and other terms can be derived in a similar manner (derive as an exercise).

For models with variable viscosity, the respective residuals and coefficients in Eqs. (15.32)–(15.34) become

$$\Delta R_{i,j,l}^{x\text{-}Stokes} = R_{i,j,l}^{x\text{-}Stokes} - \frac{\partial \sigma'_{xx}}{\partial x} - \frac{\partial \sigma_{xy}}{\partial y} - \frac{\partial \sigma_{xz}}{\partial z} + \frac{\partial P}{\partial x}, \tag{15.44}$$

$$\Delta R_{i,j,l}^{y\text{-}Stokes} = R_{i,j,l}^{y\text{-}Stokes} - \frac{\partial \sigma'_{yy}}{\partial y} - \frac{\partial \sigma_{yx}}{\partial x} - \frac{\partial \sigma_{yz}}{\partial z} + \frac{\partial P}{\partial y}, \tag{15.45}$$

$$\Delta R_{i,j,l}^{z\text{-}Stokes} = R_{i,j,l}^{z\text{-}Stokes} - \frac{\partial \sigma'_{zz}}{\partial z} - \frac{\partial \sigma_{zx}}{\partial x} - \frac{\partial \sigma_{zy}}{\partial y} + \frac{\partial P}{\partial z}, \tag{15.46}$$

$$\frac{\partial \sigma'_{xx}}{\partial x} = 2\eta_{n(i-1,j,l-1)} \frac{v_{x(i,j+1,l)} - v_{x(i,j,l)}}{\Delta x^2}$$
$$- 2\eta_{n(i-1,j-1,l-1)} \frac{v_{x(i,j,l)} - v_{x(i,j-1,l)}}{\Delta x^2}, \quad \text{(Fig. 15.5)} \tag{15.47}$$

$$\frac{\partial \sigma_{xy}}{\partial y} = \eta_{xy(i,j,l-1)} \left(\frac{v_{x(i+1,j,l)} - v_{x(i,j,l)}}{\Delta y^2} + \frac{v_{y(i,j+1,l)} - v_{y(i,j,l)}}{\Delta y \Delta x} \right)$$

$$- \eta_{xy(i-1,j,l-1)} \left(\frac{v_{x(i,j,l)} - v_{x(i-1,j,l)}}{\Delta y^2} + \frac{v_{y(i-1,j+1,l)} - v_{y(i-1,j,l)}}{\Delta y \Delta x} \right),$$

$$\text{(Fig. 15.5)} \qquad \text{(15.48)}$$

$$\frac{\partial \sigma_{xz}}{\partial z} = \eta_{xz(i-1,j,l)} \left(\frac{v_{x(i,j,l+1)} - v_{x(i,j,l)}}{\Delta z^2} + \frac{v_{z(i,j+1,l)} - v_{z(i,j,l)}}{\Delta z \Delta x} \right)$$

$$- \eta_{xz(i-1,j,l-1)} \left(\frac{v_{x(i,j,l)} - v_{x(i,j,l-1)}}{\Delta z^2} + \frac{v_{z(i,j+1,l-1)} - v_{z(i,j,l-1)}}{\Delta z \Delta x} \right),$$

$$\text{(Fig. 15.5)} \qquad \text{(15.49)}$$

$$C_{v_x(i,j,l)} = -2 \frac{\eta_{n(i-1,j,l-1)} + \eta_{n(i-1,j-1,l-1)}}{\Delta x^2} - \frac{\eta_{xy(i,j,l-1)} + \eta_{xy(i-1,j,l-1)}}{\Delta y^2}$$

$$- \frac{\eta_{xz(i-1,j,l)} + \eta_{xz(i-1,j,l-1)}}{\Delta z^2}, \qquad \text{(15.50)}$$

and other terms can be derived similarly (derive as an exercise).

The methodology of using a multigrid approach for 3D models is the same as in 2D cases, with the single difference that trilinear (Eqs. (15.14)–(15.16)) and not bilinear interpolation schemes should be used to construct the restriction and prolongation operations. Example implementations of the above algorithm for the case of constant and variable viscosity are given respectively in the programs, **Stokes_Continuity3D_Multigrid.m**, **Variable_viscosity3D_Multigrid.m** and **Variable_viscosity3D_MultiMultigrid.m** associated with this chapter. As can be seen from Figs. 15.11, 15.12, multigrid allows us to obtain high-accuracy 3D mechanical solutions at various viscosity contrasts. In the case of large viscosity contrasts, a computer accuracy solution can often be obtained with a 'multi-multigrid' approach (Fig. 15.12) using repetitive cycles of gradual increase in a computational viscosity contrast, as described in Chapter 14.

Elastic stress rotation. 3D numerical models including elasticity should take into account the elastic stress rotation (Chapter 12). One possible way is to use the general form of *Jaumann stress rate* (see Eqs. (12.29)–(12.36)), which allows us to compute the rate of change caused by rotation for various stress components

$$\dot{\sigma}'_{ij(Jaumann)} = \sigma'_{ik}\omega_{kj} - \omega_{ik}\sigma'_{kj}, \qquad \text{(15.51)}$$

where $\dot{\sigma}'_{ij(Jaumann)}$ is rate of change for the σ'_{ij} deviatoric stress components, repeated index k indicates summation and ω_{kj}, ω_{ik} are components of the anti-symmetric rotation rate tensor (Eq. (12.29)). Using Eq. (15.51) in 3D for e.g. σ'_{xx}, the deviatoric stress component gives (see Eq. (12.31))

$$\dot{\sigma}'_{xx(Jaumann)} = 2\sigma_{xy}\omega_{yx} + 2\sigma_{xz}\omega_{zx}. \qquad \text{(15.52)}$$

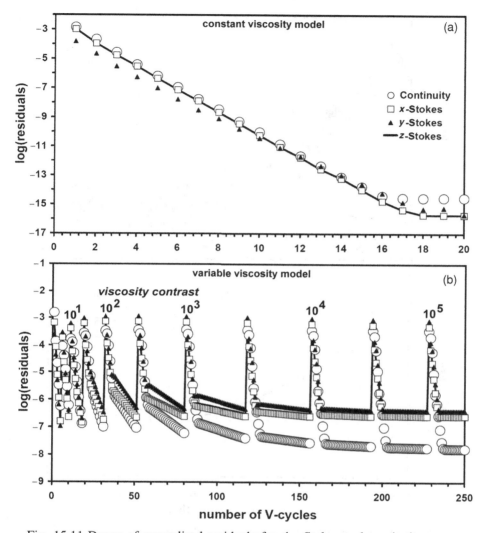

Fig. 15.11 Decay of normalised residuals for the Stokes and continuity equations versus number of multigrid V-cycles for 3D models with constant (a) and variable (b) viscosity. A 5-level multigrid with resolution $49 \times 49 \times 49$ nodes on the finest level is used with relaxation parameters $\theta_{relaxation}^{continuity} = 0.3$ and $\theta_{relaxation}^{Stokes} = 0.9$. Numerical setup: rectangular block having higher density (and viscosity in (b) by factor 10^5) sinks in lower density fluid. Iterations start from a hydrostatic pressure field, zero velocities and no viscosity contrast. Residuals in (a) stabilise at computer accuracy. Spikes in the solutions in (b) are caused by an increase in viscosity contrast by the factor of 3.333 after reaching given level of tolerance (10^{-6}) for the sum of residuals. Results are obtained with the programs **Stokes_Continuity3D_Multigrid.m** (a) and **Variable_viscosity3D_Multigrid.m** (b).

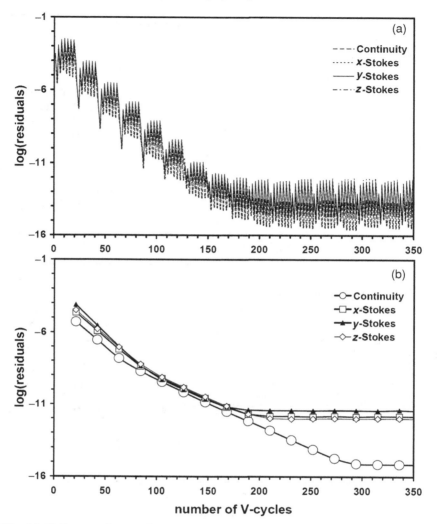

Fig. 15.12 Decay of normalised residuals for the Stokes and continuity equations with the number of multigrid V-cycles for a 3D model with variable viscosity in the case of a 'multi-multigrid' approach using repetitive cycles of gradual increase in computational viscosity contrast. Residuals stabilise at computer accuracy level. 5-level multigrid with resolution $49 \times 49 \times 49$ points on the finest level is used with relaxation parameters $\theta_{relaxation}^{continuity} = 0.75$ and $\theta_{relaxation}^{Stokes} = 1.25$. Numerical setup: rectangular block having higher density and viscosity (by factor 10^6) which sinks in fluid with lower density and viscosity. Iterations start from a hydrostatic pressure field and zero velocities. (a) Decay of local residuals computed with current corrections and right-hand side *within* each cycle (steps) of gradual viscosity contrast increase (spikes). (b) Decay of global residuals computed *after* each cycle of gradual viscosity contrast increase. Spikes in (a) are caused by an increase in viscosity contrast by the factor of 10 every 3 multigrid cycles. Results are obtained with the program **Variable_viscosity3D_MultiMultigrid.m**.

Similar derivations can be made for other stress components (see Eqs. (12.32)–(12.36)). Like in 2D models (Chapter 13, Eqs. (13.35)–(13.36)), the numerical implementation of stress rotation is done by re-computing elastic stress components stored at markers according to the first-order accurate scheme

$$\sigma'_{ij(rotated)} = \sigma'_{ij(m)} + \Delta t_m \times \dot{\sigma}'_{ij(Jaumann)} = \sigma'_{ij(m)} + \Delta t_m(\sigma'_{ik}\omega_{kj} - \omega_{ik}\sigma'_{kj}), \quad (15.53)$$

where $\sigma'_{ij(m)}$ is the deviatoric stress component for a given marker, Δt_m is the marker displacement time step (Chapter 13) and ω_{kj} the rotation rate components that are defined at the same nodal points and computed with similar FD schemes as respective $\dot{\varepsilon}_{kj} = \dfrac{1}{2}\left(\dfrac{\partial v_k}{\partial x_j} + \dfrac{\partial v_j}{\partial x_k}\right)$ strain rate components (see Figs. 15.1, 15.3) and then interpolated to markers using standard interpolation formula (Eq. (15.15), Fig. 15.8).

Numerical algorithms. Numerical algorithms for the thermomechanical codes in 3D, do not differ from algorithms described in Chapters 11, 13 for 2D codes with the exception that direct solvers for various equations should rather be substituted by the iterative ones described above.

Programming exercises and homework

Exercise 15.1
Program solving the 3D temperature equation on a regular $51 \times 51 \times 51$ grid based on Gauss–Seidel iteration (Eqs. (15.17)–(15.22)). The model setup corresponds to hot rectangular block ($20 \times 20 \times 20$ km, $T = 1500$ K, $\rho = 3100$ kg/m^3, $C_P = 1500$ J/K/kg, $k = 1$ W/m/K) which is located in a colder medium ($T = 1000$ K, $\rho = 3000$ kg/m^3, $C_P = 1000$ J/K/kg, $k = 3$ W/m/K). The block is located in the middle of the model, which is $100 \times 100 \times 100$ km in size. Boundary conditions are insulating at all boundaries. Use the relaxation parameter $\theta^{temperature}_{relaxation} = 1.25$ in Eq. (15.17). An example is in **Temperature3D_Gauss_Seidel.m**.

Exercise 15.2
Generalise Exercise 14.1 to 3D. Solve the Poisson equation for the case of a spherical planetary body embedded in a mass less-like medium (Eqs. (15.23)–(15.29), Fig. 14.7). The number of resolution levels $= 4$, the resolution on the coarsest (last) grid $= 7 \times 7 \times 7$ nodal points, and the factor of increase in resolution between the levels $= 2$. All other model parameters and material properties are the same as in the 2D exercise. Use a ghost node approach along the spherical boundary surface for the gravity potential in the same way as in 2D (Fig. 14.7, Eqs. (14.13)–(14.16)), i.e. independently in x, y and z directions. An example is in **Poisson3D_Multigrid_planet_arbitrary.m**.

Exercise 15.3

Generalise Exercises 14.2 and 14.3 to 3D. Solve the Stokes and continuity equations for both constant and variable viscosity cases using a pressure–velocity formulation and ghost node approach (Eqs. (15.30)–(15.50), Figs. 15.1–15.5). Model parameters: model size $= 100 \times 100 \times 100$ km, block size $= 20 \times 20 \times 20$ km. Material properties are the same as in 2D cases. Boundary conditions $=$ free slip at all boundaries, number of resolution levels $= 4$, resolution of the basic grid on the coarsest (last) level $= 4 \times 4 \times 4$ nodal points, factor of increase in resolution between the levels $= 2$, relaxation coefficients for Gauss–Seidel iterations $\theta_{relaxation}^{Stokes} = 0.9$ and $\theta_{relaxation}^{continuity} = 0.3$, number of smoothing iterations on the finest (basic) level $= 5$, factor of increase in the number of iterations with the level coarsening $= 2$. For the variable viscosity case, use a gradual increase in the viscosity contrast by a factor $10^{1/2}$ (Eq. (14.57)), to reach an accurate solution (Fig. 15.11). Examples are in **Stokes_Continuity3D_Multigrid.m** and **Variable_viscosity3D_Multigrid.m**.

16

Numerical benchmarks

Theory: Numerical benchmarks: testing of numerical codes for various problems. Examples of thermomechanical benchmarks.
Exercises: Programming of models for various numerical benchmarks.

16.1 Code benchmarking: why should we spend time on it?

Benchmarking of a numerical code means comparing the numerical solution obtained with solving the system of linear equations with (i) analytical solutions (ii) results of physical (analogue) experiments (iii) numerical results from other (well-established) codes and (iv) general physical considerations. Benchmarking of newly created numerical tools is sometimes very tedious, but an absolutely necessary stage of code development as its purpose is to test the code's robustness in a broad range of situations relevant to geodynamic modelling applications. For instance, if you plan to model with your code shear heating processes in deforming rocks – make sure that your code provides the correct temperature changes related to mechanical energy dissipation; if you model subduction – make sure that your code handles correctly large viscosity contrasts and has no notable numerical diffusion of the temperature field and composition; if you intend to model self-gravitating planetary bodies – make sure that your code computes correct gravity field etc. We should not be lazy and limit ourselves to one or two common benchmarks, such as the Rayleigh–Taylor instability and convection with constant viscosity hoping that everything else will work automatically. No, it will not! Therefore, test your code on a broad range of challenging cases (several of which are discussed below), to explore its limitations and be creative in inventing and calibrating new numerical benchmarks. Then, in the end, you will be really proud of your 'numerical child'. Many of the analytical solutions that can be used for testing of thermomechanical codes *for geodynamically relevant situations* can be taken from the textbook of Turcotte and Schubert (2002), which we will also use for constraining some of our numerical benchmarks.

Below we discuss details of several important benchmarks which test various aspects of thermomechanical codes. These calibrating tests aim to verify the efficacy of numerical solutions for a variety of circumstances relevant to geodynamics. These will include:

(a) sharply discontinuous viscosity distribution (test 1 and 2);
(b) strain rate dependent viscosity (test 3);
(c) non-steady development of temperature field (test 4);
(d) shear heating for temperature dependent viscosity (test 5);
(e) advection of a sharp temperature front (test 6);
(f) heat conduction for temperature-dependent thermal conductivity (test 7);
(g) thermal convection with constant and variable viscosity (tests 8);
(h) elastic stress build-up and advection (test 9 and 10);
(i) localisation of visco-elasto-plastic deformation (test 11).

Of course this list is incomplete and many additional benchmarks exist, or can be invented. Yet, performing the discussed benchmarks, we will at least get some confidence that we created a state-of-the-art numerical geodynamic modelling tool which correctly reproduces a number of challenging geodynamic models.

16.2 Test 1. Rayleigh–Taylor instability benchmark

This is a typical analytical solution based benchmark. To test correctness of the numerical velocity solution for gravity driven flows, in the case of sharply heterogeneous density and viscosity fields, one can use a two-layer Rayleigh–Taylor instability model (e.g. Ramberg, 1968) with a no-slip condition on the top and at the bottom and symmetry conditions along the vertical walls (Fig. 16.1(a)). An initial sinusoidal disturbance of the boundary between the upper (η_1, ρ_1) and the lower (η_2, ρ_2) layers of thicknesses h_1 and h_2, respectively, has a small initial amplitude (ΔA) and a wavelength (λ). Under this condition, the velocity of the diapiric growth (v_y) is given by the relation (Ramberg, 1968)

$$\frac{v_y}{\Delta A} = -K \frac{\rho_1 - \rho_2}{2\eta_2} h_2 g,$$

$$K = \frac{-d_{12}}{c_{11} j_{22} - d_{12} i_{21}},$$

$$c_{11} = \frac{\eta_1 2\phi_1^2}{\eta_2 \left(\cosh 2\phi_1 - 1 - 2\phi_1^2\right)} - \frac{2\phi_2^2}{\cosh 2\phi_2 - 1 - 2\phi_2^2}$$

$$d_{12} = \frac{\eta_1 \left(\sinh 2\phi_1 - 2\phi_1\right)}{\eta_2 \left(\cosh 2\phi_1 - 1 - 2\phi_1^2\right)} + \frac{\sinh 2\phi_2 - 2\phi_2}{\cosh 2\phi_2 - 1 - 2\phi_2^2},$$

(16.1)

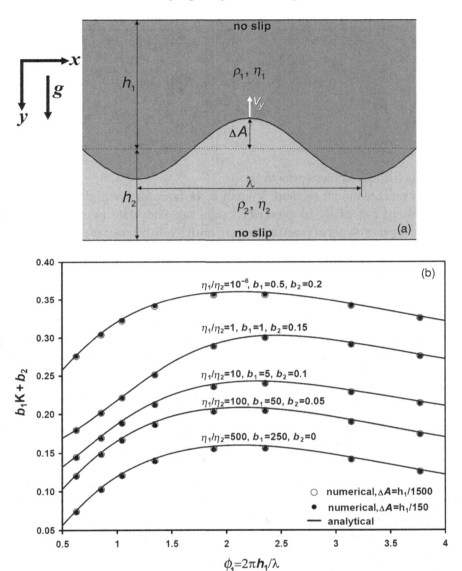

Fig. 16.1 Rayleigh–Taylor instability benchmark. (a) Initial setup. (b) Comparison of numerical (symbols) and analytical (lines, Eq. (16.1)) solutions for the case of two layers with equal thicknesses ($h_1 = h_2$). Arbitrary scaling coefficients b_1 and b_2 are used in (b) for plotting results computed at variable viscosity contrasts on the same diagram. Growth factor K for numerical cases is computed from velocity field based on Eq. (16.1). Numerical results are calculated at resolution 51×51 nodes and 250×250 markers with the code **Variable_viscosity_Ramberg.m**.

$$i_{21} = \frac{\eta_1 \phi_2 \left(\sinh 2\phi_1 + 2\phi_1\right)}{\eta_2 \left(\cosh 2\phi_1 - 1 - 2\phi_1^2\right)} + \frac{\phi_2 \left(\sinh 2\phi_2 + 2\phi_2\right)}{\cosh 2\phi_2 - 1 - 2\phi_2^2},$$

$$j_{22} = \frac{\eta_1 2\phi_1^2 \phi_2}{\eta_2 \left(\cosh 2\phi_1 - 1 - 2\phi_1^2\right)} - \frac{2\phi_2^3}{\cosh 2\phi_2 - 1 - 2\phi_2^2},$$

$$\phi_1 = \frac{2\pi h_1}{\lambda},$$

$$\phi_2 = \frac{2\pi h_2}{\lambda},$$

where K is a dimensionless growth factor.

With a marker-in-cell method, a small layer boundary perturbation (less than one grid step) can easily be prescribed on a sub-grid scale by small sinusoidal (vertical) displacements of markers that are initially distributed regularly inside the numerical grid

$$\Delta A_m = \cos\left(2\pi \frac{x_m - 0.5L}{\lambda}\right) \Delta A, \tag{16.2}$$

where x_m and ΔA_m are horizontal coordinate and vertical displacement for a given marker m and L is the horizontal width of the numerical model. For proper constraining the numerical models, the relationship $L = 2\lambda$ can be used together with free slip (i.e. horizontal symmetry) conditions on two vertical boundaries.

Figure (16.1(b)) compares numerical and analytical solutions for the growth rate of the instability estimated for two layers of equal thickness of (i.e. $h_1 = h_2$) at different values of ΔA, λ and η_1/η_2. Good accuracy at large variations of the disturbance wavelength and layer viscosity contrasts ($\eta_1/\eta_2 = 10^{-6}$–5×10^2) suggests that the tested numerical code is capable of correctly modelling the velocity fields for gravity driven flows across a boundary with sharp changes in density and viscosity. Even very small perturbations of the horizontal boundary are properly captured by variations in relative position of markers via a bilinear density interpolation procedure from markers to nodes. An example of the numerical setup for conducting the Rayleigh–Taylor instability benchmark is given by the code **Variable_viscosity_Ramberg.m**.

16.3 Test 2. Falling block benchmark

This is a typical example of a benchmark which is based on general physical considerations. I personally like it very much, since it is simple to implement but creates challenging conditions to be handled numerically. In the case of an isolated rigid object, sinking in a low viscosity surrounding, the velocity of the object mainly depends on the viscosity of its surrounding (weakest medium). As

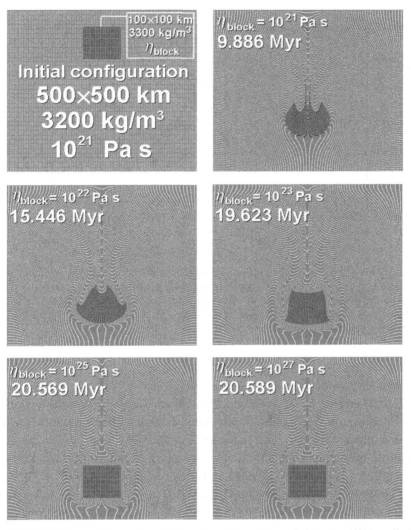

Fig. 16.2 Initial conditions (top left) and results of the numerical experiments for the falling block benchmark performed by Gerya and Yuen (2003a). Boundary conditions: free slip at all boundaries. Black and white dots represent positions of markers for the block and the medium, respectively. Grid resolution of the model is 51×51 nodes, 22 500 markers.

was discussed in Chapter 14, this situation differs from modelling the Rayleigh–Taylor instability where the strong layer is attached to the model boundaries and the velocity field is defined by the strongest medium. According to our physical intuition, (i) deformation of the block should vanish with increasing viscosity contrast (Fig. 16.2) and (ii) the sinking velocity at high viscosity contrasts should

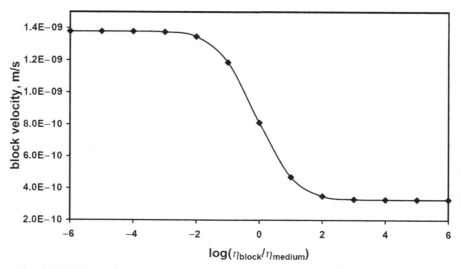

Fig. 16.3 Velocity of the rectangular block sinking in a viscous medium as a function of viscosity contrast between the block and the background medium. The model setup corresponds to Fig. 16.2. (top left). Numerical results are calculated at a resolution of 51×51 nodes and 250×250 markers with the code **Variable_viscosity_block.m**.

be independent of the absolute value of the viscosity of the block but should solely depend on that of the surrounding medium (Fig. 16.3). In the case of using finite differences for solving the momentum equations, they should be formulated in a stress-conserving way (see Chapter 7). This test also proves the accurate conservation properties of a numerical procedure in terms of *preserving the block edges geometry* (Fig. 16.2) at large deformation and high (10^2–10^6) viscosity contrast between the stiffer block and the weak surroundings. An example of the numerical setup for conducting the falling block benchmark is given by the code **Variable_viscosity_block.m**.

16.4 Test 3. Channel flow with a non-Newtonian rheology

This test can be conducted to check the numerical solution of the momentum and continuity equations for flows with a strongly strain-rate/stress-dependent rheology, which is characteristic of dislocation creep (Ranalli, 1995). The computation is carried out for vertical flow of a non-Newtonian (with a power-law index n) viscous medium in a section of an infinite vertical channel (Fig. 5.2) of width L in the absence of gravity. Boundary conditions are taken as follows: a given constant vertical pressure gradient, $\dfrac{\partial P}{\partial y}$ along the channel and no-slip conditions at the walls. The viscosity of the non-Newtonian flow is defined by the following rheological equation formulated in term of second stress and strain rate invariants

$$2\dot{\varepsilon}_{II} = C_1 \left(\sigma_{II}\right)^n, \tag{16.3}$$

where C_1 is a material constant in $Pa^{-n} \cdot s^{-1}$. Equation (16.3) can be reformulated in terms of effective viscosity as a function of second strain rate invariant

$$\eta_{eff} = \frac{\sigma_{II}}{2\dot{\varepsilon}_{II}} = C_1^{-1/n} \left(2\dot{\varepsilon}_{II}\right)^{1/n-1}. \tag{16.4}$$

Analytical solutions for the velocity and viscosity profiles across the channel are given by (Turcotte and Schubert, 2002; Gerya and Yuen, 2003a)

$$v_y = \frac{C_1}{n+1} \left(-\frac{\partial P}{\partial y}\right)^n \left[\left(\frac{L}{2}\right)^{n+1} - \left(x - \frac{L}{2}\right)^{n+1}\right], \tag{16.5}$$

$$\eta_{eff} = \frac{\sigma_{II}}{2\dot{\varepsilon}_{II}} = \frac{\sigma_{yx}}{2\dot{\varepsilon}_{yx}} = \frac{1}{C_1}\left(-\frac{\partial P}{\partial y}\right)^{1-n}\left(x - \frac{L}{2}\right)^{1-n}, \tag{16.6}$$

$$\frac{\partial P}{\partial y} = \frac{P_{end} - P_{beg}}{H}, \tag{16.7}$$

where P_{beg} and P_{end} are pressures at the beginning ($y=0$) and at the end ($y=H$) of the channel section of height H, respectively. Figure 16.4 compares analytical and numerical (2D) solutions based on conservative finite differences with marker-in-cell techniques obtained with the code **Variable_viscosity_channel.m**. Numerical and analytical solutions overlap, implying high accuracy of the numerical method for modelling flows with strong lateral variations in viscosity caused by the non-Newtonian rheology. *Open channel boundary conditions* at the top and at the bottom imply an infinite vertical channel with constant vertical pressure gradients. These boundary conditions are programmed by defining P_{beg} and P_{end} in the first and the last row of pressure nodes, respectively, and prescribing $\dfrac{\partial v_x}{\partial y} = 0$, $\dfrac{\partial v_y}{\partial y} = 0$ at the upper and lower boundary of the model. Note that the vertical length of the channel section H used in Eq. (16.7) for computing pressure gradient corresponds to the distance between the first and the last row of pressure nodes and not to the vertical length of the 2D model.

16.5 Test 4. Non-steady temperature distribution in a Newtonian channel

Here we describe another channel flow based benchmark. This one can be performed to test the numerical accuracy of solving the time-dependent (non-steady)

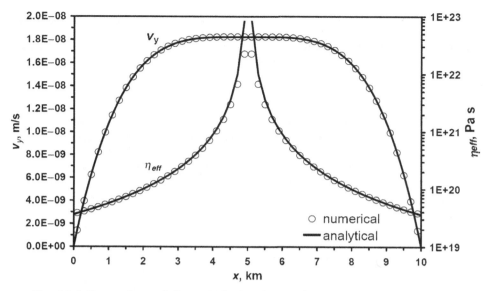

Fig. 16.4 Comparison of the analytical and numerical solutions for the velocity and viscosity profiles across a channel with non-Newtonian flow rheology given by Eq. (16.3). Numerical results are calculated at the resolution of 51×21 nodes and 250×100 markers with the code **Variable_viscosity_channel.m** associated with this chapter. Model parameters: $L = 10$ km, $H = 9.5$ km, $n = 3$, $C_1 = 10^{-37}$ $Pa^{-3} \cdot s^{-1}$, $P_{beg} = 10^9$ Pa, $P_{end} = 0$.

temperature equations in cases when heat advection is coupled with heat diffusion. The model corresponds to the vertical flow of a heat-conductive medium of constant viscosity η in a channel, in the absence of gravity. Boundary conditions are: a given constant vertical pressure gradient, $\dfrac{\partial P}{\partial y}$ along the channel, non-slip conditions and $T = const = T_o(y)$ and $\dfrac{\partial T}{\partial y} = const = \dfrac{\partial T_o(y)}{\partial y}$ at the walls (Exercise 9.2). The initial conditions for the temperature distribution inside the model are $T = T_o(y)$, $\dfrac{\partial T}{\partial y} = \dfrac{\partial T_o(y)}{\partial y}$ and $\dfrac{\partial T}{\partial x} = 0$. The horizontal steady-state profile for vertical velocities, v_y, is defined by the equation which can be derived either from Eq. (16.5) with $n = 1$ and $C_1 = \dfrac{1}{\eta}$ or from Eq. (5.30) with $g_y = 0$

$$v_y = -\frac{1}{2\eta} \left(\frac{\partial P}{\partial y} \right) (Lx - x^2). \tag{16.8}$$

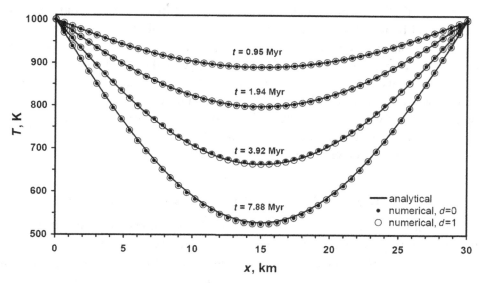

Fig. 16.5 Comparison of the analytical and numerical solutions for temperature profiles across the channel with constant viscosity. Numerical results are calculated at resolution 51×11 nodes and 250×50 markers with the code **Constant_viscosity_channel_T.m**. Model parameters: $L = 30$ km, $H = 11.25$ km, $\eta = 10^{19}$ Pa · s, $P_{beg} = 10^5$ Pa, $P_{end} = 0$, $\partial T/\partial y = 40$ K/km, $T(y = 0) = 1000$ K.

The corresponding temperature changes in the channel with time are then given by the following series expansion (Gerya and Yuen, 2003a)

$$\Delta T(x, t) = \sum_{m=1}^{\infty} F_m E_{mt} \sin \left[\pi \, (2m - 1) \frac{x}{L} \right],$$

$$F_m = -8\xi \frac{L^2}{[\pi \, (2m - 1)]^3},$$

$$E_{mt} = L^2 \frac{1 - \exp \left\{ -\dfrac{\kappa t}{L^2} [\pi \, (2m - 1)]^2 \right\}}{\kappa \, [\pi \, (2m - 1)]^2}, \qquad (16.9)$$

$$\kappa = \frac{k}{\rho C_P},$$

$$\xi = -\frac{1}{2\eta} \left(\frac{\partial P}{\partial y} \right) \left(\frac{\partial T_o(y)}{\partial y} \right),$$

where $\Delta T(x, t)$ is the temperature change as a function of the horizontal coordinate x and time t; κ is a constant thermal diffusivity in m²/s. Equation (16.9) does not account for shear heating: in this numerical test it is considered as negligible. Figure 16.5 compares the analytical solution from Eq. (16.9) with the numerical one obtained with the 2D thermomechanical code **Constant_viscosity_channel_T.m.** Mechanical boundary conditions for the infinite vertical channel are the same as for the previous benchmark. A constant temperature gradient is used as a boundary condition for temperature nodes located at the upper and lower thermal boundaries implying infinity of the thermal profile in the vertical direction. Figure 16.5 shows that numerical and analytical results coincide well for calculations performed both with ($d = 1$) and without ($d = 0$) numerical subgrid diffusion (Eqs. (10.15)–(10.19)) implying robustness of the coupled thermomechanical solution for the case of non-steady heat conduction associated with heat advection.

16.6 Test 5. Couette flow with viscous heating

This benchmark is designed to verify the numerical solution of the coupled momentum and temperature equations for flows with temperature-dependent rheology in the situation of strong shear heating (viscous dissipation). The analytical model setup corresponds to a vertical *Couette flow* (simple shear deformation in a laterally limited planar zone of width L) in the absence of gravity. Boundary conditions are taken as follows: zero vertical pressure gradient, $\dfrac{\partial P}{\partial y} = 0$ along the flow, $v_y = 0$, $T = T_0$ and $\sigma_{yx} = const = \sigma_{yx1}$, $\dfrac{\partial T}{\partial x} = 0$ at the left and right walls, respectively. Viscosity of the flow is given by the following rheological equation (Turcotte and Schubert, 2002)

$$\eta = A \exp\left[\frac{E_a}{RT_0}\left(1 - \frac{T - T_0}{T_0}\right)\right]$$

where E_a is the activation energy, R is gas constant and A is pre-exponential rheological constant, which depends on the material. The analytical solution for steady temperature distribution $T(x)$ inside the flow is given by the relation (Turcotte and Schubert, 2002)

$$x = \frac{L}{B}\ln\left[\frac{(D + B)(C - B)}{(D - B)(C + B)}\right], \tag{16.10}$$

$$B = \ln\left[\frac{1 + \left(1 - \dfrac{2Br}{B^2}\right)^2}{1 - \left(1 - \dfrac{2Br}{B^2}\right)^2}\right], \tag{16.11}$$

$$C = \{2\,[\phi_1 - \phi(x)]\,Br\}^{1/2}, \tag{16.12}$$

$$D = [2\,(\phi_1 - 1)\,Br]^{1/2}, \tag{16.13}$$

$$\phi(x) = \exp [\theta(x)], \tag{16.14}$$

$$\theta(x) = \frac{E_a [T(x) - T_0]}{RT_0^2}, \tag{16.15}$$

$$\phi_1 = \frac{B^2}{2Br}, \tag{16.16}$$

$$\phi_1 = \exp (\theta_1), \tag{16.17}$$

$$\theta_1 = \frac{E_a (T_1 - T_0)}{RT_0^2}, \tag{16.18}$$

$$Br = \frac{(\sigma_{yx1} L)^2 E_a}{kART_0^2} \exp \left(-\frac{E_a}{RT_0}\right), \tag{16.19}$$

where Br is the non-dimensional Brinkman number, θ the non-dimensional temperature change, σ_{yx1} the shear stress that remains constant within the flow, k the thermal conductivity of the flow medium, T_1 the temperature at the right wall of the flow (i.e. maximal temperature). Solving the non-linear Equations (16.10)–(16.11) analytically for given values of k, L, A, E_a, T_0 and σ_{yx1} is non-trivial and the solution for T_1 is not unique at a given value of σ_{yx1}. Rather than defining σ_{yx1}, non-negative values of B can be chosen and then the Brinkman number and shear stress in the channel can be computed from Eq. (16.11) and Eq. (16.19), respectively, as

$$Br = \frac{B^2}{2} \left[1 - \left(\frac{\exp B - 1}{\exp B + 1}\right)^2\right], \tag{16.20}$$

$$\sigma_{yx1} = \left[Br \frac{kART_0^2}{L^2 E_a} \exp \left(\frac{E_a}{RT_0}\right)\right]^{1/2}. \tag{16.21}$$

Other unknown parameters can be computed from B and Br by using Equations (16.10)–(16.18). Based on such calculations, the dependence of maximal non-dimensional temperature change in the channel θ_1 from the Brinkman number Br can be computed (Fig. 16.6(a)).

To test the Couette flow solution numerically, the constant vertical velocity boundary condition $v_y = v_{y1}$ should be applied at the right boundary instead of $\sigma_{yx} = \sigma_{yx1}$ used in the analytical model. The upper and lower boundary conditions are the same as in the Tests 3 and 4 taken that $P_{beg} = P_{end} = 0$ and $\frac{\partial T}{\partial y} = 0$. This modification will ensure uniqueness of the numerical thermomechanical solution which becomes steady-state in a finite number of time steps. The value of the parameter B should be computed iteratively with Equations (16.16)–(16.18) and (16.20)

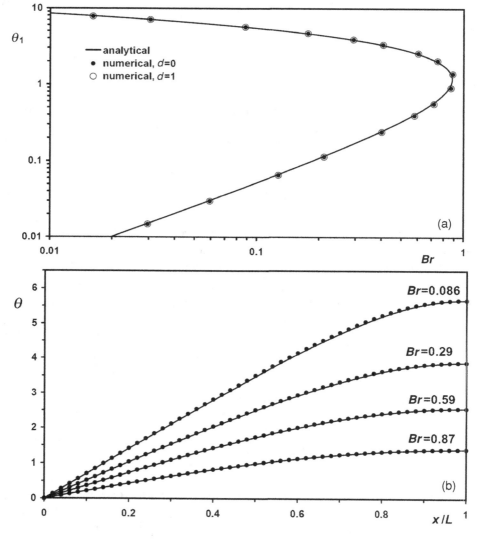

Fig. 16.6 Comparison of analytical and numerical results for a case of steady-state Couette flow with temperature-dependent viscosity of the medium and shear heating. (a) Maximal temperature change within the flow θ_1 (Eq. 16.18) versus Brinkman number Br (Eq. 16.19), (b) Distribution of temperature changes θ (Eq. 16.15) across the flow at different Brinkman number. Numerical results are calculated at a resolution of 51×11 nodes and 250×50 markers with the code **Variable_viscosity_Couette_T.m** associated with this chapter. Model parameters: $L = 30$ km, $H = 11.25$ km, $A = 10^{15}$ Pa · s, $E_a = 150$ kJ/mol, $k = 2$ W/m/K, $T_0 = 1000$ K.

from the steady-state temperature (T_1) at the right boundary (see this type of computation in the end of the program example **Variable_viscosity_Couette_T.m**). Other parameters can again be computed with B from Equations (16.10)–(16.21). Figure 16.6 shows that numerical and analytical results coincide well, implying that the numerical solution holds for thermomechanical effects of shear heating in case of strongly variable temperature-dependent viscosity.

16.7 Test 6. Advection of sharp temperature fronts

The verification of the ability to advect sharp temperature fronts is fundamental in numerical tests of various advection algorithms. The geodynamic relevance of this test is obvious when modelling rapidly moving subducting and detached slabs is envisaged. Numerical solutions for this type of benchmark (see e.g. Chapter 8) are typically calculated in 2D for the solid body rotation of a two-dimensional temperature wave of an arbitrary shape. One can, for example, perform such a test for a square wave with width L and thermal amplitude $\Delta T_o = 500$ K. The results of the test obtained with our finite-difference and marker-in-cell techniques are shown in Fig. 16.7 for a regularly spaced grid of moderate resolution (51×51 nodes, 250×250 markers). If heat conduction is insignificant, (Fig. 16.7(a)) the adopted marker-in-cell advection scheme is obviously not numerically diffusive, even for many revolutions, as long as after each complete revolution the initial positions of markers (with the corresponding values of initially prescribed temperature field which is negligibly affected by the heat diffusion) are reproduced well with the fourth-order Runge–Kutta integration scheme (see code **Solid_Body_Rotation_T.m**). In the case of significant heat conduction (Fig. 16.7(b)), the final temperature distribution does not depend noticeably on the number of revolutions. This point suggests good conservation properties of the adopted numerical scheme when advecting diffusing temperature fronts. Introducing numerical subgrid diffusion (Chapter 10) only negligibly affects the temperature when heat conduction is significant (Fig. 16.7(b)). Obviously, this numerical diffusion, which gives a small addition to the physical diffusion, exerts little influence in the case of negligible heat conduction (Fig. 16.7(a)). Generally, the tested method of solving the temperature equation using markers works very well in the two distinct regimes of advection for both non-diffusive (Fig. 16.7(a)) and diffusive (Fig. 16.7(b)) sharp temperature fronts.

16.8 Test 7. Channel flow with variable thermal conductivity

This analytical benchmark can be conducted to verify the accuracy of a thermomechanical code in the case of strong variations in temperature-dependent

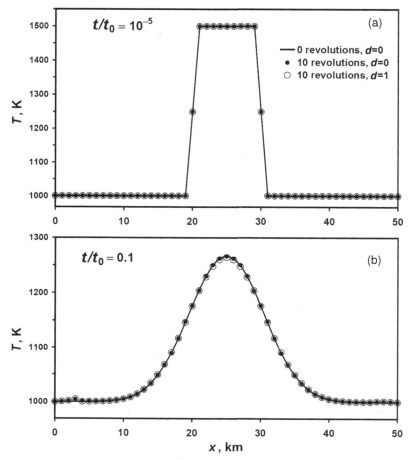

Fig. 16.7 The results of test of numerical solution for the solid body rotation of square temperature wave. The figure shows the horizontal profiles across the wave at different time t after given number of revolutions. (a) and (b) Results of numerical experiments at different characteristic thermal diffusion timescale $t_0 = \rho C_P L^2 / k$, where $L = 10$ km is the initial length of the temperature wave and d is numerical subgrid diffusion parameter (see Eq. 10.16 in Chapter 10). Numerical results are calculated at resolution 51×51 nodes and 250×250 markers with the code Solid_Body_Rotation_T.m. Model parameters: size $= 50 \times 50$ km, $C_P = 1000$ J/kg/K, $\rho = 3000$ kg/m^3, $k = 2 \times 10^{-6}$ and 0.02 W/m/K for (a) and (b), respectively, $T_{medium} = 1000$ K, $T_{wave} = 1500$ K. Small temperature perturbation at the left boundary in (b) is a boundary effect (i.e. a trace of the rotating wave interacting with the model thermal boundary; run program **Solid_Body_Rotation_T.m** in order to see this trace in 2D).

thermal conductivity which are relevant to many geodynamic situations that involve large variations in temperature (mantle convection, lithospheric processes, etc.). For this purpose, we can again use vertical Newtonian channel flow (as in Test 4 but without temperature gradients along the channel) with a velocity distribution defined by Equation (16.8) and shear heating, which provides a strong heat-source term in the temperature equation (Chapter 9). The thermal conductivity is taken to be decreasing with temperature, which is very characteristic (e.g. Hoffmeister, 1999) for *lattice conductivity* defined by *phonons* in crystal lattice

$$k = \frac{k_0}{1 + b(T - T_0)/T_0},\tag{16.22}$$

where T_0 is a constant temperature applied at the walls of the channel; k_0 is thermal conductivity at T_0; b is a dimensionless coefficient.

The steady temperature profiles across the channel $T(x)$ are then defined by equation (Gerya and Yuen, 2003a)

$$T(x) = T_0 \frac{C(x) + b - 1}{b},$$

$$C(x) = \exp\left\{\frac{L^4 b}{192 k_0 T_0 \eta} \left(\frac{\partial P}{\partial y}\right)^2 \left[1 - \left(2\frac{x}{L} - 1\right)^4\right]\right\},\tag{16.23}$$

where L is the channel width, η is a constant viscosity of the medium and $\dfrac{\partial P}{\partial y}$ is the pressure gradient along the channel (Eq. 16.7).

Figure 16.8 compares the analytical solution for both temperature and thermal conductivity profiles with the numerical solutions obtained with the 2D thermomechanical code, **Variable_conductivity_channel.m**. This figure demonstrates the high accuracy of the numerical solution, suggesting that the adopted conservative FD scheme correctly computes heat transport in the case of strong variations in thermal conductivity (factor of 4 variation across the channel for the given case, Fig. 16.8).

16.9 Test 8. Thermal convection with constant and variable viscosity

This benchmark can be conducted to test the ability of the code to model mantle convection. Blankenbach *et al.* (1989) tested several 2D mantle convection models with a broad variety of numerical techniques and reported steady-state values for a number of model parameters to which the numerical solution should converge with increasing grid resolution.

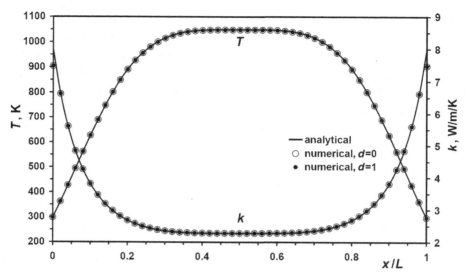

Fig. 16.8 Comparison of the analytical (Eq. (16.23)) and numerical solutions for the steady temperature and thermal conductivity profiles across a channel with constant viscosity and strong shear heating. Numerical results are calculated at resolution 51×11 nodes and 250×50 markers with the code **Variable_conductivity_channel.m.** Model parameters: $L = 30$ km, $H = 11.25$ km, $\eta = 10^{19}$ Pa·s, $P_{beg} = 3 \times 10^7$ Pa, $P_{end} = 0$, $T_0 = 298$ K, $k_0 = 8$ W/m/K, $b = 1$.

Table 16.1 represents the physical parameters for five steady-state convection models with both constant (models 1a, 1b, 1c) and variable (models 2a, 2b) temperature- and depth-dependent viscosity. Convection is studied in a rectangular box of height H and width L ($H = L = 1000$ km for all models with except of model 2b). The boundary conditions are free-slip along all boundaries, a specified temperature on the top (T_{top}) and at the bottom (T_{bottom}) and thermal insulation ($\partial T / \partial x = 0$) along the left and right walls. The difference between T_{top} and T_{bottom} in all experiments is 1000 K. The following formulation for temperature- and depth-dependent viscosity of the mantle is used

$$\eta = \eta_0 \exp \left(-b \frac{T - T_{top}}{T_{bottom} - T_{top}} + c \frac{y}{H} \right), \tag{16.24}$$

where η_0 is viscosity at the top of the model (i.e. at $T = T_{top}$ and $y = 0$); b and c are coefficients establishing dependencies of viscosity with temperature and depth, respectively ($b = 0$ and $c = 0$ in constant viscosity tests 1a, 1b and 1c). Density in all models depends linearly on temperature

$$\rho = \rho_0 [1 - \alpha (T - T_{top})],$$

Table 16.1 *Parameters of mantle convection benchmarks (Blankenbach et al., 1989)*

Test	1a	1b	1c	2a	2b
Gravitational acceleration, g (m/s^2)	10	10	10	10	10
Model height, H (km)	1000	1000	1000	1000	1000
Model width, L (km)	1000	1000	1000	1000	2500
Temperature at the top, T_{top} (K)	273	273	273	273	273
Temperature at the bottom, T_{bottom} (K)	1273	1273	1273	1273	1273
Thermal conductivity, k (W/m/K)	5	5	5	5	5
Heat capacity, C_P (J/kg)	1250	1250	1250	1250	1250
Standard density, ρ_0 (kg/m^3)	4000	4000	4000	4000	4000
Thermal expansion, α (1/K)	2.5×10^{-5}	2.5×10^{-5}	2.5×10^{-5}	2.5×10^{-5}	2.5×10^{-5}
Flow law parameters: η_0 (Pa s)	10^{23}	10^{22}	10^{21}	10^{23}	10^{23}
b	0	0	0	$\ln(1000)$	$\ln(16384)$
c	0	0	0	0	$\ln(64)$
Nusselt number, $Nu = \dfrac{H}{T_{bottom}\,L} \displaystyle\int_{x=0}^{L} \left(\dfrac{\partial T}{\partial y} \right)_{top} dx$	4.8844	10.534	21.972	10.066	6.9229
Non-dimensional root mean square (rms) velocity, $v_{rms} = \dfrac{H \rho_0 C_P}{k} \sqrt{\dfrac{1}{HL} \displaystyle\int_{x=0}^{L}\int_{y=0}^{H} (v_x^2 + v_y^2)\,dy\,dx}$	42.865	193.21	833.99	480.43	171.76
Non-dimensional temperature gradients, $q_{corner} = \dfrac{H}{T_{bottom} - T_{top}} \left(\dfrac{\partial T}{\partial y} \right)_{corner}$					
q_1 (top-left corner, above upwelling)	8.0593	19.079	45.964	17.531	18.484
q_2 (top-right corner, above downwelling)	0.5888	0.7228	0.8772	1.0085	0.1774
q_3 (bottom-right corner, below downwelling)	8.0593	19.079	45.964	26.809	14.168
q_4 (bottom-left corner, below upwelling)	0.5888	0.7228	0.8772	0.4974	0.6177
Local minimum along the central vertical temperature profile: $T_c = \dfrac{T - T_{top}}{T_{bottom} - T_{top}}$	0.4222	0.4284	0.4322	0.7405	0.3970
$Z_c = \dfrac{H - y}{H}$	0.2249	0.1118	0.0577	0.0623	0.1906
Local maximum along the central vertical temperature profile: $T_c = \dfrac{T - T_{top}}{T_{bottom} - T_{top}}$	0.5778	0.5716	0.5678	0.8323	0.5758
$Z_c = \dfrac{H - y}{H}$	0.7751	0.8882	0.9423	0.8243	0.7837

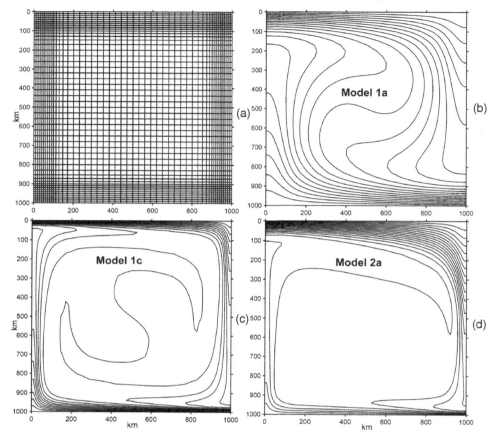

Fig. 16.9 Irregularly (10–30 km) spaced grid (a) and steady-state temperature structures (b)–(d) for the three mantle convection benchmarks from Table 16.1. Numerical results are computed at a resolution of 51×51 nodes and 200×200 randomly distributed markers with the code **Variable_viscosity_convection_irregular_grid.m.** Solid lines in (b)–(d) represent isotherms between T_{top} and T_{bottom} with an interval of 50 K.

where $\rho_0 = 4000 \text{ kg/m}^3$ is the standard density and $\alpha = 2.5 \times 10^{-5}$ 1/K is thermal expansion coefficient.

Despite this relatively simple setup, obtaining an accurate steady-state solution for mantle convection models is quite challenging. This is mainly due to (i) many (typically several thousands) time steps required to obtain a steady-state solution and (ii) a strong localisation of thermal upwellings and downwellings along the walls (e.g. Fig. 16.9(c)) in models with low mantle viscosity (or more precisely with high Rayleigh number $Ra = \dfrac{\rho_0 \alpha \left(T_{bottom} - T_{top}\right) g H^3 C_P}{\eta k}$, where g is the

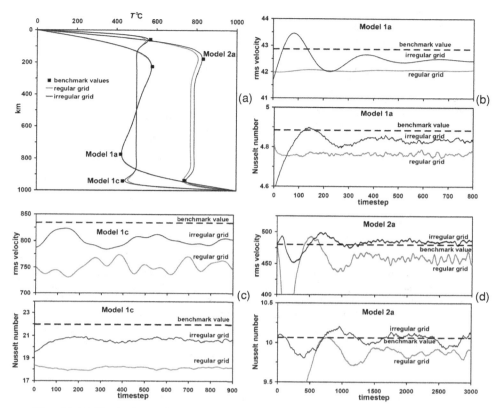

Fig. 16.10 Vertical steady-state temperature profiles in the centre of the model (a) and near-steady-state variations of root mean square (rms) velocity and Nusselt number (b)–(d) for the three mantle convection benchmarks from Table 16.1. Dashed lines in (b)–(d) show the benchmark values for respective parameters from Table 16.1. Solid lines show the numerical results calculated at resolution 51 × 51 nodes and 200 × 200 randomly distributed markers with the code **Variable_viscosity_convection_irregular_grid.m.**

gravitational acceleration, C_P the heat capacity and k is thermal conductivity). The problem of localisation can be overcome by either using high resolution of the entire model or (more efficiently), by using an irregularly spaced grid which is denser at the model walls (Fig. 16.9(a)). The steady-state thermal structures computed for some of the models of Table 16.1 are shown in Figure 16.9(b)(c)(d). Figure 16.10 presents the results of the mantle convection benchmark for these models obtained with the program **Variable_viscosity_convection_irregular_grid.m** associated with this chapter. As can be seen at the same model resolution of 51 × 51 nodes and 40 000 markers, models with irregularly spaced grid show results that are much closer to the benchmark values. Therefore the use of irregularly spaced grids, can

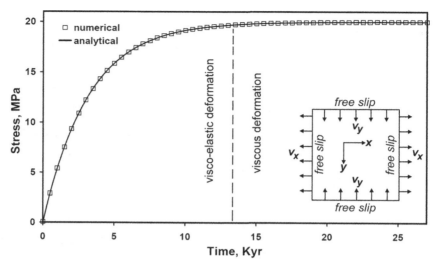

Fig. 16.11 Comparison of numerical (symbols) and analytical (solid line) solutions for the case of visco-elastic stress build-up due to pure shear (x-y direction) with constant normal strain rate and in the absence of gravity. Numerical and analytical (Eq. (16.25)) solutions are compared for $\dot{\varepsilon}_{xx} = 10^{-14}$ s^{-1}, $\eta = 10^{21}$ Pa s and $\mu = 10^{10}$ Pa. Panel with numerical setup is shown in the right part of the diagram. Numerical results are calculated at resolution 51×51 nodes and 200×200 markers with the code **Stress_buildup.m.**

in many cases significantly increase the accuracy of a numerical solution without a notable increase in computational costs.

16.10 Test 9. Stress build-up in a visco-elastic Maxwell body

This test can be performed to verify the 2D numerical solutions for the case of a deforming visco-elastic Maxwell body (Exercise 12.1). In case of uniform pure shear, deformation of an initially un-stressed, incompressible visco-elastic medium with a constant strain rate $\dot{\varepsilon}_{xx}$ elastic deviatoric stress σ'_{xx} grows with time t according to the equation

$$\sigma'_{xx} = 2\dot{\varepsilon}_{xx}\eta\,[1 - \exp(-t\mu/\eta)], \tag{16.25}$$

where t is the time from the beginning of deformation and η and μ are the constant viscosity and shear modulus of the medium, respectively. Based on Eq. (16.25), one can perform a numerical test of stress build-up shown in Fig. 16.11. The numerical experiment is designed on a rectangular model (cf. panel in Fig. 16.11) by prescribing constant outward directed velocity v_x along the vertical boundaries and inward directed velocity v_y for the horizontal boundaries of the model computed as

$$v_x = \frac{1}{2}\dot{\varepsilon}L_x,$$

$$v_y = \frac{1}{2}\dot{\varepsilon}L_y,$$

where $\dot{\varepsilon}$ is prescribed deviatoric strain rate, and L_x and L_y correspond to horizontal and vertical dimensions of the model, respectively. At each time step, all deviatoric stress components are interpolated from markers (either regularly or randomly distributed) to nodes and stress increments are then interpolated back to markers (Fig. 13.1 in Chapter 13) after numerically solving the momentum and continuity equations for the entire model domain. Figure 16.11 is computed with the code **Stress_buildup.m** and demonstrates the high accuracy of the numerical solution, which overlaps with the analytical one, hence properly describing the transition from the dominant elastic regime to the prevailing viscous deformation.

16.11 Test 10. Recovery of the original shape of an elastic slab

This benchmark can be performed to test the 2D visco-elastic numerical solutions in terms of proper advection and conservation of elastic stresses. Figure 16.12 shows the results of a numerical experiment for the recovery of the original shape of an elastic slab surrounded by a low-density, much lower viscosity and much higher shear modulus medium.

The initially un-stressed slab is attached to the left wall of the box and is spontaneously deformed within 20 Kyr under a purely vertical gravity field ($g_y = 10$ m/s^2, $g_x = 0$). The slab deformation is purely elastic due to the large Maxwell time (3 170 000 Kyr) of slab material compared to the total deformation time (20 000 Kyr). In contrast, the low-viscosity medium is subjected to irreversible, purely viscous deformation since its Maxwell time (3.17×10^{-10} Kyr) is negligible compared to the deformation time. The degree of elastic deformation in the slab is large (Fig. 16.12(b)) and the stresses stored on markers are, therefore, subjected to significant advection and rotation under both simple shear and pure shear deformation. After gravity is 'switched off' (i.e. after $g_x = g_y = 0$ condition is set), the slab starts to unbend and finally fully recovers its original shape (Fig. 16.12(c)). In contrast, the low-density medium does not recover its original configuration since the viscous deformation is irreversible (see perturbations of the checkerboard pattern in the weak medium around the slab corners).

For the model shown in Fig. 16.12, the deformation rate is time-step independent and is fully determined by the viscosity of the low-density medium which acts as a stronger material (note upbending of the lower-right edge of the slab in response

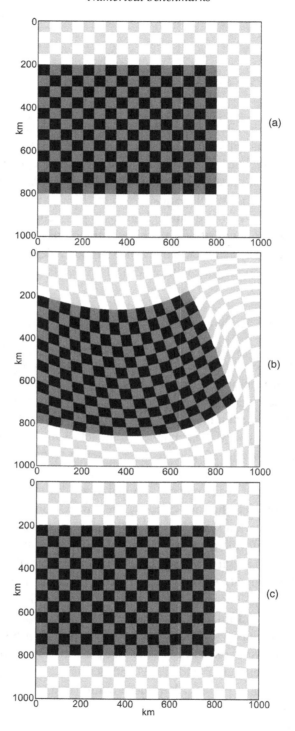

to the flow of the low-density medium around the slab). This relationship is caused by the low shear modulus of the slab (10×10^{10} Pa) compared to that of the low-density medium (10×10^{20} Pa). In contrast, in Figure 12.2 from Chapter 12, another situation is shown (Gerya and Yuen, 2007) where shear moduli of both materials are the same and the low-density medium acts as a weak material. The character of slab deformation changes correspondingly (dominant simple shear deformation and no significant upbending). In this case, however, the deformation rate is time-step dependent which does not preclude, indeed, testing the slab shape recovery (Fig. 12.2).

16.12 Test 11. Numerical sandbox benchmark

Finally, let us consider the comparison of numerical results with physical (ana-logue) sandbox experiments. Numerical modelling of sandbox experiments poses significant computational challenges because the numerical code must be able to (1) calculate large strains along spontaneously forming narrow shear zones, (2) represent complex boundary conditions, including frictional boundaries and free surfaces and (3) include a complex rheology involving both viscous and frictional/plastic materials. These challenges reflect directly, the state-of-the-art requirements for numerical modelling of large-scale tectonic processes. A numer-ical sandbox benchmark was described by Buiter *et al.* (2006) in which the results of analogue and numerical experiments for both shortening (Fig. 16.13) and exten-sion settings were compared. The shortening experiments were conducted with the use of a mobile wall moving leftward at a velocity of 2.5 cm/hour (Fig. 16.13(a)). The original cross-section is composed of sand (density $\rho = 1560$ kg/m^3, cohesion $C = 10$ Pa, an initial internal friction angle of $\varphi_{initial} = 36°$ which linearly changes to the stable value of $\varphi_{stable} = 31°$ with strain increasing from 0 to 1) and includes a

---

Fig. 16.12 Results of a numerical experiment for the recovery of the original shape of a visco-elastic slab (black, dark grey, $\rho = 4000$ kg/m^3, $\eta = 10^{27}$ Pa s and $\mu = 10^{10}$ Pa) embedded in a weak visco-elastic medium (light grey, white, $\rho = 1$ kg/m^3, $\eta = 10^{21}$ Pa s and $\mu = 10^{20}$ Pa). (a) Initial configuration, (b) config-uration after 20 Kyr of deformation under constant vertical gravity field ($g_x = 0$, $g_y = 10$ m/s^2, (c) configuration achieved within 9980 Kyr of spontaneous defor-mation after switching off gravity (i.e. after $g_x = g_z = 0$ condition is applied at 20 Kyr). Boundary conditions: no slip at the left boundary and free slip at all other boundaries. Numerical results are calculated at a resolution 51×51 nodes and 200×200 markers with the code **Slab_deformation.m** associated with this chap-ter. Note the irreversible viscous deformation of the weak surrounding medium, which is visible in its perturbed checkerboard structure close to slab corners in (c).

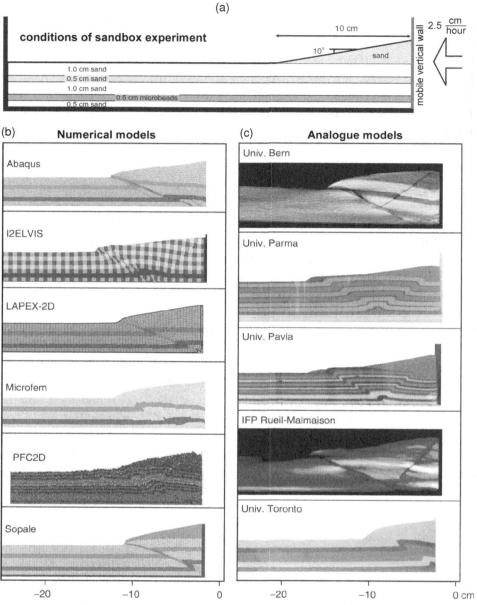

Fig. 16.13 Setup of a shortening experiment (a) and comparison of numerical (b) and analogue (c) models (at ~2 cm of shortening) performed by Buiter *et al.* (2006). (a) Horizontal layers of 'sand' (which have the same properties and differ in colour only) with an embedded layer of weaker 'microbeads' are shortened through a mobile wall on the right-hand side which is pushed leftwards. (b),(c) Names of participating numerical codes (b) and analogue labs (c) are given in respective model boxes.

0.5 cm thick weak layer of microbeads ($\rho = 1480$ kg/m^3, $C = 10$ Pa, $\varphi_{initial} = 22°$, $\varphi_{stable} = 20°$). In the right part, the model includes a 10 cm wide surface wedge composed of sand. Boundary friction on all sandbox walls is lowered ($C = 0$, $\varphi_{initial} = 19°$, $\varphi_{stable} = 19°$). Boundary conditions corresponding to the mobile wall can be implemented in a number of ways. One option is to include a rigid (highly viscous) mobile wall and prescribe constant velocity conditions ($v_x = -2.5$ cm/hour, $v_y = 0$) on Eulerian nodes located inside this wall (Fig. 16.14(a)). This can be done in combination with a weak layer included in the model, which simulates air and shifts behind the wall as it moves. In order to ensure that the wall does not leave nodes with prescribed velocity, it can be thickened from behind, by accreting displaced air markers (Fig. 16.14(b)). It should be pointed out that the implementation of the mobile wall condition may notably affect the results of numerical experiments: for example a backthrust that forms in most of analogue experiments is absent in many numerical models where a mobile wall condition was implemented by prescribing a shortening velocity directly on the right model boundary (cf. Fig. 16.13(b) and 16.13(c)). The numerical and analogue models share many similarities (Buiter *et al.*, 2006):

(1) Shortening is accommodated by an in-sequence forward propagation of thrusts (Fig. 16.14(c) also see Fig. 12.6 in Chapter 12).
(2) The first-formed thrust roots at the base of the mobile wall (Fig. 16.14(c)).
(3) By 2 cm of displacement an active thrust has formed in all models (Figs. 16.13(b)(c), 16.14(c)).
(4) The location where the first-formed forward thrust reaches the surface is influenced by the surface wedge in almost all of the experiments (Figs. 16.13(b)(c), 16.14(c)).

It should be pointed out, however, that details of shear zone patterns formed in individual analogue and numerical models are strongly variable. Such variations are an inherent feature of plastic deformation and reproducing the exact pattern of shear zones should not be considered as the benchmarking goal. More importantly, with this benchmark a numerical code should rather demonstrate its ability to hold for large deformation, for strong strain localisation along spontaneously forming narrow (1–2 grid cell wide) shear zones and for reproducing the general structural pattern of both forward and backward faults formed in analogue experiments. Figure 16.14 show the results of the numerical sandbox experiments obtained with the code **Sandbox_shortening_ratio.m.** The difference between the numerical and analogue models occurred on the same order as the differences between analogue models from different laboratories (cf. Figs. 16.13(b)(c), 16.14(b)(c)). The implemented numerical approach of plasticity treatment (Chapters 12 and 13) allows for

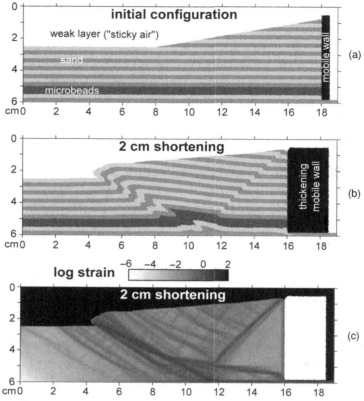

Fig. 16.14 Initial setup (a) and results (b),(c) of the numerical experiment for the shortening benchmark (Fig. 16.13(a)). The numerical model employs a visco-elasto-plastic rheology with the following material properties: sand (light grey, grey) – $\rho = 1560$ kg/m^3, $C = 10$ Pa, $\varphi_{initial} = 36°$, $\varphi_{stable} = 31°$, $\eta = 10^9$ Pa s, $\mu = 10^6$ Pa; microbeads (dark grey) – $\rho = 1480$ kg/m^3, $C = 10$ Pa, $\varphi_{initial} = 22°$, $\varphi_{stable} = 20°$, $\eta = 10^9$ Pa s, $\mu = 10^6$ Pa; weak layer ('sticky air', white) – $\rho = 1$ kg/m^3, $\eta = 10^2$ Pa s, $\mu = 10^6$ Pa; mobile wall (black) – $\rho = 1520$ kg/m^3, $\eta = 10^{12}$ Pa s, $\mu = 10^{16}$ Pa. Boundary conditions: no slip at the left and bottom boundaries and free slip on all other boundaries. Boundary friction is implemented by prescribing $\varphi_{initial} = \varphi_{stable} = 19°$ for sand and microbeads located within 2 mm near the lower and left boundaries and near the mobile wall. Shortening condition ($v_x = -2.5$ cm/hour, $v_y = 0$) is prescribed on the Eulerian nodes located inside the mobile wall. Note that the mobile wall is separated from the bottom by 2 mm thick layer of sand and is thickening from the right by converting markers of the displaced 'sticky air'. Numerical results are calculated at resolution of 191×61 nodes with 182 400 randomly distributed markers by using the code **Sandbox_ shortening_ratio.m**.

spontaneous onset of narrow shear zones, which forms a sequence of forward and backward faults like in analogue experiments.

16.13 Possible further benchmarks

Obviously, the potential number of benchmarks for testing numerical codes is infinite and not all of them are described in the present chapter. A few additional references for further numerical benchmarking problems are listed below:

- 2D analytical solutions for mantle thermal convection (Hager and O'Connell, 1981; Revenaugh and Parsons, 1987);
- 2D thermochemical convection (van Keken *et al.*, 1997);
- 2D buoyancy driven flows for strongly varying viscosity in the horizontal and vertical directions (Zhong, 1996; Moresi *et al.*, 1996);
- 2D flow around deformable elliptic inclusions (Schmid and Podladchikov, 2003; Deubelbeiss and Kaus, 2008);
- 2D visco-elastic Rayleigh–Taylor instability. (Kaus and Becker, 2007);
- 2D thermomechanical corner flows in subduction zones (van Keken *et al.*, 2008);
- 2D spontaneous subduction with a free surface (Schmeling *et al.*, 2008);
- 3D mantle convection in Cartesian geometry (Busse *et al.*, 1994);
- 3D mantle convection in spherical geometry (Zhong *et al.*, 2008);
- 3D infinitesimal and finite amplitude folding instability (Kaus and Schmalholz, 2006).

Programming exercises and homework

Exercise 16.1

Program an external MATLAB function for the 2D pressure–velocity Stokes + continuity variable viscosity solver for a regular staggered grid with external velocity nodes (Figs. 7.17, 14.8) based on the ghost node approach (Eqs. (14.39)–(14.42)) and respective global indexing of unknowns as discussed in Chapter 7 (Fig. 7.17). Implement this solver into your viscous thermomechanical code (programming exercise for Chapter 11). The advantage of using external velocity nodes is better resolving flows near the model boundaries. Modify the marker–node interpolation routines for the new grid. Particularly, shear stress $\sigma_{xy(i,j)}$ and strain rate $\dot{\varepsilon}_{xy(i,j)}$ will now be defined in all (and not only internal) basic nodes. With this new viscous code, perform falling block benchmark and compare results with Figs. 16.2 and 16.3. Model setup corresponds to Fig. 16.2. An example is in **Variable_viscosity_block.m**.

Exercise 16.2

Implement the ghost node based solver from the previous example into your visco-elastic thermomechanical code (Exercise 13.1). You will only have to modify viscosity and right-hand-side arrays given to this solver and it will solve the visco-elastic problems as well. Do not forget to modify marker–node interpolation routines for the new grid. With this new code perform visco-elastic slab bending benchmark and compare results with (Fig. 16.12). The model setup corresponds to Fig. 16.12. An example is in **Slab_deformation.m.** By the way, **Slab_deformation.m** can also employ irregularly spaced grid based on bisection algorithm (Fig. 8.10, Exercise 8.3) – think about how to implement these features in your code as well.

17

Design of 2D numerical geodynamic models

> **Theory:** Warning message! What is numerical modelling all about?
> Rock properties for numerical geodynamic models. Design of numer-
> ical models for different geodynamic processes: visco-elasto-plastic
> slab bending, retreating subduction, lithospheric extension, collision,
> slab detachment, intrusion emplacement, mantle convection with phase
> changes, core formation.
> **Exercises:** Designing numerical models for studying extension of the
> continental lithosphere.

17.1 Warning message!

Several robust visco-elasto-plastic thermomechanical codes are provided with this chapter and one can 'play' with them by changing the model geometry and resolution, as well as the material properties and boundary conditions. There is nothing wrong with that and everyone is welcome to do it. Just be aware that numerical geodynamic modelling is not *'pressing the button and automatically obtaining results'* but *knowing in depth what you and your code are doing*. So, *don't play a lottery* by starting your numerical career by immediately using these codes as research tools. Before doing this, study carefully this rather short book and make sure to correctly complete all the exercises to learn about the *advantages and limitations* of the numerical modelling techniques used in the provided codes. Otherwise, there is a big risk that your 'automatically obtained results' appearing after 'pressing the button' will be EXTREMELY WRONG ...

17.2 What is numerical modelling all about?

Having continuously studied programming and numerical modelling techniques, we might get the impression that writing a good thermomechanical code is the

269

main thing that guarantees success in numerical geodynamic modelling. Thinking that is a big mistake! Writing an extremely reliable code DOES NOT automatically imply that you will become a successful modeller... We have to learn how to use our codes in the most efficient way, how to construct robust numerical models of various geodynamic and planetary processes, how to visualise and investigate these models and how to compare them to nature. In short: *designing thoughtful and realistic numerical models is at least as important as writing efficient numerical codes*. In order to help us with this issue, several examples of numerical models of various geodynamic and planetary processes are presented with the design and technical details of the numerical experiments. The choice of these examples is, of course, subjective and mainly based on my scientific and aesthetical preferences, but this is what we have to live with.

What is numerical modelling about? Is it concerned with reproducing geological reality or investigating virtual ones? None of the two and both! We are not (either unfortunately or fortunately...) working in experimental physics where the conditions of experiments are 'relatively well' defined and known. Geological objects and systems are too complicated and their physical conditions are, in many cases, too poorly known to build fully deterministic numerical models. On the other hand, numerical modelling allows one to obtain some physical knowledge about such complex systems by studying systematically, simpler end-member cases. And this is normal! Like in experimental petrology, rather simple systems like MgO-Al_2O_3-SiO_2 (MAS) or CaO-FeO-MgO-Al_2O_3-SiO_2 (CFMAS) are often studied instead of natural rocks that are composed of at least 10–13 major oxides (Na_2O-K_2O-TiO_2-CaO-FeO-Fe_2O_3-MgO-MnO-Al_2O_3-SiO_2-H_2O-P_2O_5-CO_2). However, this simplification does not preclude the broad applicability of experimental results to natural rocks. Likewise, numerical modelling *is not a tool for fitting models to nature*, but instead *a research instrument to understand how nature works*.

17.3 Material properties

The choice of material properties in numerical models is very important. How models are set up and which prediction they allow, crucially depends on this choice. Indeed, there is a large uncertainty in material properties that are strongly variable in nature. In addition, many physical properties of natural rocks (e.g. gross-scale rheology) are poorly constrained. Therefore, some subjectivity is always present in defining model parameters. Tables 17.1 and 17.2 summarise the material properties that are used in the following examples. The choice of properties is based on widely accepted geodynamic literature (such as Turcotte and Schubert, 2002; Ranalli, 1995) and (unavoidably) on the author's personal experience of building

Table 17.1 *Rheological flow laws* used in numerical experiments (from compilation by Ranalli, 1995)*

Material	A_D $MPa^{-n} s^{-1}$	n	E_a $kJ\ mol^{-1}$
olivine (dry)	2.5×10^4	3.5	532
olivine (wet)	2.0×10^3	4.0	471
rock salt	6.3	5.3	102
quartz	1.0×10^{-3}	2	167
plagioclase An_{75}	3.3×10^{-4}	3.2	238
orthopyroxene	3.2×10^{-1}	2.4	293
clinopyroxene	15.7	2.6	335
granite	1.8×10^{-9}	3.2	123
granite (wet)	2.0×10^{-4}	1.9	137
quartzite	6.7×10^{-6}	2.4	156
quartzite (wet)	3.2×10^{-4}	2.3	154
quartz diorite	1.3×10^{-3}	2.4	219
diabase	2.0×10^{-4}	3.4	260
anorthosite	3.2×10^{-4}	3.2	238
felsic granulite	8.0×10^{-3}	3.1	243
mafic granulite	1.4×10^4	4.2	445

$$* \ \dot{\varepsilon}_{II} = A_D \, (\sigma_{II})^n \exp\left(-\frac{E_a}{RT}\right)$$

'realistic geodynamic models' (to be honest this is a highly ambiguous term and the judgment between 'realistic' and 'non-realistic' is often affected by aesthetic preferences which are, in turn, strongly defined by cartoons provided in textbooks and geological literature . . .).

17.4 Visco-elasto-plastic slab bending

Modelling of slab bending is very important in geodynamics since this process is always associated with subduction and is related to the structural and seismic features in the trench area (e.g. Ranero *et al.*, 2003, 2005). Of special interest is bending-related faulting of the incoming plate, which creates a pervasive tectonic fabric that cuts across the crust, penetrating deep into the mantle (Ranero *et al.*, 2003, 2005). Faulting is active across the entire oceanic trench slope, thereby promoting hydration of the cold crust and upper mantle surrounding these deep active faults, which may in turn cause seismic anisotropy of subducting slabs (Faccenda *et al.*, 2008a). The along-strike length and depth of penetration of

Table 17.2 *Physical properties of rocks* used in numerical experiments*

Material	ρ_0, kg/m³	k, W/(m K) (at T_K, P_{MPa})	$T_{solidus}$, K (at P_{MPa})	$T_{liquidus}$, K (at P_{MPa})	Q_L, kJ/kg	H_r, μW/m³	Flow law	μ, GPa
Sediments	2700 (solid) 2400 (molten)	$[0.64 + 807/(T + 77)]$	$889 + 17900/(P + 54) + 20200/(P + 54)^2$ at $P < 1200$ MPa, $831 + 0.06P$ at $P > 1200$ MPa	$1262 + 0.09P$	300	0.5–5	wet quartzite	10
Upper continental crust	2700 (solid) 2400 (molten)	$[0.64 + 807/(T + 77)]$	$889 + 17900/(P + 54) + 20200/(P + 54)^2$ at $P < 1200$ MPa, $831 + 0.06P$ at $P > 1200$ MPa	$1262 + 0.09P$	300	0.5–5	wet quartzite	10
Lower continental crust	2800–3000 (solid) 2500–2700 (molten)	$[1.18 + 474/(T + 77)]$	$973 - 70400/(P + 354) + 77800000/(P + 354)^2$ at $P < 1600$ MPa, $935 + 0.0035P + 0.0000062P^2$ at $P > 1600$ MPa	$1423 + 0.105P$	380	0.25–0.5	plagioclase An$_{75}$	25
Upper oceanic crust (basalt)	3000–3500 (solid) 2700 (molten)	$[1.18 + 474/(T + 77)]$	$973 - 70400/(P + 354) + 77800000/(P + 354)^2$ at $P < 1600$ MPa, $935 + 0.0035P + 0.0000062P^2$ at $P > 1600$ MPa	$1423 + 0.105P$	380	0.25	wet quartzite	25

	Density (kg m⁻³)	Thermal conductivity					Rheology	
Lower oceanic crust (gabbro)	3000–3500 (solid) 2700 (molten)	$[1.18 + 474/(T + 77)]$	$973 - 70400/(P + 354) + 77800000/(P + 354)^2$ at $P < 1600$ MPa, $935 + 0.0035P + 0.0000062P^2$ at $P > 1600$ MPa	$1423 + 0.105P$	380	0.25	plagioclase An$_{75}$	25
lithosphere – asthenosphere dry mantle	3300 (solid) 2700 (molten)	$[0.73 + 1293/(T + 77)] \times (1 + 0.00004P)$	$1394 + 0.132899P - 0.000005104P^2$ at $P <10000$ MPa $2212 + 0.030819 (P - 10000)$ at $P > 10000$ MPa	$2073 + 0.114P$	400	0.022	dry olivine	67
lithosphere – asthenosphere hydrated mantle	3000–3300 (solid) 2700 (molten)	$[0.73 + 1293/(T + 77)] \times (1 + 0.00004P)$	$1240 + 49800/(P + 323)$ at $P < 2400$ MPa, $1266 - 0.0118P + 0.0000035P^2$ at $P > 2400$ MPa	$2073 + 0.114P$	400	0.022	wet olivine	67
References**	1, 2	3,9	4	4,5,6,7,8	1, 2	1	Table 17.1	1

* other properties (for all rock types): $C_P = 1000$ J kg⁻¹ K⁻¹, $\alpha = 3 \times 10^{-5}$ K⁻¹, $\beta = 1 \times 10^{-11}$ Pa⁻¹

** 1 = (Turcotte and Schubert, 2002); 2 = (Bittner and Schmeling, 1995); 3 = (Clauser and Huenges, 1995); 4 = (Schmidt and Poli, 1998); 5 = (Hess, 1989); 6 = (Hirschmann, 2000); 7 = (Johannes, 1985); 8 = (Poli and Schmidt, 2001); 9 = Hofmeister (1999).

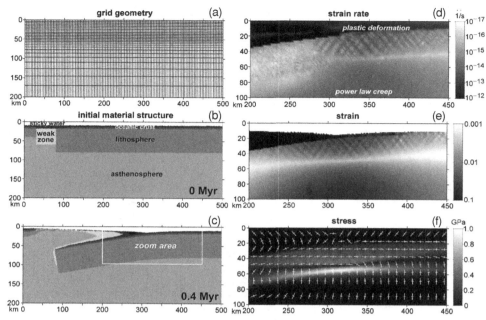

Fig. 17.1 Numerical grid (a) initial conditions (b) and results (c)–(f) of numerical experiment for visco-elasto-plastic slab bending during spontaneously retreating subduction. Model resolution is 251×51 nodal points with 100 000 randomly distributed markers. Grid resolution (a) is non-uniform in the vertical direction (each second grid line is shown). Cooling ages of the left and right plates in (b) are 1 Myr and 70 Myr respectively. (d), (e) and (f) shows numerical results for the zoomed area outlined in (c). Note that plastic deformation along faults in (d) is deactivated in the subducted portion of the slab. White crosses in (f) show the orientation of the principal stress axes; long and short branches of the crosses show extension and shortening directions, respectively. Lithospheric and asthenospheric mantle in (b) and (c) do not differ in properties (dry mantle, Table 17.2), different colours for them are used for better visualisation of slab bending. Results are computed with the code **Subducting_slab_bending.m**.

these faults are similar to the dimensions of the rupture area of intermediate-depth earthquakes.

The numerical setup for bending of a subducting slab is rather simple (Fig. 17.1(a)(b)), but requires relatively high resolution (at least 2×2 km in the slab bending area to adequately resolve the bending-related normal faults) and realistic pressure-, temperature- and stress-dependent visco-elasto-plastic rheology. One way to investigate spontaneous slab subduction and bending consists of using an initial setup for subduction initiation across a pre-existing transform fault (Hall *et al.*, 2003). The experiment begins with two plates of different ages, juxtaposed along a transform fault (cf. light grey weak zone in Fig. 17.1(b)) with low plastic strength $(\sin(\varphi) = 0)$, which creates favourable conditions for spontaneous initiation of subduction and concurrent slab bending. The vertical thermal structure of the plates is computed according to the cooling of a semi-infinite half-space (Turcotte and Schubert, 2002)

$$T(d) = T_1 + (T_0 - T_1)\left(1 - erf\left(\frac{d}{2\sqrt{\kappa\tau}}\right)\right),$$ (17.1)

where $T_0 = 273$ K is the temperature at the surface for both plates, $T_1 = 1700$ K is the temperature at the bottom of the model, d is the depth in meters below the surface, κ is thermal diffusivity (10^{-6} m^2/s) and τ is the age in seconds of the plates.

Bending is driven by strong negative buoyancy of the older plate while the weak fault allows initial displacement, which results in the spontaneous retreating subduction. To ensure self-sustaining, one-sided subduction, the weak, hydrated upper oceanic crust (basalts, sediments) is present atop the slab providing stable lubrication against the moving and cooling overriding plate (e.g., Sobolev and Babeyko, 2005; Gerya *et al.*, 2008a). On the other hand, a weak upper layer present above the crust ('sticky water', $\eta = 10^{18}$ Pa s, $\rho = 1000$ kg/m^3) provides a free-surface-like condition which is essential for a natural slab bending to occur. The validity of the weak layer approach to approximate the free surface has recently been tested and proven (Schmeling *et al.*, 2008) with the use of a large variety of numerical techniques (including our methodology based on conservative finite-differences and marker-in-cell techniques) and comparison with analogue models. The thickness of the weak layer should be at least 4–5 grid cells and its viscosity should be at least 100 times less than that of the underlying lithosphere. For regional models like the one presented here, optimal parameters for the weak layer are 10–15 km and 10^{18}–10^{19} Pa s (larger thickness and lower viscosity of this layer may require shorter time steps to avoid oscillations of numerical solution for velocity and pressure fields in the weak layer).

Figure 17.1 shows the results of a numerical experiment for spontaneous bending of a retreating subducting plate obtained with the code **Subducting_ slab_bending.m**. The deformation pattern in the bending slab is distinct (Fig. 17.1(d)): the top of the slab is subjected to intense plastic deformation with localised faults while the bottom of the slab deforms in a ductile way (i.e. by the temperature- and stress-activated dislocation creep, cf. Table 17.1) with enhancement of the deformation (see dark zone in the lower part of Fig. 17.1(d)) due to high stresses (see, light zones in Fig. 17.1(f)) in the bending area. The plastic faulting and ductile deformation fields are characterised by extension and compression in a horizontal direction, respectively (see orientation of stress principal axes in Fig. 17.1(f)). These two fields are clearly separated by the narrow, non-deforming middle plane of the slab (see light zone inside the slab in Fig. 17.1(e)) which is characterised by small deviatoric stresses (see dark zone inside the slab in Fig. 17.1(f)). The penetration depth of faults (10–50 km) (Fig. 17.1(e)) is in agreement with the observational constraints (Ranero *et al.*, 2003, 2005). Results of the experiment show that a slab with a free upper surface can easily be bent by its own weight, thereby triggering spontaneous retreating subduction. Bending is facilitated by (i) lowered pressure in the extension region, which favours deep penetration of faults and (ii) by large stresses in the compression region, which produces a local lowering of the slab viscosity due to the power-law nature of ductile creep.

17.5 Retreating oceanic subduction

This model is comparable to the previous one, but the model size is much larger to allow a longer slab retreat and deeper penetration. To avoid a significant increase in a number of grid points, one can employ a non-uniformly spaced grid with a high-resolution area that moves together with the trench (Gerya *et al.*, 2008a; Nikolaeva *et al.*, 2008). The grid spacing increases gradually away from the area of high resolution by a constant factor F at every nodal point (cf. vertical resolution in the lower part of Fig. 17.1(a)). In order to compute this incremental factor, the following formula is solved iteratively

$$F = \left(1 + \frac{D}{b}\left(1 - \frac{1}{F}\right)\right)^{1/N}, \tag{17.2}$$

where D is the distance that should be covered by N non-uniform grid steps and b is the grid spacing in the high resolution area (i.e. grid spacing from which the incremental increase should start). In order to avoid sharp changes in the numerical solution, the grid modification can be done at every time step. Re-meshing has no major effect on the algorithm since in our marker-in-cell approach, relative positions of markers and nodes change at every time step anyway. Also, nodal values of physical parameters (including temperature changes) are only used for updating properties at the moving markers. Therefore, an Eulerian node has no 'memory' and can be shifted between two time steps. In addition, since the model is much deeper than the previous one and pressure varies significantly from the top to the bottom, we should take into account the activation volume of dislocation creep.

The activation volume V_a for olivine creep (this creep is assumed to represent the mantle rheology, Tables 17.1, 17.2) varies from 0 (wet olivine) to 17 cm^3 (dry olivine) (e.g. Ranalli, 1995). Intermediate values of V_a are possible as water content in the mantle varies. Subduction model development can be notably affected by this parameter (e.g. Mishin *et al.*, 2008) and, therefore, investigating some range of activation volume variations is required to understand how the model works. The same applies to the asthenospheric mantle temperature, the plastic strength of the upper oceanic crust (lubricating layer), plate ages, initial weak zone thickness etc. This generally means that *investigating the model parameter space* (to some degree, of course, since the parameter space of a model with 10 variable parameters has at least 2^{10} permutations and consequently at least 1024 numerical experiments have to be run to investigate it in a systematic manner . . .) is a necessary component of any numerical geodynamic study. Robust conclusions should only be based on the observations obtained from several/many models.

Figure 17.2 shows the evolving numerical grid and the results of a subduction experiment. The subducting slab spontaneously retreats toward the right side of the box and the high-resolution area of the grid follows the trench area (cf. change in position of solid triangle in Fig. 17.2(a) and (b)). The slab steepens and the intensity of visco-elasto-plastic bending (Fig. 17.1) increases with time. The rate of trench retreat is very fast during the first 1.5 Myr (around 30 cm/year, Fig. 17.3) but it drops to a few cm/year as soon as the lower tip of the slab moves toward the lower boundary of the model box, and penetrates into the deeper mantle that has a larger effective viscosity (influence of the chosen $V_a = 10\,\text{cm}^3$). Generally, in models like this one, trench retreat is mainly controlled by the density contrast between the slab and the asthenosphere and by the viscous resistance of the asthenospheric mantle, which in turn depends on its rheology and temperature. For example, the rate of trench retreat decreases if a higher activation volume and/or a lower temperature for the asthenosphere are used (try to experiment using the program **Subduction.m**).

The overall model behaviour is realistic and captures several important features of retreating intra-oceanic subduction zones. Slab bending is spontaneous and slab deep angle naturally increases with depth (Fig. 17.2(b)). An accretion prism and wedge-like subduction channel form spontaneously at the plate interface (cf. zoom-in in Fig. 17.2(a)(b)). The upper (weak) part of the subducted oceanic crust partly detaches from the slab and circulates in the subduction channel, which provides a pathway for the exhumation of high-pressure tectonic melanges to the surface (e.g. Cloos, 1982; Gerya *et al.*, 2002, Gorczyk *et al.*, 2007a). The overriding plate grows in length due to the trench retreat, and cools with time (cf. isotherms in Figs. 17.2 and 17.3). The plate growth process is accommodated by the spontaneous backarc spreading where the hot mantle approaches the surface (cf. diamonds in Fig. 17.3). The model topography also behaves in a realistic manner and shows a pronounced minimum in the trench area as well as a visible maximum above the spreading centre (Fig. 17.3). Obviously, no topography maximum is formed in the middle of the overriding plate (i.e. in the area where a natural magmatic arc would grow) since our model does not account for magmatism and crustal growth. This aspect

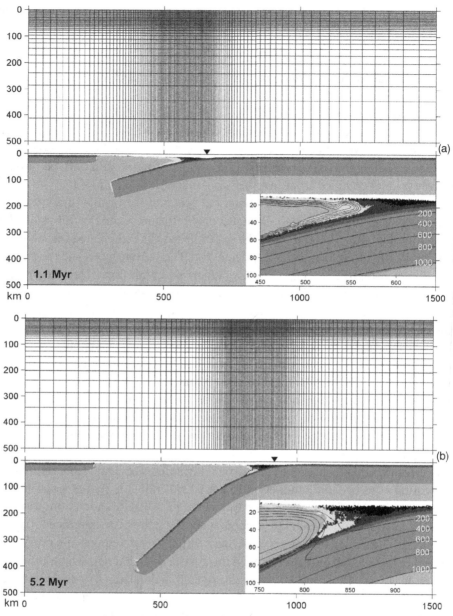

Fig. 17.2 An evolving irregularly spaced numerical grid (each second grid line is shown) and results (lithological field, see Fig. 17.1b for shade code) of the numerical experiment for spontaneous retreating subduction. The model resolution is 251 × 61 nodal points with 750 000 randomly distributed markers. Activation volume of mantle used in this experiment is 10 cm³. The asthenospheric mantle temperature at the bottom of the model is 1850 K and decreases upward according to an adiabatic gradient of 0.5 K/km. The high-resolution area of the grid moves together with the retreating trench (solid triangles). Inserts show a zoom-in of the moving trench area with a spontaneously evolving accretion prism and deep subduction channel in which the subducted oceanic crust is circulating. Black labelled lines on the inserts are isotherms in °C. Results are computed with the code **Subduction.m**.

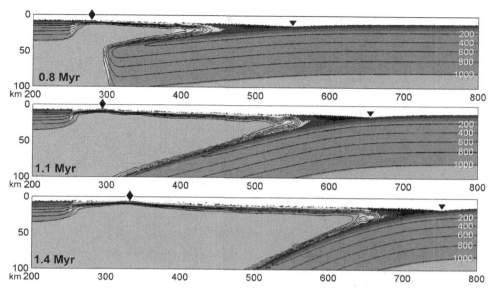

Fig. 17.3 Time evolution of the trench (triangle) and backarc spreading centre (diamond) associated with the growth and cooling of the overriding plate during a retreating subduction experiment (Fig. 17.2).

however, can be improved by using more sophisticated and realistic models, which involve mantle wedge hydration and melting, and related melt extraction and crustal growth (e.g. Nikolaeva *et al.*, 2008).

17.6 Lithospheric extension

Lithospheric extension is an important geodynamic process, for example at mid-ocean ridges, backarc (Fig. 17.3) and intra-arc extension zones, passive and active rifting, continental break-up, formation of sedimentary basins, etc. (e.g. Turcotte and Schubert, 2002). Realistic modelling of such settings poses computational challenges (e.g. Burov and Poliakov, 2001) since extension of the lithosphere and the underlying mantle is associated with intense and simultaneous viscous and brittle/plastic (faulting) deformations, as well as with notable changes in topography. Modelling requires sufficiently high (at least 2×2 km) resolution in the extension area where strongly localised deformation along faults takes place. One way to address these challenges consists in using an evolving non-uniformly spaced numerical grid (as we used for modelling retreating subduction, Fig. 17.2) combined with a variable model size which evolves with time in response to imposed bulk extension. This approach suggests re-meshing at every time step, which can be easily done with our marker-in-cell algorithm.

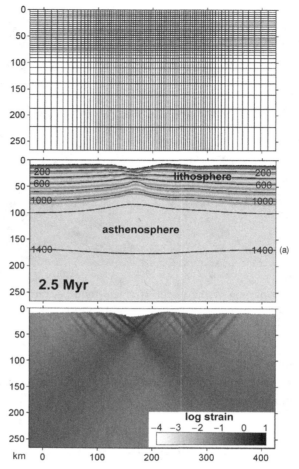

Fig. 17.4 Evolving irregularly spaced numerical grid (each second grid line is shown), lithological field (see Fig. 17.1(b) for colour code) and finite strain for two stages of an oceanic lithosphere extension experiment. Model resolution is 161 × 61 nodal points with 120 000 randomly distributed markers. Activation volume of the mantle used in this experiment is 10 cm^3. No strain weakening is used for plastic deformation. Initial asthenospheric mantle temperature at the bottom of the model is 1750 K and decreases upward according to an adiabatic gradient of 0.5 K/km. Non-compositional layering of the mantle lithosphere is shown for visualising deformation. Black labelled lines are isotherms in °C. Results are computed with the code **Extension.m**.

Figure 17.4 presents the results of modelling extension of a 70 Myr old oceanic lithosphere under an imposed constant extension rate ($v_{extension} = 2$ cm/year). The initial model is 400 × 300 km in size and includes both lithospheric and asthenospheric domains as well as a weak top layer (sticky water) that allows a natural thermomechanical evolution of both upper and lower boundaries of the lithosphere.

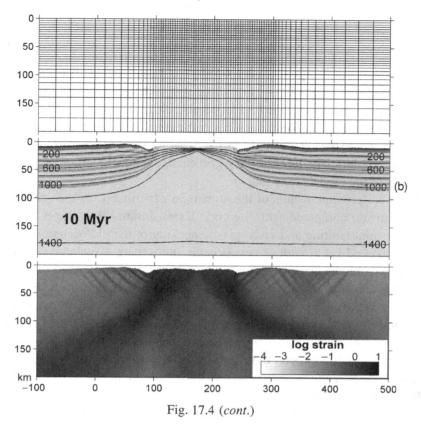

Fig. 17.4 (*cont.*)

As in previous examples, the lower lithosphere boundary is not prescribed and forms spontaneously as a rheological boundary between colder and stronger parts of the mantle and underlying hotter and weaker (asthenospheric) region. Extension is prescribed symmetrically as constant horizontal outward velocity boundary conditions at two sides of the model

$$v_{outward} = \frac{1}{2} v_{extension}. \qquad (17.3)$$

In order to ensure mass conservation in the computational model, a vertical inward velocity which changes at every time step, is prescribed along the lower model boundary

$$v_{inward(t)} = \frac{H_{(t)}}{L_{(t)}} v_{extension}, \qquad (17.4)$$

where $L_{(t)}$ and $H_{(t)}$ are width and height of the model, respectively. The model width and height change with time as

$$L_{(t)} = L_0 + t \times v_{extension} \qquad (17.5)$$

and

$$H_{(t)} = L_0 \times H_0/L_{(t)}, \tag{17.6}$$

where $L_0 = 400$ km is initial model width, t is time from the beginning of the experiment and $H_0 = 300$ km is the initial model height. The resolution of the numerical grid contains 161×61 nodal points, with $120\,000$ randomly distributed markers. The dense 2×2 km part of the grid covers the central 200×80 km area of the model where the extension is localised (not necessarily exactly in the middle, of course). Resolution in the remaining grid changes (Eq. (17.2)) at every time step in response to the model stretching.

Figure 17.4 presents results of the numerical experiment for the oceanic lithosphere extension computed with the code **Extension.m**. Two stages are clearly seen: (1) normal faulting and visco-plastic necking of the old oceanic lithosphere (Fig. 17.4(a)) and (2) growth of new oceanic lithosphere associated with the formation of a spreading centre. Extensional deformation associated with normal faulting (Fig. 17.4(a)) initially starts in a 200–300 km wide area of the lithosphere. It is followed by a spontaneous and gradual (within 1.5–2.5 Myr) focusing of the deformation in the necking area where a new spreading centre forms and the hot asthenospheric mantle comes close to the surface (see distribution of isotherms in Fig. 17.4(b)). The topography development is characteristic: the topographic low develops in the spreading centre and two elevated regions form on the rift flanks (e.g. Burov and Cloetingh, 1997; Burov and Poliakov, 2001). However, our model topography is assumed to be submarine and contains no erosion or sedimentation processes which may notably affect lithospheric extension (e.g. Burov and Cloetingh, 1997; Burov and Poliakov, 2001) if taken into account. Like in the retreating subduction model, this experiment does not take into account decompression melting of the mantle or the formation of new oceanic crust in the spreading centre, which again points toward the need for developing more sophisticated and realistic models.

17.7 Continental collision

Continental collision is another very 'popular' geodynamic setting that has been widely addressed by numerical modelling. In that case, much care should be taken to address erosion and sedimentation processes which play major roles in orogeny associated with colliding continents (e.g. Willett, 1999; Beaumont *et al.*, 2001). According to numerical studies, variations in erosion/sedimentation rates during subduction and collision may significantly affect crustal mass flux and consequently alter crustal deformation and the behaviour of the crust–mantle interface (e.g. Willett, 1999; Beaumont *et al.*, 2001; Pysklywec, 2006; Gerya *et al.*, 2008b). One possibility to model these processes is to use *an internal evolving erosion/sedimentation surface* (Gerya and Yuen, 2003b) that separates the top boundary of the lithosphere from the overlaying sticky water/air layer with vertically stratified density (either 'air', 1 kg/m^3, for $y < y_{water}$ or 'water', 1000 kg/m^3, for $y > y_{water}$, where y_{water} is the water level adopted in the model). This surface evolves according to the transport equation solved at the Eulerian coordinates at each time step (Gerya and Yuen, 2003b):

$$\frac{\partial y_{es}}{\partial t} = v_y - v_x \frac{\partial y_{es}}{\partial x} + v_e - v_s, \tag{17.7}$$

where y_{es} is the vertical position of the surface as a function of the horizontal distance x; v_y and v_x are the vertical and horizontal components of the material velocity vector at the surface; v_s and v_e are sedimentation and erosion rates, respectively.

Erosion and sedimentation rates in Eq. (17.7) can be computed in various ways. The simplest is to use both slope and elevation independent *large-scale erosion and sedimentation rates* (e.g. Vance *et al.*, 2003; Gerya and Yuen, 2003b; Gerya *et al.*, 2008b) which correspond to the relation:

$$v_s = 0 \text{ mm/a}, v_e = v_{e0} \quad \text{when} \quad y < y_{water},$$
$$v_s = v_{s0} \text{ mm/a}, v_e = 0 \quad \text{when} \quad y > y_{water},$$

where v_{e0} and v_{s0} are imposed as constant erosion and sedimentation rates. Another possibility is to use more sophisticated models of surface processes that combine downhill diffusion erosion and fluvial erosion (Kooi and Beaumont, 1994; Burov and Cloetingh, 1997; Braun and Sambridge, 1997; Beaumont *et al.*, 2001; Burov *et al.*, 2001). In the case of short-distance, downhill diffusion erosion (e.g., Kooi and Beaumont, 1994; Burov and Cloetingh, 1997), Eq. (17.7) can be modified to

$$\frac{\partial y_{es}}{\partial t} = v_y - v_x \frac{\partial y_{es}}{\partial x} + \frac{\partial}{\partial x} \left(K_s \frac{\partial y_{es}}{\partial x} \right), \tag{17.8}$$

where K_s is the effective 'topography diffusion' coefficient, which is highly variable (0–10^5 m^2/year, e.g. Kooi and Beaumont, 1994; Burov and Cloetingh, 1997). One possibility to solve the transport equations (17.7) and (17.8) consists in using a 1D Eulerian advection methods such as upwind differences or the FCT method discussed in Chapter 8.

The oceanic lithosphere initially present between the two continental plates is another important component of continental collision models. Prescribing this lithosphere (e.g. Faccenda *et al.*, 2008b; Gerya *et al.*, 2008b; Warren *et al.*, 2008) allows a more natural beginning and more faithfully captures the initial stages of the

collision process. During these stages, the model behaviour changes very rapidly
due to the arrival of the positively buoyant continental crust in the subduction zone.
Obviously, experiments of post-subduction collision should allow deep subduction
of the oceanic slab and spontaneous bending of plates, which require using a
sufficiently wide (>500 km) and deep (>200 km) model (e.g. Burov *et al.*, 2001;
Faccenda *et al.*, 2008b; Gerya *et al.*, 2008b; Warren *et al.*, 2008). One possibility
to fit these requirements consists of using models with a variable size domain
(e.g. Burov *et al.*, 2001), like the one we explored for lithospheric extension.
Another option is to use setups with a constant model size and a permeable lower
boundary (Burg and Gerya, 2005; Faccenda *et al.*, 2008b; Gerya *et al.*, 2008b)
where infinity-like thermal and mechanical boundary conditions (Chapters 7 and
10) are prescribed.

   Figure 17.5 shows the initial setup and results of a post-subduction collision
experiment conducted with the code **Collision.m**. The 1000×300 km model
(Fig. 17.5(a)) uses a non-uniform 201×61 rectangular grid with a constant high
resolution of 2×2 km in the 300×60 km area of the subduction/collision zone.
Coarser resolution around this zone changes at every time step in response to the
model shortening (leftward) and thickening (downward), which accommodate the
convergence whose velocity is prescribed at the right model boundary. Boundary
conditions are similar to the lithospheric extension example (Eq. (17.3)–(17.6))
but the model is shortening rather than extending and the left boundary does not
change position with time. In order to compensate for the thickening of the sticky
air/water layer on the top of the model, the water level changes at every time step
as

$$y_{water(t)} = y_{water0} L_0 / L_{(t)}, \tag{17.9}$$

where $y_{water0} = 7.5$ km is the initial water level.

   The initial material setup (Figure 17.5a) implies early oceanic–continental subduction with two continental sections (each 400 km wide) and a relatively short
(200 km) intermediate oceanic plate, which is 40 Myr old. The continental crust is
35 km thick, with the upper and lower crustal layers (Table 17.2) of equal thickness.
The nucleated subduction zone at the left ocean/continent boundary is prescribed
as a 4–15 km wide weak zone cutting across the entire mantle lithosphere and
reaching a depth of 90 km. The weak zone is prescribed as a wet brittle/plastic fault
within mantle rocks, characterised by wet olivine rheological parameters and a low
plastic strength of 1 MPa (i.e. assuming a high pore fluid pressure and $\sin(\varphi) = 0$).
During subduction, the pre-defined weak zone is spontaneously replaced by weak
subducted crustal rocks, thereby preserving the decoupling along the interface.
Obviously, the subduction zone is prescribed in a rather arbitrary way but this is

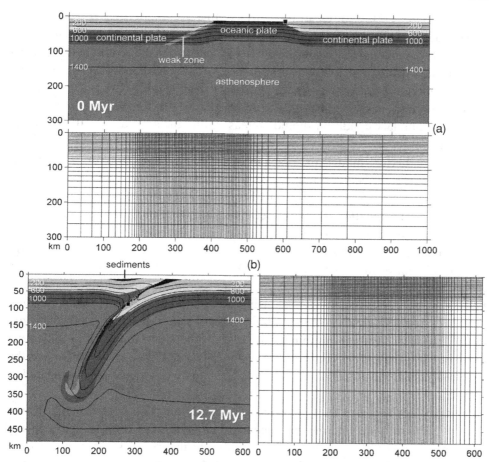

Fig. 17.5 Lithological field and evolving irregularly spaced numerical grid (each second grid line is shown) for the initial model setup (a) and culminate stage (b) of the continental collision experiment. Model resolution is 201 × 61 nodal points with 300 000 randomly distributed markers. Activation volume of the mantle used in this experiment is 10 cm³. The initial asthenospheric mantle temperature at the bottom of the model is 1750 K and decreases upward according to an adiabatic gradient of 0.5 K/km. Convergence of 3 cm/yr is prescribed from the right. Black labelled lines are isotherms in °C. Black square shows position of a representative marker from the deeply subducting sedimentary rock unit for which *P–T*-time path is shown in Fig. 17.8. Results are computed with the code **Collision.m**.

what we have to live with since the issue of subduction initiation is very controversial and over 10(!) conceptual models of subduction initiation have already been suggested in the literature (discussion by Ueda *et al.*, 2008).

The initial thermal structure of the continental lithosphere is laterally uniform and corresponds roughly to the usual continental geotherms (e.g., Turcotte and Schubert, 2002): 0 °C at the surface linearly increasing to 1380 °C at 100 km depth. The temperature structure of the oceanic plate corresponds to its age (Eq. (17.1)). A gradual linear transition from the oceanic to the continental geotherm is prescribed within a 50 km wide area at the two ocean/continent boundaries. The initial temperature gradient in the asthenospheric mantle is 0.5 °C/km. Due to thickening in response to shortening of the model with time, the temperature imposed for the lower boundary condition should increase with time according to the adiabatic gradient.

In this numerical model, the mechanisms driving subduction are a combined 'plate push' (prescribed constant convergence velocity at the right boundary) and 'slab pull' (temperature induced density contrast between the subducted lithosphere and surrounding mantle, see above retreating subduction model example). This type of boundary condition is widely applied in numerical models of subduction and collision (e.g., Burov *et al.*, 2001; Burg and Gerya, 2005; Faccenda *et al.*, 2008b; Warren *et al.*, 2008), assuming that in the globally confined system of plates, the 'external push' imposed on a plate (coming from a different slab) can be significant. A spontaneously increasing slab pull mainly regulates slab bending dynamics and delamination of the slab from the overriding plate (e.g., Gerya *et al.*, 2008b). Indeed, similar (and in relation) to the subduction initiation problem, the issue of choosing proper convergence conditions is not fully resolved yet and the external push at the initial stages of convergence may be exaggerated compared to nature.

Figure 17.5(b) shows the final stage of an experiment for continental collision with a 200 km wide intermediate oceanic plate (30 Myr cooling age) moving leftward at a constant velocity of 3 cm/yr due to the model shortening imposed from the right boundary. Results are computed with the code **Collision.m**. The development of the continental collision zone is associated with deep (>100 km) subduction of the continental crust underneath the orogen (Fig. 17.5(b)). This feature appears in many numerical models of continental collision (e.g. Burov *et al.*, 2001; Gerya *et al.*, 2008b; Warren *et al.*, 2008) including those driven by slab pull rather than by a prescribed convergence velocity (e.g. Faccenda *et al.*, 2008b; Baumann *et al.*, 2009). This fits findings of *ultrahigh-pressure (UHP) rocks* (e.g. Chopin, 2003; Liou *et al.*, 2004) that contain *metamorphic diamonds* (e.g., Rosen *et al.*, 1972; Sobolev and Shatsky, 1990; Dobrzhinetskaya *et al.*, 1995; Massonne, 1999) and coesite (e.g., Chopin, 1984; Smith, 1984) in Phanerozoic collision belts. The topography development predicted with our simplified model (Fig. 17.6) also appears realistic and demonstrates the growth of a positive topography (up to 2500 m above the water level, Eq. (17.9)) after the beginning of collision at around 5–7 Myr. Growth of the elevated region is associated with erosion and deposition of sediments on both sides of the 'orogen' (see black material in Fig. 17.5(b)).

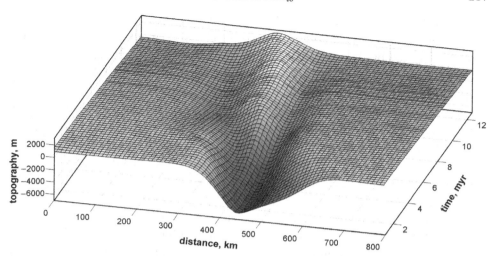

Fig. 17.6 Development of topography with time (relative to the water level, Eq. (17.9)) for the model shown in Fig. 17.5. $K_s = 0.00003$ m^2/s (1000 m^2/year) is used in Eq. (17.8) which corresponds to approximately 1 mm/year gross-scale erosion/sedimentation rates. Results are computed with the code **Collision.m**.

17.8 Slab breakoff

Slab breakoff (also called *slab detachment*) is an interesting 'hidden' process happening at depth within the mantle (ideal process for modellers indeed . . .). It was initially hypothesised on the basis of gaps in hypocentre distribution and tomographic images of subducted slabs (Isacks and Molnar, 1969; Barazangi *et al.*, 1973; Pascal *et al.*, 1973; Chung and Kanamori, 1976; Fuchs *et al.*, 1979) and is supported by both theoretical considerations (Sacks and Secor, 1990; von Blanckenburg and Davies, 1995; Davies and von Blanckenburg, 1995) and detailed seismic tomography (Spakman *et al.*, 1988; Wortel and Spakman, 1992, 2000; Xu *et al.*, 2000; Levin *et al.*, 2002). Slab detachment is often attributed to a decrease in subduction rate subsequent to continental collision (e.g., Davies and von Blanckenburg, 1995; Wong A Ton and Wortel, 1997), an effect caused by the buoyancy of the continental lithosphere introduced into the subduction zone (Fig. 17.5(b)). In addition to multiple geophysical, geological and geochemical investigations (see Andrews and Billen, 2009 and references therein), analytical models, laboratory and numerical experiments have been undertaken to characterise the breakoff processes (e.g., Davies and von Blanckenburg, 1995; Yoshioka and Wortel, 1995; Wong A Ton and Wortel, 1997; Yoshioka *et al.*, 1995; Buiter *et al.*, 2002; Chemenda *et al.*, 2000; Faccenna *et al.*, 2006; Gerya *et al.*, 2004a; Andrews and Billen, 2009; Faccenda *et al.*, 2008b; Zlotnik *et al.*, 2008; Baumann *et al.*, 2009).

Recent thermomechanical models indicate two detachment modes (Andrews and Billen, 2009): (1) deep *viscous breakoff*, which is characteristic of strong slabs and is controlled by thermal relaxation (heating) of the slab and subsequent thermomechanical necking in dislocation creep regime (Gerya *et al.*, 2004a; Faccenda *et al.*, 2008b; Zlotnik *et al.*, 2008; Baumann *et al.*, 2009), and (2) relatively fast, shallow *plastic breakoff* which is characteristic of weaker slabs and is controlled by plastic necking of the slab (Andrews and Billen, 2009; Mishin *et al.*, 2008; Ueda *et al.*, 2008). It was demonstrated that the time before the onset of viscous (but not plastic) detachment increases with the slab age, indicating that detachment time is controlled by the thickness and integrated stiffness of the thermally relaxing slabs (Gerya *et al.*, 2004a; Andrews and Billen, 2009).

Breakoff can be modelled in a sufficiently self-consistent way starting, for example, from the configuration obtained in the continental collision experiment (Fig. 17.5(b)). We can essentially use the same code and stop convergence (either sharply or gradually) after the continental crust of the incoming plate reaches asthenospheric depths, assuming that the crustal buoyancy can potentially block further subduction. Even more consistent breakoff models use a spontaneous convergence of plates (driven by the slab pull) allowed after some period of forced convergence creating sufficient slab pull but before the actual beginning of collision (Faccenda *et al.*, 2008b; Baumann *et al.*, 2009). In this case, plates should be detached from the model walls to permit horizontal movements.

Figure 17.7 shows results of a breakoff experiment performed with the code **Collision_and_breakoff.m**. It uses the first approach and sharply stops model shortening (and obviously its thickening as well, otherwise mass conservation condition in the model will be violated which would be really bad . . .) at 12.7 Myr (Fig. 17.5(b)). The model is then free to evolve spontaneously. In the beginning, dominating processes are downward bending (steepening) and thermal relaxation of the slab, as well as the buoyant escape of previously subducted continental crust toward shallower depths (see the movement of a black square in Fig. 17.7(a)(b)(c) and the respective *P–T*-time path in Fig. 17.8). This stage lasts over 15 Myr (from 12.7 Myr to 27.8 Myr, cf. Figs. 17.5(b) and 17.7(b)). After the strength of the slab interior is lowered by the temperature increase, a self-accelerating (due to feedbacks from stress concentration and shear heating, Gerya *et al.*, 2004a), thermomechanical *necking* is activated and leads to rapid (within <1 Myr) detachment of the slab (Fig. 17.7(b)(c)). This necking is driven by thermally activated, stress-sensitive dislocation creep. The depth of breakoff is relatively shallow (around 140 km), but this model feature is sensitive to many model parameters and results may widely vary, ranging from 50 to 500 km (Gerya *et al.*, 2004a; Faccenda *et al.*, 2008b; Mishin *et al.*, 2008; Ueda *et al.*, 2008; Andrews and Billen, 2009; Baumann *et al.*, 2009). Finally, the detached slab rapidly sinks and rotates in a coherent manner (as

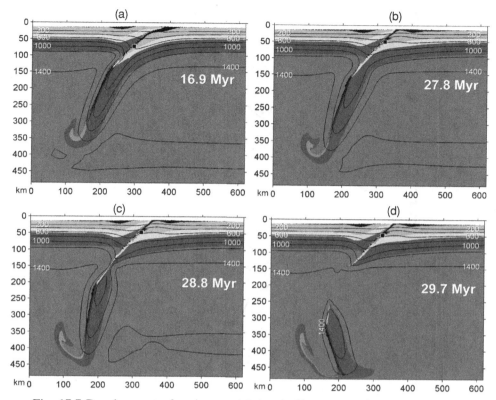

Fig. 17.7 Development of a viscous slab breakoff process. Initial conditions as well as model parameters correspond to Fig. 17.5(b). Boundary conditions are free slip on all model boundaries. Black labelled lines are isotherms in °C. Four stages of breakoff are shown: (a) downward bending (steepening) and thermal relaxation of the slab, (b) beginning of the thermomechanical necking process, (c) slab detachment from the upper part of the plate, (d) rapid sinking and coherent rotation of the slab. Black square shows position of a representative marker from deeply subducted sedimentary rock unit for which P–T-time path is shown in Fig. 17.8. Time in figures is shown from the beginning of the collision experiment (Fig. 17.5). Results are computed with the code **Collision_and_breakoff.m**.

a rigid body), interacting with the lower model boundary which is an artifact of the impermeable lower boundary condition. Rigid slab rotation is a realistic process resulting from slab interaction with the underlying and surrounding mantle (particularly near the 670 km deep spinel-to-perovskite transition), which is confirmed by numerical experiments with larger and deeper models that employ mantle phase transitions (Mishin *et al.*, 2008; Baumann *et al.*, 2009).

Figure 17.8 show a P–T-time path of a marker representing a sedimentary rock unit deeply subducted during collision and exhumed toward the surface

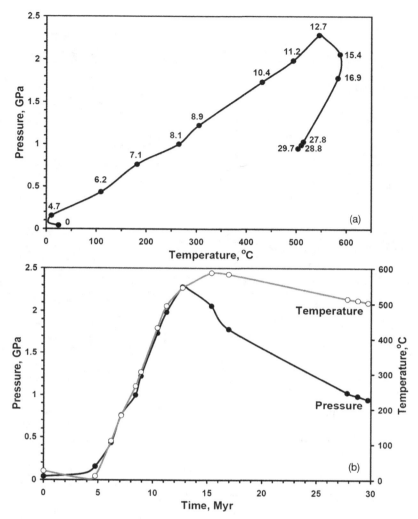

Fig. 17.8 Representative *P–T*-time path for the sedimentary rock unit deeply sub-ducted during the continental collision (see small black square in Fig. 17.5). This unit exhumes toward the surface during thermal relaxation and detachment of the slab (see small black square in Fig. 17.7). Numbers along the *P–T* path in (a) show time in Myr from the beginning of the collision experiment (Fig. 17.5).

during thermal relaxation and detachment of the slab (see small black square in Figs. 17.5, 17.7). The possibility of computing *P–T*-time trajectories of various rock units represent another very useful feature of the marker-in-cell approach. The synthetic *P–T*-time paths can be further compared to *P–T* paths of natural rock complexes derived from petrological data. The comparison of *P–T* paths poses strong constraints for testing numerical models on the basis of natural data which is now broadly used in numerical modelling of various geodynamic processes (e.g. Gerya *et al.*, 2000, 2002, 2008b; Jamieson *et al.*, 2002; Gerya and Maresch, 2004; Stoeckhert and Gerya, 2005; Gerya and Stoeckhert, 2006; Gorczyk *et al.*, 2007b; Faccenda *et al.*, 2008b; Warren *et al.*, 2008).

17.9 Intrusion emplacement into the crust

James Hutton conceived the idea of plutonism already in the late eighteenth century (e.g. Ellenberger, 1994) but the formation of large magma bodies in the crust, the plutons, still eludes full understanding. Research on the topic has, however, been much focused on the emplacement of granitic magma (e.g. Pitcher, 1979; Ramberg, 1981; Petford *et al.*, 2000). Thermomechanical modelling of magma intrusion is not yet very 'popular' (e.g. Burov *et al.*, 2003; Gerya and Burg, 2007; Burg *et al.*, 2009) and is numerically challenging because it involves simultaneous and intense deformation of materials with very contrasting rheological properties. To see the contrast, consider that typical crustal rocks are visco-elasto-plastic whilst the intruding magma is a low viscosity, complex fluid/crystal mixture. Indeed, the finite-differences+marker-in-cell (*FDM+MIC*) numerical methodology discussed in this book is appropriate for such type of experiments and has already been used for thermomechanical modelling of mafic-ultramafic intrusion emplacement (Gerya and Burg, 2007; Burg *et al.*, 2009).

In the case of trans-lithospheric emplacement (i.e. intrusion of a magmatic body from sub-lithospheric depths into the crust) which is the process assumed for many mafic-ultramafic bodies found in cratons and within magmatic arcs (e.g. Burg *et al.*, 2009), the following model design can be used (Fig. 17.9(a)(b) 0 Kyr). The model domain should obviously include both lithospheric and asthenospheric regions. The envisaged intrusion emplacement area has to be well resolved (grid spacing should be several times smaller than the modelled intrusion size which typically implies a resolution of 0.5–1 km or better, Gerya and Burg, 2007). Non-uniform grids similar to those we used for the slab bending model (Fig. 17.1(a)) and for the lithospheric extension/collision models (Figs. 17.4, 17.5) are appropriate. The initial thermal structure of the lithosphere, with a 25 km thick crust corresponding to a magmatic arc, is represented by a relatively hot sectioned geotherm (0 °C on the top of the crust, 627 °C at the Moho and 1317 °C at 93 km depth). An adiabatic gradient of 0.5 K/km is used in the asthenospheric mantle. Dense layering in the crust is used to better visualise the deformation. Boundary conditions are free slip on all boundaries. No far field shortening/extension is introduced to affect the intrusion process.

In this model, we assume that magma is coming from a *sub-lithospheric magmatic source region (SMSR)* (Gerya and Burg, 2007). The SMSR and the magmatic

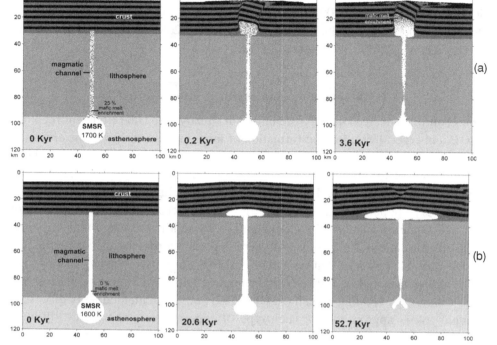

Fig. 17.9 Dynamics of intrusion emplacement into the crust in the case of (a) hot magma with lowered (10^{14} Pa s) viscosity and (b) colder magma with higher (10^{16} Pa s) viscosity. Model resolution is 201×61 nodal points with 192 000 randomly distributed markers. The grid is similar to Fig. 17.1(a) and has a non-uniform vertical resolution of 1.0–6.4 km and uniform horizontal spacing of 0.5 km. Asthenospheric mantle temperature at the bottom of the model is 1600 K and decreases upward according to adiabatic gradient of 0.5 K/km. Moho temperature is 900 K for both models. Results are computed with the code **Intrusion_emplacement.m**.

channel across the lithospheric mantle are then prescribed initially. The SMSR and the median channel enriched in mafic melt have an initially uniform magmatic temperature, which can be varied within reasonable limits (e.g. between 1200 and 1500 °C). This is the presumed temperature range at the head of a partially molten, hydrous thermal-chemical mantle plumes which can represent the SMSR (e.g. Gerya and Yuen, 2003b; Gerya *et al.*, 2004b; Castro and Gerya, 2008). We emphasise that the modelled SMSR is only a thermally and chemically distinct region of the hydrated, partially molten mantle rocks and not a chamber of fully molten magma. Partially molten material has a much lower viscosity than the surrounding dry mantle (e.g. Pinkerton and Stevenson, 1992) and can move through the magmatic channel as a melt/crystal mixture. In that sense, the SMSR is equivalent to a magma reservoir. According to the melting model described below, this hydrated region initially has 10–30% melt fraction (depending on the assumed initial temperature) which varies with depth. Though the bulk composition of the SMSR in the model is ultramafic, the melt composition within the SMSR depends on the degree of melting and thus varies between mafic and ultramafic.

By implementing a pre-existing hot and thus weak (with plastic strength of 1 MPa) magmatic channel, we implicitly accept the general assumption that magma rises from the source area as a result of any lithospheric perturbation. Natural hot channels may originate at an early magmatic stage due to the rapid and localised upward percolation of hot mobile fluids/melts, which are differentiation products at the top of the SMSR. Respectively, enrichment (up to 25%) by mafic melt is prescribed for the channel. Mechanisms of localised upward fluid/melt transport include hydrofracture (e.g., Clemens and Mawer, 1992), diffusion (e.g., Scambelluri and Philippot, 2001), porous flow (e.g., Scott and Stevenson, 1986; Connolly and Podladchikov, 1998; Vasilyev *et al.*, 1998; Ricard *et al.*, 2001) and reactive flow (e.g. Spiegelman and Kelemen, 2003). An important point is that fluid percolation diminishes the plastic strength of rocks through an increase in pore fluid pressure, which in turn allows massive amounts of partially molten rocks to ascend through the lithosphere. Propagation of magmatic rocks from the channel into the crust develops in a spontaneous manner and is mainly controlled by the crustal rheology and density.

According to geological observations, fault tectonics, and hence plastic deformation of the crust, play a significant role during plutonic emplacement resulting from fluid/melt percolation along forming fracture zones (e.g. Clemens and Mawer, 1992). The plastic yield strength of rocks under fluid-present conditions, strongly depends on the ratio between the solid (P_{solid}) and fluid (P_{fluid}) pressure Eqs. (12.41)–(12.43) (Chapter 12). During intrusion, the most intensive percolation of magmatic fluids is expected to follow the pattern of fractured rocks along spontaneously propagating fault zones. Fluid supply will then increase the pore fluid pressure along the fault zones, thus lowering the plastic yield strength of the fractured rocks. This will in turn further localise deformation along the weakening fault zones. Under such circumstances, the plastic strength of fractured rocks will be inversely correlated with the amount of continuous plastic deformation experienced by the rocks. In order to model this process in a simplified way, the fluid pressure factor λ for a given model rock increases with plastic strain experienced by the rock

$$\lambda = 1 - (1 - \lambda_0)\left(1 - \frac{\gamma_{plastic}}{\gamma_{cr}}\right) \text{ when } \gamma_{plastic} < \gamma_{cr} \text{ and } \lambda = 1 \text{ when } \gamma_{plastic} \geq \gamma_{cr},$$

$$(17.11a)$$

or in terms of the effective friction angle φ (as in the sandbox benchmark of Chapter 16),

$$\sin(\varphi) = \sin(\varphi_0)\left(1 - \frac{\gamma_{plastic}}{\gamma_{cr}}\right) \text{ when } \gamma_{plastic} < \gamma_{cr} \text{ and } \sin(\varphi)=0 \text{ when } \gamma_{plastic} \geq \gamma_{cr}$$

$$(17.11\text{b})$$

$$\sin(\varphi_0) = \sin(\varphi_{dry})(1 - \lambda_0), \qquad (17.12)$$

$$\sin(\varphi) = \sin(\varphi_{dry})(1 - \lambda), \qquad (17.13)$$

$$\gamma_{plastic} = \int_t \left(\frac{1}{2}\dot{\varepsilon}_{ij(plastic)}\dot{\varepsilon}_{ij(plastic)}\right)^{1/2} dt, \qquad (17.14)$$

where λ_0 and φ_0 are the initial (before plastic yielding) pore fluid pressure factor and effective friction coefficient characteristic for the rock, respectively; γ_{cr} is the critical plastic strain required for weakening the rock by percolating fluid (i.e. the strain necessary for reaching condition $P_{fluid} = P_{solid}$, $\lambda = 1$). In our example, we used $\gamma_{cr} = 0.1$ and $\sin(\varphi_0)$ equal to 0.2 and 0.6 for the crustal and mantle rocks, respectively.

For simplicity, the effective viscosity η of partially molten rocks ($M > 0.1$) is assigned by a low constant value of 10^{16} Pa s. In real melt-crystal aggregates, this viscosity is strongly and non-linearly dependent on the melt fraction and can be calculated, for example, by using the formula (Bittner and Schmeling, 1995; Pinkerton and Stevenson, 1992):

$$\eta = \eta_0 \exp\left[2.5 + (1 - M)\left(\frac{1 - M}{M}\right)^{0.48}\right], \qquad (17.15)$$

where η_0 is an empirical parameter depending on rock composition. According to Bittner and Schmeling (1995), $\eta_0 = 10^{13}$ Pa s can be taken for partially molten mafic rocks (i.e., $1 \times 10^{14} \leq \eta \leq 2 \times 10^{15}$ Pa s for $0.1 \leq M \leq 1$) and $\eta_0 = 5 \times 10^{14}$ Pa s (i.e., $6 \times 10^{15} \leq \eta \leq 8 \times 10^{16}$ Pa s for $0.1 \leq M \leq 1$) can be adopted for felsic rocks. Also, an empirical equation has been calibrated by Caricchi *et al.*, (2008) which describe the non-Newtonian strain-rate-dependent rheology of partially molten felsic rocks.

Compared to previous numerical examples, the new feature that we have to introduce is a treatment of the melting/crystallisation processes. Crystallisation of the intruding magma and, to some extent, partial melting of host rocks are two important and coeval processes during plutonism (e.g. Marsh, 1982; Best and Christiansen, 2001) because they affect the density and the rheology of both intruding and intruded rocks, respectively. The numerical models presented allow the gradual crystallisation of magma and partial melting of the crust (e.g. Bittner and Schmeling, 1995) in the pressure–temperature domain between the wet solidus and dry liquidus of corresponding rocks (Table 17.2). As a first approximation, the volumetric fraction of melt M at constant pressure is assumed to increase linearly with temperature according to the relations (Gerya and Yuen, 2003b; Burg and Gerya, 2005):

$$M = 0 \quad \text{at} \quad T \le T_{solidus}, \tag{17.16a}$$

$$M = \frac{(T - T_{solidus})}{(T_{liquidus} - T_{solidus})} \quad \text{at} \quad T_{solidus} < T < T_{liquidus}, \tag{17.16b}$$

$$M = 1 \quad \text{at} \quad T \ge T_{liquidus}, \tag{17.16c}$$

where $T_{solidus}$ and $T_{liquidus}$ are the solidus and liquidus temperatures of the considered rock, respectively (Table 17.2).

The effective density, ρ_{eff}, of partially molten rocks is then calculated from:

$$\rho_{eff} = \rho_{solid} \left(1 - M + M \frac{\rho_{0molten}}{\rho_{0solid}} \right), \tag{17.17}$$

where ρ_{0solid} and $\rho_{0molten}$ are the standard densities of solid and molten rock, respectively (Table 17.2) and ρ_{solid} is the density of solid rocks at given P and T computed according to Eq. (2.4b) (Chapter 2) based on the thermal expansion and compressibility coefficients (Table 17.2).

The effect of latent heating due to equilibrium melting/crystallisation is included implicitly by increasing the effective heat capacity (C_{Peff}) and the thermal expansion (α_{eff}) of the partially crystallised/molten rocks ($0 < M < 1$), calculated as (Burg and Gerya, 2005):

$$C_{Peff} = C_P + Q_L \left(\frac{\partial M}{\partial T} \right)_{P=const}, \tag{17.18a}$$

$$\alpha_{eff} = \alpha + \rho \frac{Q_L}{T} \left(\frac{\partial M}{\partial P} \right)_{T=const}, \tag{17.18b}$$

where C_P is the heat capacity of the solid rock and Q_L is the latent heat of melting of the rock (Table 17.2). Pressure and temperature derivatives of the melt fraction M can be obtained numerically by calling an external melt fraction computation routine (see MATLAB function **Melt_fraction.m** used by the code **Intrusion_emplacement.m**).

Figure 17.9 displays results of an intrusion experiment performed with the code **Intrusion_emplacement.m**. Two emplacement regimes are compared: rapid emplacement of hot, low viscosity magma with higher degree of melting (Fig. 17.9(a)) and slower emplacement of colder, higher viscosity magma with lower degree of melting (Fig. 17.9(b)). In the first case, emplacement is mainly controlled by the plastic deformation (faulting) of the crust. Magma rises rapidly toward the surface (Fig. 17.9(a), 3.6 Kyr), despite the fact that its density is notably higher (by 150–300 kg/m^3) than that of the crustal host rocks (though it is lower than the density of the non-molten lithospheric mantle). Extrusion of hot, partially molten rocks through the magmatic channel is primarily driven by the density contrast between these partially molten rocks and the mantle lithosphere. To minimise gravitational energy, intrusive rocks of intermediate density should tend to pool along the crust/mantle boundary (i.e. at the neutral buoyancy level). However, this tendency is only realised if the magma viscosity is relatively high and its emplacement, therefore, is relatively slow such that viscous deformation of the lower crust can accommodate the intrusion emplacement along the Moho (Fig. 17.9(b)). In the first model, however, (Fig.17.9(a)) the magma viscosity is low and its emplacement is, therefore, fast and can only be accommodated by plastic deformation. Since plastic strength of the crust rapidly decreases with decreasing depth (more precisely with decreasing dynamic pressure), the intrusion propagates upwards to the upper crust where emplacement requires less mechanical work. The restraining gravitational energy produced by the arrival of a dense intrusion in the less dense crust is compensated and overcome by the positive buoyancy of partially molten rocks in the magmatic channel. This phenomenon is comparable to the penetration of a diapir head into lower density rocks (e.g., Ramberg, 1981). This respective intrusion mechanism is also called trans-lithospheric mantle diapirism (Burg *et al.*, 2009). It is worth emphasising that the gravitational balance controls the height of the column of molten rock, but not the volume of magmatic rock below and above the Moho. This is expressed in the mechanical equilibrium relation:

$$h_{Channel}(\rho_{Mantle} - \rho_{Magma}) = h_{Intrusion}(\rho_{Magma} - \rho_{Crust}), \qquad (17.19)$$

where $h_{Channel}$ is the height of the column of magmatic rock in the channel below the Moho, $h_{Intrusion}$ is the height of magmatic rock above the Moho, ρ_{Mantle}, ρ_{Magma} and ρ_{Crust} are the density of the mantle lithosphere, the intruding magma and the crust, respectively. Since the width of the channel below the Moho is limited by the rigidity of the cold mantle lithosphere, the volume of magma intruding into the crust can be much larger than the volume of rock remaining in the channel (Fig. 17.9(a), 3.6 Kyr).

17.10 Mantle convection with phase changes

As we discussed in the introduction, thermomechanical modelling of mantle convection has a rich history dating back to the early 1970s (e.g. Richter, 1978; Schubert, 1992; Bercovici, 2007). Not surprisingly, it is one of the most advanced fields of geodynamic modelling in terms of both technical and conceptual progresses (e.g., Yuen *et al.*, 2000). Realistic modelling of terrestrial and planetary convection is a challenging topic (e.g., Hansen and Yuen, 1988; Larsen *et al.*, 1995; Yuen *et al.*, 2000; Zhong *et al.*, 2007; Tackley, 2008) and requires the application of sophisticated 3D numerical codes working with spherical geometries at high grid resolution which can almost exclusively only be preformed by parallel computing on 'big machines'. Indeed, one important aspect of modelling mantle convection, which is of interest for this chapter, is the incorporation of solid-state phase transitions into such numerical models.

Solid-state phase transitions are crucial phenomena in the Earth's mantle. Major phase transitions include olivine–spinel at 410 km depth and spinel–perovskite at 670 km depth. These transitions are associated with significant changes in mantle density and seismic wave speeds (Turcotte and Schubert, 2002). It was also suggested recently, that the so-called D″ (D-double-prime) discontinuity near the core–mantle boundary is related to perovskite–post-perovskite phase transition (Oganov and Ono, 2004; Murakami *et al.*, 2004). Phase transitions affect the dynamics of mantle convection due to (1) density changes and (2) latent heating (Richter, 1973; Schubert *et al.*, 1975; Christensen and Yuen, 1985; Tackley, 1993; Zhong and Gurnis, 1994).

Phase changes are traditionally included in mantle convection models (e.g., Richter, 1973; Schubert *et al.*, 1975; Christensen and Yuen, 1985; Tackley, 1993; Zhong and Gurnis, 1994) by programming each transition individually (i.e. similarly to what we did with melting reactions in the previous example). However, for realistic mantle compositions, the amount of various phase transitions is larger than only three (Fig. 17.10) and these phase transitions involve several minerals of variable composition (so called *solid solutions*, Table 17.3) which makes the traditional approach quite inconvenient. An alternative method has been developed recently based on Gibbs free energy minimisation (Chapter 2). This method was initially applied for crustal- and lithospheric-scale thermal (Petrini *et al.*, 2001; Gerya *et al.*, 2001; Kaus *et al.*, 2005) and thermomechanical (Gerya *et al.*, 2004c, 2006; Yamato *et al.*, 2008) models and then expanded to mantle convection models (Tackley, 2008).

The idea of this *petrological-thermomechanical* method is relatively simple (Gerya *et al.*, 2004c; 2006): (i) phase diagrams (*P–T pseudosections*) and related density (ρ) and enthalpy (H) maps (see programming exercise 2.3; Chapter 2) are first computed for the necessary rock compositions in a relevant range of $P–T$ conditions and (ii) these maps are then used in thermomechanical experiments for computing density (ρ), effective heat capacity incorporating latent heat (C_{Peff}) and energetic effects (both adiabatic and latent heating) for isothermal (de)compression (H_P) for material points (markers) based on standard thermodynamic formulas and

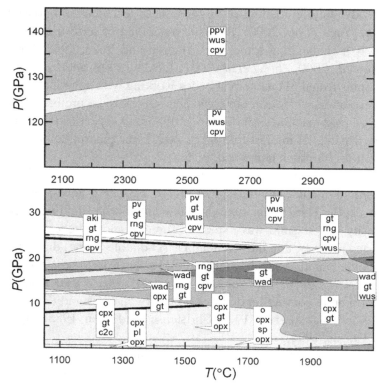

Fig. 17.10 Phase relations for the CaO-FeO-MgO-Al$_2$O$_3$-SiO$_2$ pyrolite model (see Table 17.3 for notations of minerals) computed (Mishin *et al.*, 2008) with the Gibbs free energy minimisation program Perple_X (Connolly, 2005). To permit the resolution of phase relations the diagram is split to exclude the large depth interval between the transition zone and core–mantle boundary in which the model does not predict phase transformations. Composition for the pyrolite model is 3.87 wt% CaO, 8.11 wt% FeO, 3.61 wt% Al$_2$O$_3$, 38.59 wt% MgO and 45.82 wt% SiO2.

numerical differentiation in P–T space (Fig. 17.11)

$$\rho = \rho_{i,j}\left(1 - \frac{\Delta T_m}{\Delta T}\right)\left(1 - \frac{\Delta P_m}{\Delta P}\right) + \rho_{i+1,j}\left(1 - \frac{\Delta T_m}{\Delta T}\right)\frac{\Delta P_m}{\Delta P}$$

$$+ \rho_{i,j+1}\frac{\Delta T_m}{\Delta T}\left(1 - \frac{\Delta P_m}{\Delta P}\right) + \rho_{i+1,j+1}\frac{\Delta T_m}{\Delta T}\frac{\Delta P_m}{\Delta P}, \tag{17.20}$$

$$Cp_{eff} = \left(\frac{\partial H}{\partial T}\right)_{P=const} = \frac{H_{i,j+1} - H_{i,j}}{\Delta T}\left(1 - \frac{\Delta P_m}{\Delta P}\right) + \frac{H_{i+1,j+1} - H_{i+1,j}}{\Delta T}\left(\frac{\Delta P_m}{\Delta P}\right), \tag{17.21}$$

Table 17.3 *Phase notation and formulae of minerals for the*
CaO-FeO-MgO-Al$_2$O$_3$-SiO$_2$ pyrolite model (Fig. 17.9)

Symbol	Phase	Formula*
aki	akimotoite	$Mg_xFe_{1-x-y}Al_{2y}Si_{1-y}O_3$, $x + y \leq 1$
c2c	pyroxene	$[Mg_xFe_{1-x}]_4Si_4O_{12}$
cpv	Ca–perovskite	$CaSiO_3$
cpx	clinopyroxene	$Ca_{2y}Mg_{4-2x-2y}Fe_{2x}Si_4O_{12}$
gt	garnet	$Fe_{3x}Ca_{3y}Mg_{3(1-x+y+z/3)}Al_{2-2z}Si_{3+z}O_{12}$, $x + y \leq 1$
o	olivine	$[Mg_xFe_{1-x}]_2SiO_4$
opx	orthopyroxene	$[Mg_xFe_{1-x}]_{4-2y}Al_{4(1-y)}Si_4O_{12}$
ppv	post–perovskite	$Mg_xFe_{1-x-y}Al_{2y}Si_{1-y}O_3$, $x + y \leq 1$
pv	perovskite	$Mg_xFe_{1-x-y}Al_{2y}Si_{1-y}O_3$, $x + y \leq 1$
rng	ringwoodite	$[Mg_xFe_{1-x}]_2SiO_4$
sp	spinel	$Mg_xFe_{1-x}Al_2O_3$
wad	waddsleyite	$[Mg_xFe_{1-x}]_2SiO_4$
wus	magnesiowuestite	$Mg_xFe_{1-x}O$

* Unless otherwise noted, the compositional variables w, x, y, and z may
vary between zero and unity and are determined as a function of pressure
and temperature by free-energy minimisation with Perple_X program (Connolly,
2005).

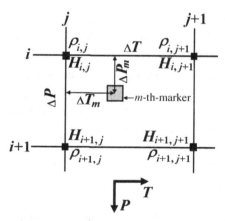

Fig. 17.11 Stencil associated with the *P–T* grid used for the interpolation
of physical properties from enthalpy and density look-up tables, to the
markers.

$$\frac{H_P}{DP/Dt} = \left(1 - \rho\frac{\partial H}{\partial P}\right)_{T=const} = 1 - \frac{H_{i+1,j} - H_{i,j}}{\Delta P}\left(1 - \frac{\Delta T_m}{\Delta T}\right)\frac{\rho_{i+1,j} + \rho_{i,j}}{2}$$

$$- \frac{H_{i+1,j+1} - H_{i,j+1}}{\Delta P}\left(\frac{\Delta T_m}{\Delta T}\right)\frac{\rho_{i+1,j+1} + \rho_{i,j+1}}{2}. \tag{17.22}$$

The temperature equation (9.8) is respectively modified as

$$\rho C_{Peff}\frac{DT}{Dt} = -\frac{\partial q_i}{\partial x_i} + H_r + H_s + H_P. \tag{17.23}$$

The stable mineralogy and physical properties for the mantle used in our example are computed (Mishin *et al.*, 2008) with Perple_X (Connolly, 2005) by a free energy minimisation approach. For this purpose the Mie-Grueneisen formulation of Stixrude and Bukowinski (1990) was adopted with the parameterisation of Stixrude and Lithgow-Bertelloni (2005) augmented for lower mantle phases as described by Khan *et al.* (2006). This parameterisation limits the chemical model to the CaO-FeO-MgO-Al$_2$O$_3$-SiO$_2$ (Table 17.3). The mantle rheology is based on the dry olivine flow law (Table 17.1) with an activation volume of 5 cm^3, which allows us to mimic 1.5–3 order of magnitude increase in viscosity for the lower mantle. This range is often used in mantle convection models (e.g. Tackley, 2000). Olivine is obviously not stable in the deep mantle below the olivine–spinel transition and, therefore, our rheological choice is rather arbitrary. This is related to the limited availability of experimentally calibrated flow laws applicable to the deep mantle such that simplified temperature- and depth-dependent rheological models are traditionally used in numerical mantle convection studies (e.g., Richter, 1973; Schubert *et al.*, 1975; Christensen and Yuen, 1985; Zhong and Gurnis, 1994; Tackley, 1993, 2000, 2008).

Figure 17.12 shows several stages of a mantle convection modelled with the code **Mantle_convection.m**. The model design is simple: a square 3000 × 3000 km box with free slip boundaries, constant temperature conditions applied at the top and in the bottom and no heat flux condition across the vertical walls. In this model we do not intend to model self-consistent plate generation (e.g. Tackley, 2000) and therefore use uniform, relatively coarse grid resolution. Mantle convection is characterised by a semi-layered structure with a strongly convecting upper part (above 670 km depth) and a much more slowly deforming lower mantle with several plumes penetrating from the core–mantle boundary (Figs. 17.12, 17.13(b)). Density changes (Fig. 17.13(d)) and thermal effects (Fig. 17.13(c)) of various phase transitions (Fig. 17.10) are captured with our petrological-thermomechanical numerical approach without programming them individually. This approach makes coding simpler and easily allows changes to be made in the case of testing different models for mantle composition.

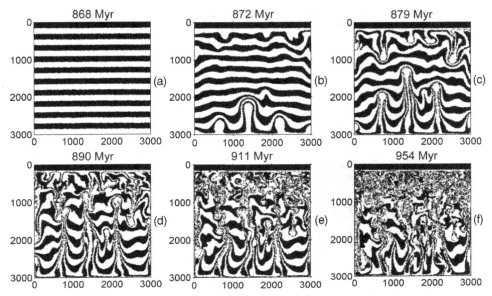

Fig. 17.12 Development of mantle convection processes with phase changes (Fig. 17.10) computed with the petrological-thermomecanical code **Mantle_convection.m** associated with this chapter. Model resolution is 51×51 nodal points with 40 000 randomly distributed markers. Grid resolution is uniform in both directions. The model is shown from an arbitrary stage of 868 Myr when non-compositional layering is superimposed for visualising deformation onto a pre-computed non-steady thermal structure. Note, moderate deformation of the lower mantle by localised upwellings and downwellings (thermal plumes) which contrasts with the intense chaotic mixing of the upper mantle.

17.11 Deformation of self-gravitating planetary body

Lastly, let us discuss some planetary-scale applications of numerical geodynamic modelling which are becoming more and more widespread in relation to the problem of planetary accretion and core formation processes (e.g. Stevenson, 1981). One important group of such applications concerns the internal deformation of an inhomogeneous, self-gravitating body. Numerical modelling of deformation of such a body was already discussed in Chapter 11 with the use of a 'spherical-Cartesian' approach (Honda *et al.*, 1993; Gerya and Yuen, 2007; Lin *et al.*, 2009). This approach is relatively simple and does not require major rewriting of the Cartesian code used for the previous examples. All that is required is to add the numerical solution of the Poisson equation before solving the momentum and continuity equations (Fig. 11.3). Gravitational acceleration components are then computed from the gravity potential locally and are used in the right-hand side of the momentum equation (Eqs. (11.18), (11.19)). As a boundary condition for gravity potential, one can use constant gravitational potential value (e.g. $\Phi = 0$)

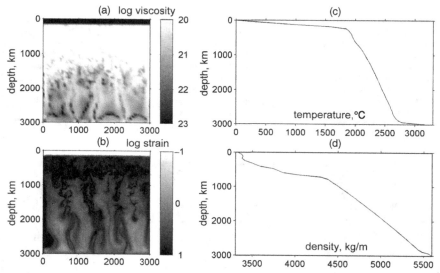

Fig. 17.13 Viscosity (a) and strain (b) maps and horizontally averaged temperature (c) and density (d) profiles for the computed mantle convection model at 954 Myr (Fig. 17.12(f)). Note, 1–2 order of magnitude viscosity contrast between the upper and lower mantle obtained with our simplified rheological model based on dry olivine flow law with activation volume of 5 cm³. Mantle deformation in (b) is semi-layered and is strongly affected by the spinel–perovskite transition at around 670 km depth due to its negative Clapeyron slope (Fig. 17.10) which creates difficulties for both upwellings and downwellings for penetrating this boundary. Cold lithosphere at the top of the model remains undeformed (stagnant lid regime, e.g. Tackley, 2000) due to the prescribed high plastic strength ($\sin(\varphi) = 0.6$) which corresponds to a dry mantle. Plate bending processes (Fig. 17.1) in this lithosphere are not modelled due to the imposed free slip upper boundary condition without a weak layer (in contrast to free-surface condition formed by sticky air/water used in all previous examples) as well as due to the very coarse grid resolution employed here (60 × 60 km in contrast to 2 × 2 km needed for properly resolving slab bending processes).

applied at a circular surface located at a distance from the planet (Fig. 14.7, Eq. (14.12)). The chosen value of the potential along the surface is arbitrary since it does not affect the resulting gravitational acceleration field (given by derivatives of the potential). The use of such a boundary condition is based on the fact that with growing distance from the planet, both the gravitational acceleration and gravity potential tend to become solely a function of the planetary mass (m_p) and the distance to its centre (d)

$$g = G\frac{m_p}{d^2}, \tag{17.24}$$

$$\Phi = const - G\frac{m_p}{d}. \tag{17.25}$$

However, it should be mentioned that forcing the potential to be uniform outside the planet at a certain distance from the planetary centre also affects the gravitational field inside the planet (especially its tangential component). Therefore, the gravity potential boundary should preferably be located at a significant distance from the planetary surface which should be comparable to the planetary radius (Lin *et al.*, 2009).

The initial setup for the numerical experiment on gravitational redistribution of metal and silicate in a Mars-sized body (3000 km in radius) is shown in Figure 17.14 (0.03 Myr). This setup is based on the numerical study of Golabek *et al.* (2008a) who investigated rheological controls on the terrestrial core formation mechanism. The initial temperature is 273 K at the surface and 1200 K at 100 km depth and then rises linearly to 1500 K in the centre of the planet. This initial profile is rather arbitrary and mimics to some degree the effects of various planetary heating processes: (i) decay of short-lived radioactive isotopes (MacPherson *et al.*, 1995) such as ^{26}Al (half-life time is 0.73 Ma) and ^{60}Fe (half-life time is 1.5 Ma), (ii) impact heating by accreted planetesimals (Davies, 1985; Melosh, 1990), (iii) impact associated gravitational unloading (Asphaug *et al.*, 2006) and (iv) adiabatic heating caused by growing pressure in the planetary interior. The planet is heterogeneous and composed of a silicate matrix and randomly distributed iron diapirs with radii varying from 50 to 100 km. Using a variable-sized diapir is related to the fact that the size distribution of the planetesimals and planetary embryos after runaway growth and the formation of Mars-sized bodies should be heterogeneous (Melosh, 1990; Tonks and Melosh, 1992; Stevenson, 2008). As in the previous example, we also apply a coupled petrological-thermomechanical approach for modelling density and thermal properties of the silicate matrix, which is assumed to have pyrolitic composition (Fig. 17.10). The rheology of silicate corresponds to dry olivine (Table 17.1) with an activation volume of 10 cm^3. Constant density (10000 kg/m^3) and lowered viscosity (10^{20} Pa s) are used for the metal.

Viscosity of the weak mass less-like medium (1 kg/m^3) surrounding the planet is also taken to be 10^{20} Pa s, which is 2–5 order of magnitude lower than that of silicate at the planetary surface. This is sufficient to provide free surface like condition (Schmeling *et al.*, 2008; Lin *et al.*, 2009). Lower viscosity will require shorter time steps which is inconvenient. Feedback from shear heating caused by gravitational energy release is also taken into account since this process may strongly affect temperature distribution inside the differentiating planet (Gerya and Yuen, 2007; Golabek *et al.*, 2008a).

Figure 17.14 shows several stages of the core formation computed with the code **Core_formation.m** associated with this chapter. Model development corresponds to so called 'decomposition mode' (Golabek *et al.*, 2008a) which is defined by

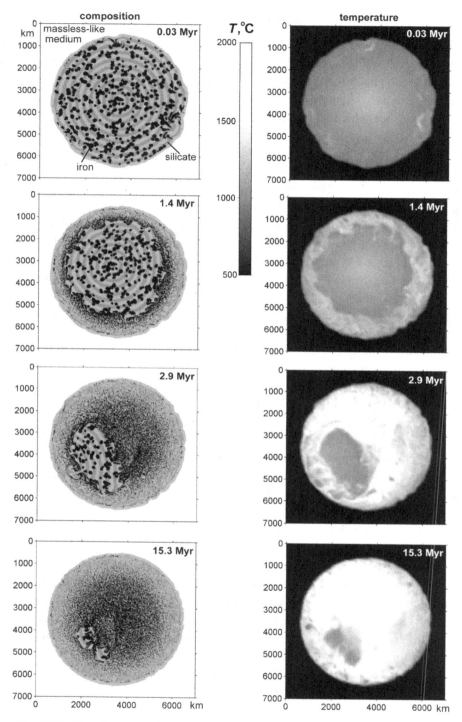

Fig. 17.14 Development of the compositional (left column) and temperature (right column) fields in the numerical experiment on core formation computed with the code **Core_formation.m** associated with this chapter. Model resolution is 101×101 nodal points with $160\,000$ randomly distributed markers.

choosing a relatively large activation volume (10 cm^3) and plastic strength (no Peierls plasticity limit is imposed) of the silicate matrix that effectively makes the pressurised planetary interior rheologically stronger than the outer shell. Therefore, in this model, the iron diapirs near the surface are activated foremost (Fig. 17.14, 0.03 Myr). Even the slightest asymmetries in the initial iron distribution lead to an earlier initiation of diapir sinking in some distinct regions where significantly larger temperatures develop due to shear heating processes. The diapir sinking releases large amounts of energy. This leads to the activation of neighbouring diapirs and finally to the formation of large iron ponds. The underlying material of the planetary interior is too strong to be deformed by stresses arising from the available iron agglomerations. Therefore, all iron diapirs in the outer region of the planet will be finally activated. This leads to a global temperature rise in the upper layers of the planet (Fig. 17.14, 1.4 Myr). Consequently, a low-viscosity shell, basically analogous to a magma ocean, is formed around the remaining highly viscous central region. The iron ponds on top of this high-viscosity sphere finally coalescence and form an iron-rich ring around the central region (Fig. 17.14, 1.4 Myr), as was suggested by Stevenson (1981) and Ida *et al.* (1987). Temperature and density dis-equilibrium around the ring leads to the degree-one instability (e.g. Ida *et al.*, 1987) resulting in the formation of advective streams of the iron-rich material around the central region. Asymmetry of these flows causes and sustains the rotation of the central sphere and the resulting shear heating aids in further decomposition (Fig. 17.14, 2.9 Myr). Iron accumulating on the one side of the planet pushes the stiff, non-differentiated planetary interior (including the passive iron diapirs) out of the central region creating a noticeable asymmetry in the planet (Fig. 17.14, 2.9 Myr). A similar scenario was proposed by Elsasser (1963) and modelled numerically in a simplified way (Honda *et al.*, 1993; Lin *et al.*, 2009), but under the assumption of a cold central region. This translation favours the decompression-related decomposition of the 'exhuming' interiors when the material approaches shallower depths. Decompression causes (via activation volume) a viscosity reduction in the regions of the translated central sphere closer to the planet's surface. The silicates released at the leading side, rise due to their low density as Rayleigh–Taylor instabilities upwards into the high-temperature zone (Fig. 17.14, 2.9 Myr). This causes further iron release and makes the process of interior 'decomposition' self-sustaining, which finally results in the formation of an iron core (Fig. 17.14, 15.3 Myr). Obviously, this scenario is non-unique and several others core formation modes may develop instead, depending on variations in planetary size, temperature structure and material properties, and particularly the effective silicate rheology (Golabek *et al.*, 2008a).

Programming exercise and homework

Exercise 17.1

Program a model for extension of a continental lithosphere based on the provided oceanic lithosphere extension model (Fig. 17.4). Take the initial compositional and thermal structure of the continental lithosphere to be the same as in the continental collision example (Fig. 17.5(a)). Use the codes **Extension.m** and **Collision.m** associated with this chapter. Alternatively, you may also want to program a similar model using a uniform grid resolution by using your own visco-elasto-plastic codes created during exercises for Chapter 13.

Epilogue: outlook

Theory: Where are we now? Where to go further? Current and future directions of numerical geodynamic modelling development: 3D, MPI, OpenMP, PETSc, AMR, FEM, FVM, GPU/Cell-based computing, interactive computing, realistic physics, visualisation challenges etc.
Exercises: No more exercises and no more homework!

Where are we now?

Where are we after reading 17 chapters + Introduction and performing (all?) analytical and programming exercises and homework? We are still only at the very beginning of the wonderful, rapidly developing, world of numerical geodynamic modelling. The method we learned is based on a finite-difference method and marker-in-cell techniques (FDM+MIC). Although this holds for many geodynamic situations ('all in one' tool, Gerya and Yuen, 2007), its universality is not absolute and other approaches may be better suited for some numerical problems. The examples provided in this book are written in a very explicit way, with the purpose of facilitating learning and understanding, rather than to produce massive and fast numerical results. The resolution of the numerical examples is also moderate so that corresponding experiments can be completed in a reasonable amount of time (several seconds to several days) on an ordinary laptop which is, by the way, not exactly the main tool of geodynamic modellers (apart from, sometimes, for code developments and testing, for looking at results computed on *big machines* and for writing papers and books like this one . . .). Once again, we are at the beginning and if we want to go further, it is worth reading the following.

Where to go further?

Not only modellers ask this question. Recently, ten research questions shaping twenty-first-century Earth Science were identified by the US National

Research Council of the National Academy of Sciences (DePaolo *et al.*, 2008, www8.nationalacademies.org/onpinews/newsitem.aspx?RecordID=12161).
They are cited below

(1) 'How did Earth and other planets form? While scientists generally agree that this solar system's sun and planets came from the same nebular cloud, they do not know enough about how Earth obtained its chemical composition to understand its evolution or why the other planets are different from each other. Although credible models of planet formation now exist, further measurements of solar system bodies and extrasolar objects could offer insight in the origin of the Earth and the solar system.'

(2) 'What happened during Earth's "dark age" (the first 500 million years)? Scientists believe that another planet collided with Earth during the late stages of its formation, creating debris that became the moon and causing Earth to melt down to its core. This period is critical to understanding planetary evolution, especially how the Earth developed its atmosphere and oceans, but scientists have little information because few rocks from this age are preserved.'

(3) 'How did life begin? The origin of life is one of the most intriguing, difficult, and enduring questions in science. The only remaining evidence of where, when, and in what form life first appeared springs from geological investigations of rocks and minerals. To help answer the question, scientists are also turning toward Mars, where the sedimentary record of early planetary history predates the oldest Earth rocks, and other star systems with planets.'

(4) 'How does Earth's interior work, and how does it affect the surface? Scientists know that the mantle and core are in constant convective motion. Core convection produces the Earth's magnetic field, which may influence surface conditions, and mantle convection causes volcanism, seafloor generation, and mountain building. However, scientists can neither precisely describe these motions, nor calculate how they were different in the past, hindering scientific understanding of the past and prediction of Earth's future surface environment.'

(5) 'Why does Earth have plate tectonics and continents? Although plate tectonics theory is well established, scientists wonder why Earth has plate tectonics and how closely it is related to other aspects of Earth, such as the abundance of water and the existence of the continents, oceans, and life. Moreover, scientists still do not know when continents first formed, how they remained preserved for billions of years, or how they are likely to evolve in the future. These are especially important questions as weathering of the continental crust plays a role in regulating Earth's climate.'

(6) 'How are Earth processes controlled by material properties? Scientists now recognise that macroscale behaviors, such as plate tectonics and mantle convection, arise from microscale properties of Earth materials, including the smallest details of their atomic structures. Understanding materials at the microscale is essential to comprehend the Earth's history and making reasonable predictions about how planetary processes may change in the future.'

(7) 'What causes climate to change – and how much can it change? Earth's surface temperature has remained within a relatively narrow range for most of the last 4 billion years, but how does it stay well-regulated in the long run, even though it can change so abruptly? Study of Earth's climate extremes through history – when climate was extremely cold or hot or changed quickly – may lead to improved climate models that could enable scientists to predict the magnitude and consequences of climate change.'

(8) 'How has life shaped Earth – and how has Earth shaped life? The exact ways in which geology and biology influence each other are still elusive. Scientists are interested in life's role in oxygenating the atmosphere and reshaping the surface through weathering and erosion. They also seek to understand how geological events caused mass extinctions and influenced the course of evolution.'

(9) 'Can earthquakes, volcanic eruptions, and their consequences be predicted? Progress has been made in estimating the probability of future earthquakes, but scientists may never be able to predict the exact time and place an earthquake will strike. Nevertheless, they continue to decipher how fault ruptures start and stop and how much shaking can be expected near large earthquakes. For volcanic eruptions, geologists are moving toward predictive capabilities, but face the challenge of developing a clear picture of the movement of magma, from its sources in the upper mantle, through Earth's crust, to the surface where it erupts.'

(10) 'How do fluid flow and transport affect the human environment? Good management of natural resources and the environment requires knowledge of the behavior of fluids, both below ground and at the surface, and scientists ultimately want to produce mathematical models that can predict the performance of these natural systems. Yet, it remains difficult to determine how subsurface fluids are distributed in heterogeneous rock and soil formations, how fast they flow, how effectively they transport dissolved and suspended materials, and how they are affected by chemical and thermal exchange with the host formations.'

A remarkable fact is that seven (!) of these questions (1, 2, 4, 5, 6, 9 and 10) are directly related to the topics addressed by numerical geodynamic modelling and finding answers to those questions is likely to depend on the future efforts of modellers. This will keep us all busy for a while! The general ideas for future technical and conceptual advances in numerical geodynamic modelling are therefore quite clear: (*i*) *fast computing of high-resolution 3D numerical problems with complex and realistic physics applicable to nature, and (ii) obtaining a rigorous understanding of geodynamic and planetary processes and the key physical parameters controlling them.* This means that modellers have to concentrate on both (i) developing efficient, fast, realistic, high-resolution thermomechanical numerical 3D codes that will become routine tools for geodynamicists and (ii) integrating results of numerical models with natural observations and producing comprehensive testable predictions in related fields.

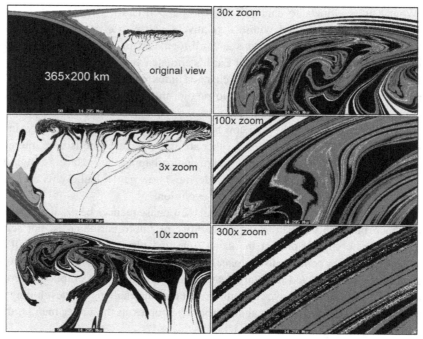

Fig. Outlook.1 High-resolution, multiple-scale visualisation of mechanical stir-ring structures related to development of a partially molten mantle wedge plume (Gorczyk *et al.*, 2006; 2007b) which spreads underneath the cold and stiff mantle lithosphere of the overriding plate (see Fig. 17.3 for evolving thermal structure of such plate). 40 billion elements (pixels) data set with spatial resolution of around 2 m is based on compressed output from the 2D numerical experiment with 10 bil-lion markers. Each pixel of the original size figure (top left frame) can be resolved as a full-size picture (bottom right frame). The experiment was computed with the use of an OpenMP-based parallelisation on the shared memory supercomputer COBALT at NCSA-Illinois.

How big are '*big numerical problems*'? At the moment, close to a billion nodal points (e.g. Cohen *et al.*, 2005; Tackley, 2008; Krotkiewski *et al.*, 2008) i.e. close to $1000^3 = 1000 \times 1000 \times 1000$ nodal points which is 8500 times larger com-pared to the $49^3 = 49 \times 49 \times 49$ that we solved with multigrid in 3D in Chapter 15 (Figs. 15.11, 15.12). Also, tens of billions of markers are explored (Gorczyk *et al.*, 2007b) in some models (Fig. Outlook.1) which is also several thousand times bigger than the 750 thousand markers explored in our largest experiment, the retreating subduction in Chapter 17 (Figs. 17.2, 17.3). Handling huge models is not a trivial task. Difficulties, as usual, do not come from the hardware side, i.e. from limited availability of corresponding 'big machines' with terabytes (i.e. thousands of gigabytes) of memory – they exist and getting access to them is suf-ficiently uncomplicated. In addition, recent hardware development trends involve using highly efficient, low-cost *GPU/Cell-based* multiprocessor units, which will greatly expand computational capabilities in the future. The key difficulties are (i) to efficiently *parallelise computations* (i.e. use in parallel many processors for the same numerical experiment) and (ii) to program efficient procedures for *compressed data storage*.

For example, the highest-resolution FDM+MIC simulations to date use 10 to 40 billion markers (Fig. Outlook.1). This output from these extremely high resolution simulations result in an uncompressed output file sizes of up to 840 GB (or more) for each time step. Similar amounts of memory are required during the runs as well, but a number of supercomputers such as BRUTUS at ETH-Zurich or COBALT at NCSA-Illinois have ample memory for these large runs. Multiplied to hundreds of time steps, which are typically needed to be stored for each experiment, and to tens of experiments typically needed for every studied numerical problem, implies that enormous storage capabilities are required. If we do not care about data compression, then supercomputers will spend 99% of their time for *IO operations* (i.e. for writing results of our experiment on hard disks)! This is not exactly what they are made for . . . The problem only appears from a certain resolution size and we would perhaps even not guess about it before. Data compression should also be sufficiently fast so that it does not take more than 10% of CPU time. Efficient and fast data compression algorithms based, for example, on wavelets (e.g. Vasiliev *et al.*, 2004) allow the reduction of storage size by a factor of 100 to 1000 (e.g. Gorczyk *et al.*, 2007b) which brings us back to efficient computing and results production.

Another data handling option is to perform *fully interactive computing* associated with no/little data storage (e.g. Damon *et al.*, 2008). The idea behind it is the following: multiprocessor hardware is currently so efficient that one can, in principle, repeat almost any kind of numerical experiment within minutes or few hours. If one spent this time looking on a screen showing progressing results of an ongoing numerical experiment, and saving limited amount of characteristic frames as image files, then the regular data storage is no longer needed. If you want to learn something more about the same experiment you simply repeat it. Of course, these two strategies of data handling may also be complementary.

Let us now go briefly through a list of things that, from a general perspective, would be useful to read, think of, learn and implement in the future. This list is obviously fragmentary, subjective and non exhaustive, but will provide some useful hints triggering further thinking.

State-of-the-art overview

A good cross-section covering the current state of the art in numerical geodynamic modelling (who is now doing what and with which numerical technique) can be retrieved from two recent special volumes of the *Physics of the Earth and Planetary Interiors* attributed to computational Earth Sciences:

- Volume 163: Computational challenges in the Earth sciences, edited by David A. Yuen and Huai Zhang in 2007.
- Volume 171: Recent advances in computational geodynamics: theory, numerics and applications, edited by Boris J.P. Kaus, myself and Daniel W. Schmid in 2008.

Several comprehensive overviews on mantle convection modelling can be found in

- *Treatise on Geophysics*, Editor-in-Chief: Gerald Schubert, *Volume 7: Mantle Dynamics*, edited by David Bercovici in 2007, Elsevier.

  Information, about existing and forthcoming computational needs for solid Earth sciences in general and for mantle convection in particular can be retrieved from the report
- *High-performance Computing Requirements for the Computational Solid Earth Sciences*, edited by Cohen and co-workers in 2005 (www.geo-prose.com/computational_ SES.html).

Efficient direct solvers

Implementation of efficient direct solvers (developed by mathematicians) to numerical codes and tuning them for geodynamic modelling problems can significantly (sometimes up to 100 times) speed up calculation in 2D and can even allow addressing 3D problems at sufficiently high resolution (e.g. Braun *et al.*, 2008). Three such direct solvers are currently used in geosciences:

1. WSMP (Watson Sparse Matrix Package, Gupta, 2000, www-users.cs.umn.edu/~agupta/ wsmp.html).
2. MUMPS (MUltifrontal Massively Parallel sparse direct Solver, Amestoy *et al.*, 2001, http://graal.ens-lyon.fr/MUMPS/).
3. PARDISO (Schenk and Gärtner, 2004, 2006, www.pardiso-project.org/).

The PARDISO solver is, for example, currently implemented in the thermomechanical 2D codes originally developed by Gerya and Yuen (2003a, 2007) and resulted in a speedup of 30 times compared to a standard Gaussian solver (Chapter 3), thereby allowing us to perform numerical experiments at high resolution of up to 800×800 grid points on a single processor (which is 50 times more grid points compared to our 101×101 grid in the core formation experiment of Fig. 17.13). Similar performances can also be reached with the MATLAB-based finite element code MILAMIN (Dabrowski, *et al.*, 2008, http://milamin.org/): MILAMIN means Million-A-Minute, i.e. one million equations are solved within one minute. This corresponds to a resolution of approximately 600×600 nodal points in case of 2D mechanical code.

Parallelisation of numerical codes

Efficient, high-resolution 2D and 3D modelling require the *parallelisation of numerical codes*. This allows them to use many processors at the same time, which proportionally speeds up numerical calculations (as long as parallelisation is efficiently implemented, see e.g. Tackley, 2008). Individual processors need to exchange information during the calculations and their activity is not fully independent and should be thoroughly correlated. This makes parallelisation a non-trivial programming task. There are several ways for code parallelisation using open source libraries such as MPI, OpenMP and PETSc (for example).

MPI – the Message Passing Interface standard (e.g. www-unix.mcs.anl.gov/mpi/, Karniadakis and Kirby, 2003). MPI is a library specification for *message passing* which defines and enables a mechanism of data exchange between different processors that are simultaneously used for the same numerical experiment. MPI was designed for high performance on both massively parallel machines and on workstation clusters. MPI works with both *distributed memory* (different parts of memory can be accessed by different processors) and *shared memory* (the entire memory can be directly accessed by each processor). Programming MPI requires a significant effort to begin but the resulting codes are very efficient.

OpenMP – Open Multi-Processing (e.g., http://openmp.org/wp/, Chapman *et al.*, 2007). The OpenMP Application Program Interface (API) supports multi-platform shared-memory parallel programming in C/C++ and Fortran on all architectures, including Unix platforms and Windows NT platforms. OpenMP is easy to learn and implement in numerical codes (typically one only has to add few lines to parallelise loops already present in their code) but its application is limited to shared memory machines.

PETSc – Portable, Extensible Toolkit for Scientific Computation (e.g. www.mcs.anl.gov/petsc/petsc-as/ and www.lifev.org/lifev/documentation/linsol/PETScManual.pdf/). PETSc, pronounced PET-see (the S is silent), – is a suite of data structures and routines for the scalable (parallel) solution of scientific applications modelled by partial differential equations. It employs the MPI standard for all message-passing communication and is quite convenient to learn and to use for programming parallel numerical codes (it is a bit similar to MATLAB, actually).

By the way, many available direct solvers already contain parallelisation which makes them even more attractive to employ in our codes.

Mesh refinement algorithms

Mesh refinement is a very suitable option when various processes should be modelled on different scales in the same numerical model (e.g. Fig. Outlook.1). In this case, the numerical grid should be preferably able to follow the regions where high numerical resolution is needed. We already discussed simplified methods of grid refinement (e.g. in the retreating subduction experiment), where we forced the high-resolution part of our grid to follow the trench area (Fig. 17.2). Indeed, the refinement ability of relatively simple meshes discussed in this book, made by inter-sections of horizontal and vertical lines, is limited. Therefore, more sophisticated approaches based on *Adaptive Mesh Refinement (AMR)* algorithms will be more appropriate for resolving many areas of complicated geometry present in the same model (e.g. Albers *et al.*, 2000; Braun *et al.*, 2008). These methods can be based on either finite differences (e.g. Albers *et al.*, 2000) or on finite elements (e.g. Braun *et al.*, 2008) and require significant work to properly derive a discrete formulation of the governing equations which is conservative, especially in the areas where res-olution changes. The *finite element methods (FEM)* (e.g., Zienkiewicz *et al.*, 2005) possess many advantages in this respect (e.g. Moresi *et al.*, 2003, 2007), as they are well suited to dealing with complex, irregular rectangular and triangular meshes and give more accurate results in cases when sharp curved material interfaces need to be followed in numerical models (e.g., Deubelbeiss and Kaus, 2008; Popov and Sobolev, 2008).

An additional method of choice for unstructured meshes is the *finite-volume method (FVM)* (Toro, 1999; LeVeque, 2002). Like the finite-difference method, FVM values are calculated at discrete nodes. 'Finite volume' refers to the small volume surrounding each node (e.g. to a cell surrounding central pressure node in the staggered grid (e.g., Fig. 15.1). In FVM, volume integrals in a partial differential equation that contain a divergence term are converted to surface integrals, using the divergence theorem. These terms are then evaluated as fluxes at the surfaces of each finite volume. Since the flux entering a given volume is identical to that leaving the adjacent volume, these methods are conservative. In fact, the conservative finite differences discussed in Chapters 7 and 10 are equivalent to applying a finite volume method on the relatively simple rectangular grids used in this book.

In the case of various marker-in-cell approaches, which are very common in geodynamic modelling, refinement may (or even should) also be applied to the unstructured grid of markers by using marker splitting and merging procedures (Moresi *et al.*, 2003).

Including complex realistic physics in numerical geodynamic models

In terms of involving more realistic and complicated physics in modelling, the current and future trends are quite obvious from the 'top ten' questions listed above:

(1) Using more realistic physical properties of rocks. Complicated visco-elasto-plastic rheologies including both diffusion and dislocation creep as well as Mohr–Coulomb, Drucker–Prager and Peierls plasticity (Ranalli, 1995; Katayama and Karato, 2008; Karato, 2008). Incorporating realistic rheology for partially molten rocks (see recent review by Caricci *et al.*, 2007). Use of compressible time-dependent forms of the continuity equation (Tackley, 2008; Gerya and Yuen, 2007)

(2) Accounting for phase transformations (including melting). Incorporating both volumetric and thermal effects of various phase transitions in numerical models (Gerya *et al.*, 2004c, 2006; Tackley, 2008). Adding kinetics of phase transitions.

(3) Address fluid and melt migration in deforming rocks. Programming coupled approaches for modelling fluid/melts generation and transport in actively deforming systems associated with many geodynamic processes in the crust and mantle (e.g. Connolly and Podladchikov, 1998; Schmeling, 2000; Katz, 2008).

(4) Accounting for geochemical processes in geodynamic models. Including modelling of geochemical processes (e.g. Sobolev *et al.*, 2005) in thermomechanical experiments (e.g. Xie and Tackley, 2004a,b). Using realistic models including fluid and melt related transport of trace elements in various geodynamic and planetary environments.

(5) Coupling of modelling of deep geodynamic processes with the Earth's surface development simulations (e.g., Kooi and Beaumont, 1994; Willett, 1999; Cloetingh *et al.*, 2007; Braun *et al.*, 2008; Kaus *et al.*, 2008b).

(6) Realistic numerical modelling of magmatic processes. Coupling of modelling of magma conduit physics and volcanic processes (e.g. Melnik and Sparks, 1999; Melnik, 2000; Papale, 1999, 2001) with magma generation and ascent (e.g. Schmelling, 2000; Katz, 2008), intrusion emplacement (e.g. Burov *et al.*, 2003, Gerya and Burg 2007; Burg *et al.*, 2009; also see Fig. 17.9), magma chamber dynamics (e.g. Oldenburg *et al.*, 1990; Bagdassarov and Fradkov, 1993; Spera *et al.*, 1995; Simakin and Botcharnikov; 2001; Bergantz, 2000; Longo *et al.*, 2006; Ruprecht *et al.*, 2008) and related hydrothermal processes (e.g., Driesner *et al.*, 2006; Driesner and Geiger, 2007).

(7) Coupling of long-term and short-term poro-visco-elasto-plastic deformation processes. Developing numerical approaches for relating long-term geodynamic processes with faulting dynamics, fluid flows, rapture processes and seismicity (e.g., Miller *et al.*, 2004; Faccenda *et al.*, 2008a; Frehner *et al.*, 2008; Pergler and Matyska, 2008; Regenauer-Lieb and Yuen, 2008; Ben-Zion, 2008).

(8) Realistic modelling of the Earth formation processes: accretion, core formation, magma ocean development, onset of mantle convection. One of the numerical challenges is in coupling of planetary accretion (typically addressed with N-body simulations, e.g. Chambers and Wetherill, 1998; Chambers, 2001), giant impacts (modelled with hydrodynamic codes, e.g. Benz *et al.*, 1986; Canup and Asphaug, 2001; Canup, 2004; Wada *et al.*, 2006; Melosh, 2008, and analytical models, e.g. Senshu *et al.*, 2002) and core formation processes (addressed with continuum mechanics approaches, Honda

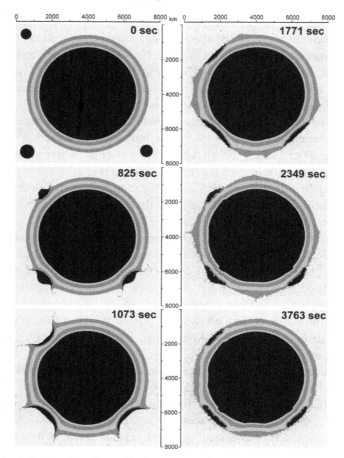

Fig. Outlook.2 Results of a preliminary numerical experiment (Gerya and Yuen, 2007) of a planetary accretion process associated with large impacts of meteorites (planetesimals) moving at relatively small speeds. The experiment is performed for a self-gravitating Mars-sized body with the use of a newly developed version of the visco-elasto-plastic code I2ELVIS (similar to the one we explored for core formation modelling in Chapter 17) which also takes inertial terms in the momentum equations into account (Chapter 5). The accretion process lasts for about one hour and is associated with large amount of ejects, large elastic waves propagating along the planetary surface and elasto-plastic deformation of the interior. Non compositional layering in the outer shell of the planet is used for visualising deformation. The grid resolution of the model is 161 × 161 nodes, 640 000 randomly distributed markers. The experiment was computed on the shared-distributed memory supercomputer BRUTUS at ETH-Zurich.

Fig. Outlook.3 Different 3D projections of thermal-chemical plumes (Fig. Outlook 1) growing atop of the subducting slab. The numerical model corresponds to a spontaneous retreating oceanic subduction (Fig. 17.2, 17.3) and explores the effects of mantle wedge hydration and melting in 3D (Zhu *et al.*, 2009). The temperature iso-surface of 1350 K is shown. Note that rising plumes are colder than the mantle wedge and move upward due to their compositional buoyancy (Gerya and Yuen, 2003b). Numerical experiments are performed with the code I3ELVIS (Gerya and Yuen, 2007) based on multigrid method (Chapters 14, 15) at the resolution $405 \times 101 \times 101$ nodes with 50 million markers. The experiment was computed on the shared-distributed memory supercomputer BRUTUS at ETH-Zurich.

et al., 1994; Golabek *et al.*, 2008a, b; Samuel and Tackley, 2008; Lin *et al.*, 2009; Ricard *et al.*, 2009; also see Fig. 17.14). A possible numerical solution of this challenging problem may be based on a combination of various types of codes into one accretion-impact-differentiation tool (Fig. Outlook.2).

3D visualisation challenges

As discussed in the introduction, modellers spend much, or even most of their time on visualising and understanding numerical models. Since these models are likely to become bigger and more complex in the future, the role of efficient visualisation technologies will be growing. Visualisation of large numerical models is a non-trivial task in 2D already (Rudolf *et al.*, 2004; Gorczyk *et al.*, 2006) since the amount of graphical information exceeds the resolutions of even the most powerful graphic screens by a factor of thousands (e.g. snapshots from 40 gigapixel database shown in Fig. Outlook.1). It is even more complicated with

3D models (e.g. Damon *et al.*, 2008; Kadlec *et al.*, 2008; Chen *et al.*, 2008) because the structure even for a single field cannot be seen at once, and requires processing of many views for proper understanding (Fig. Outlook.3, also see films **Cold_Plumes_1.mpeg** and **Cold_Plumes_2.mpeg** associated with this chapter). Future challenges in this respect are again quite obvious: geology-friendliness, ultrahigh resolution, multiple-scales, multiple-fields, interactive visualisation etc.

Conceptual warning

Discussion on the forthcoming technical advances above may give the impression that the only thing that we have to do is to write larger, more complex 3D codes which run on parallel supercomputers and have efficient ways of visualising results and compress the data. Obviously, this is only part of what needs to be done. More important is that we obtain an in-depth physical understanding of the dynamics of geological and planetary processes. This understanding should not only be based on the 'powerful numerics', but also on using *scaling laws* based on simplified theories and on comparison of predictions of such theories with the much more complex numerical simulations and with nature. If we find consistency, we have likely learned something essential, and we did not have to perform 2^{35} simulations to understand how nature works.

Conclusion

In conclusion: The future of numerical geodynamic modelling looks bright and there is a lot of exciting work and challenging research to do. Just go on!

It was fun to write all this. Thank you for reading.

Appendix

MATLAB program examples

The following resources can be found here:
www.cambridge.org/gerya

Introduction

Program 1: **Visualisation_is_important.m** (Exercise Introduction.2) – visualisation of 'sin' and 'cos' functions with 'plot', 'pcolor', 'contour' and 'surf'.

Chapter 1

Program 2: **Divergence.m** (Exercise 1.2) – computation and visualisation of velocity, divergence of velocity and time derivatives of density with 'pcolor' and 'quiver'.

Chapter 2

Program 3: **Periclase_EOS.m** (Exercise 2.2) – computation and visualisation of density, thermal expansion and compressibility for periclase (MgO) using external Gibbs free energy function *G_periclase.m*.

Program 4: **Density_map.m** (Exercise 2.3) – loading from data files (*m895_ro*, *morn_ro*) and visualising density maps for pyrolite (*m895_ro*) and MORB (*morn_ro*) and density difference between pyrolite and MORB.

Chapter 3

Program 5: **Poisson1D.m** (Exercise 3.1) – solution of 1D Poisson equation with finite differences on a regular grid using direct solver '\'.

Program 6: **Poisson2D_direct.m** (Exercise 3.2) – solution of 2D Poisson equation with finite differences on a regular grid using direct solver '\'.

Program 7: **Poisson2D_Gauss_Seidel.m** (Exercise 3.3) – solution of 2D Poisson equation with finite differences on a regular grid using Gauss–Seidel iteration.

Program 8: **Poisson2D_Jacobi.m** (Exercise 3.4) – solution of 2D Poisson equation with finite differences on a regular grid using Jacobi iteration.

Chapter 4

Program 9: **Strain_rate.m** (Exercise 4.2) – computation and visualisation of velocity field, strain rate, deviatoric strain rate, and second strain rate invariant.

Chapter 5

Program 10: **Streamfunction2D.m** (Exercise 5.2) – solution of 2D Stokes and continuity equations with finite differences on a regular grid using stream function – vorticity formulation for a medium with constant viscosity.

Chapter 6

Program 11: **Viscosity_profile.m** (Exercise 6.1) – computation and visualisation of viscosity profile across the lithosphere.

Program 12: **Viscosity_map.m** (Exercise 6.2) – computation and visualisation of viscosity map in temperature – log stress coordinates.

Program 13: **Viscosity_comparison.m** (Exercise 6.3) – computation and visualisation of viscosity maps in temperature – log stress coordinates for a combination of dislocation and diffusion creep; comparison of wet and dry olivine rheology.

Chapter 7

Program 14: **Stokes_continuity_constant_viscosity.m** (Exercise 7.1) – solution of 2D Stokes and continuity equations with finite differences on a regular grid using pressure–velocity formulation for a medium with constant viscosity.

Program 15: **Stokes_continuity_variable_viscosity.m** (Exercise 7.2) – solution of 2D Stokes and continuity equations with finite differences on a regular grid using pressure–velocity formulation for a medium with variable viscosity.

Chapter 8

Program 16: **Upwind_1D.m** (Exercise 8.1) – comparison of upwind, downwind and central differences for 1D advection of a square density wave in a constant velocity field.

Program 17: **FCT_1D.m** (Exercise 8.2) – using FCT algorithm for 1D advection of a square density wave in a constant velocity field.

Program 18: **Markers_1D.m** (Exercise 8.2) – using marker-in-cell algorithm with regular Eulerian grid for 1D advection of a square density wave in a constant velocity field.

Program 19: **Markers_1Dirregular.m** (Exercise 8.3) – using marker-in-cell algorithm with irregular Eulerian grid for 1D advection of a square density wave in a variable velocity field; using bisection algorithm.

Program 20: **Stokes_Continuity_Markers.m** (Exercise 8.4) – solution of 2D Stokes continuity and advection equations with finite-differences and marker-in-cell technique on a regular grid using pressure–velocity formulation for a medium with variable viscosity; use of the first-order accurate in space and time marker advection scheme.

Program 21: **Stokes_Continuity_Markers_Runge_Kutta.m** (Exercise 8.5) – solution of 2D Stokes continuity and advection equations with finite-differences and marker-in-cell technique on a regular grid using pressure–velocity formulation for a medium with variable viscosity; using of the fourth-order accurate in space first-order accurate in time Runge–Kutta marker advection scheme.

Chapter 9

Program 22: **Shear_heating.m** (Exercise 9.3) – solution of 2D Stokes and continuity equations with finite differences on a regular grid using pressure–velocity formulation for a medium with variable viscosity; computation and visualisation of shear heating distribution.

Program 23: **Shear_adiabatic_heating.m** (Exercise 9.4) – solution of 2D Stokes and continuity equations with finite differences on a regular grid using pressure–velocity formulation for a medium with variable viscosity; computation and visualisation of shear and adiabatic heating distribution.

Chapter 10

Program 24: **Explicit_implicit_1D.m** (Fig. 10.2) – solution of 1D temperature equation on a regular grid for a non-moving medium with constant conductivity; comparison of implicit and explicit method.

Program 25: **Explicit_Implicit2D.m** (Exercise 10.1) – solution of 2D temperature equation on a regular grid for a non-moving medium with constant conductivity; comparison of implicit and explicit method.

Program 26: **Variable_conductivity.m** (Exercise 10.2) – solution of 2D temperature equation on a regular grid for a non-moving medium with variable conductivity; comparison of implicit and explicit method.

Program 27: **Conduction_advection2D.m** (Exercise 10.3) – solution of 2D Eulerian temperature equation with advective terms on a regular grid for a moving medium with constant conductivity; use of upwind differences for advection of temperature; comparison of implicit and explicit method.

Program 28: **Variable_conductivity_advection2D.m** (Exercise 10.3) – solution of 2D Eulerian temperature equation with advective terms on a regular grid for a moving medium with variable conductivity; use of upwind differences for advection of temperature; comparison of implicit and explicit method.

Program 29: **Variable_conductivity_markers2D.m** (Exercise 10.4) – solution of 2D Lagrangian temperature equation on a regular grid with implicit finite differences for a moving medium with variable conductivity; use of marker-in-cell approach for advection of temperature.

Chapter 11

Program 30: **i2vis.m** (Exercise 11.1) – 2D thermomechanical viscous code; solution of 2D Stokes, continuity, temperature and advection equations with finite-differences and marker-in-cell technique on a regular grid using pressure–velocity formulation for a deforming incompressible medium with variable viscosity and thermal conductivity; taking into account radiogenic, shear and adiabatic heating.

Chapter 12

Program 31: **Viscoelastic_stress.m** (Exercise 12.2) – computation of visco-elastic stress build-up/relaxation with time.

Program 32: **Viscoelastoplastic_strain_rate.m** (Exercise 12.3) – computation of visco-elasto-plastic stress build-up and associated viscous, elastic and plastic strain rate evolution with time.

Program 33: **Peierls_creep.m** (Exercise 12.4) – computation and visualisation of viscosity maps in temperature – log stress coordinates for a combination of dislocation, diffusion and Peierls creep; comparison of wet and dry olivine rheology.

Chapter 13

Program 34: **Viscoelastic2D.m** (Exercise 13.1) – 2D thermomechanical visco-elastic code; solution of 2D Stokes, continuity, temperature and advection equations with finite-differences and marker-in-cell technique on a regular grid using pressure–velocity formulation for a deforming incompressible medium with variable viscosity, shear modulus and thermal conductivity; taking into account radiogenic, shear and adiabatic heating.

Program 35: **i2elvis.m** (Exercise 13.2) – 2D thermomechanical visco-elasto-plastic code; solution of 2D Stokes, continuity, temperature and advection equations with finite-differences and marker-in-cell technique on a regular grid using pressure–velocity formulation for a deforming incompressible medium with variable viscosity, shear modulus, plastic strength and thermal conductivity; taking into account radiogenic, shear and adiabatic heating.

Chapter 14

Program 36: **Gauss_Seidel_iterations_Poisson.m** (Figs. 14.1, 14.2) – solution of 2D Poisson equation with Gauss–Seidel iteration.

Program 37: **Poisson_Multigrid.m** (Fig. 14.5) – solution of 2D Poisson equation with multigrid based on V-cycle and external functions *Poisson_smoother.m, Poisson_restriction.m, Poisson_prolongation.m*; resolution between multigrid levels changes by factor of two.

Program 38: **Poisson_Multigrid_planet_arbitrary.m** – solution of 2D Poisson equation for the case of a circular planetary body embedded in a mass-less like medium with multigrid based on V-cycle and external functions *Poisson_smoother_planet.m, Poisson_restriction_planet.m, Poisson_prolongation_planet.m*; resolution between multigrid levels changes in an arbitrary way.

Program 39: **Stokes_Continuity_Multigrid.m** – solution of 2D Stokes and continuity equations for a constant viscosity medium with multigrid based on V-cycle and external functions *Stokes_Continuity_smoother.m, Stokes_Continuity_ restriction.m, Stokes_Continuity_prolongation.m*; resolution between multigrid levels changes by factor of two.

Program 40: **Variable_viscosity_MultiMultigrid_arbitrary.m** (Fig. 14.13) – solution of 2D Stokes and continuity equations for a variable viscosity medium with multi-multigrid based on V-cycle and external functions *Viscosity_restriction.m, Stokes_Continuity_viscous_smoother.m*.

Program 41: **Poisson_Multigrid_planet.m** (Exercise 14.1) – solution of 2D Poisson equation for the case of a circular planetary body embedded in a mass less-like medium with multigrid based on V-cycle and external functions *Poisson_smoother_ planet.m, Poisson_restriction_planet.m, Poisson_prolongation_planet.m*; resolution between multigrid levels changes by factor of two.

Program 42: **Constant_Viscosity_Multigrid_ghost.m** (Exercise 14.2, Fig. 14.10) – solution of 2D Stokes and continuity equations for a constant viscosity medium with multigrid based on V-cycle, ghost-node-based smoother ***Stokes_Continuity_ smoother_ghost.m*** and external functions ***Stokes_Continuity_restriction.m, Stokes_Continuity_prolongation.m***; resolution between multigrid levels changes by factor of two.

Program 43: **Variable_viscosity_Multigrid_arbitrary.m** (Exercise 14.3, Fig 14.12) – solution of 2D Stokes and continuity equations for a variable viscosity medium with multigrid based on V-cycle and external functions ***Viscosity_restriction.m, Stokes_Continuity_viscous_smoother.m, Stokes_Continuity_viscous_restriction.m, Stokes_Continuity_prolongation.m***; resolution between multigrid levels changes in an arbitrary way.

Chapter 15

Program 44: **Temperature3D_Gauss_Seidel.m** (Exercise 15.1, Fig. 15.9) – solution of 3D temperature equation on a regular grid for a non-moving medium with variable conductivity; the solution is based on Gauss–Seidel iteration with the use of external function ***Temperature3D_smoother.m.***

Program 45: **Poisson3D_Multigrid_planet_arbitrary.m** (Exercise 15.2, Fig. 15.10) – solution of 3D Poisson equation for the case of a spherical planetary body embedded in a mass-less like medium with multigrid based on V-cycle and external functions ***Poisson3D_smoother_planet.m, Poisson3D_restriction_planet.m, Poisson3D_prolongation_planet.m***; resolution between multigrid levels changes in an arbitrary way.

Program 46: **Stokes_Continuity3D_Multigrid.m** (Exercise 15.3, Fig. 15.11a) – solution of 3D Stokes and continuity equations for a constant viscosity medium with multigrid based on V-cycle and external functions ***Stokes_Continuity3D_smoother.m, Stokes_Continuity3D_restriction.m, Stokes_Continuity3D_prolongation.m***; resolution between multigrid levels changes by factor of two.

Program 47: **Variable_viscosity3D_Multigrid.m** (Exercise 15.3, Fig. 15.11b) – solution of 3D Stokes and continuity equations for a variable viscosity medium with multigrid based on V-cycle and external functions ***Viscosity_restriction3D.m, Stokes_Continuity3D_viscous_smoother.m, Stokes_Continuity3D_viscous_ restriction.m, Stokes_Continuity3D_prolongation.m***; resolution between multigrid levels changes by factor of two.

Program 48: **Variable_viscosity3D_MultiMultigrid.m** (Fig. 15.12) – solution of 3D Stokes and continuity equations for a variable viscosity medium with multi-multigrid based on V-cycle and external functions ***Viscosity_restriction3D.m, Stokes_Continuity3D_viscous_smoother.m, Stokes_Continuity3D_viscous_ restriction.m, Stokes_Continuity3D_prolongation.m***; resolution between multigrid levels changes by factor of two.

Chapter 16

Program 49: **Variable_viscosity_Ramberg.m** (Fig. 16.1) – mechanical benchmark for a two-layer Rayleigh–Taylor problem; solution of 2D Stokes, continuity and advection equations with finite-differences and marker-in-cell technique using external function *Stokes_Continuity_solver_ghost.m*.

Program 50: **Variable_viscosity_block.m** (Fig. 16.3, Exercise 16.1) – mechanical benchmark for a falling square block; solution of 2D Stokes, continuity and advection equations with finite-differences and marker-in-cell technique using external function *Stokes_Continuity_solver_ghost.m*.

Program 51: **Variable_viscosity_channel.m** (Fig. 16.4) – mechanical benchmark for a channel flow with a non-Newtonian rheology; solution of 2D Stokes, continuity and advection equations with finite-differences and marker-in-cell technique using external function *Stokes_Continuity_solver_channel.m*.

Program 52: **Constant_viscosity_channel_T.m** (Fig. 16.5) – thermomechanical benchmark for a non-steady temperature distribution in a Newtonian channel; solution of 2D Stokes, continuity, temperature and advection equations with finite-differences and marker-in-cell technique using external functions *Stokes_Continuity_solver_channel.m, Temperature_solver.m*.

Program 53: **Variable_viscosity_Couette_T.m** (Fig. 16.6) – thermomechanical benchmark for a steady Couette flow with viscous heating and temperature-dependent viscosity; solution of 2D Stokes, continuity, temperature and advection equations with finite-differences and marker-in-cell technique using external functions *Stokes_Continuity_solver_Couette.m, Temperature_solver.m*.

Program 54: **Solid_Body_Rotation_T.m** (Fig. 16.7) – thermal benchmark for advection and diffusion of sharp temperature fronts in a prescribed rigid-body rotation velocity field; solution of 2D temperature and advection equations with finite-differences and marker-in-cell technique using external function *Temperature_solver.m*.

Program 55: **Variable_conductivity_channel.m** (Fig. 16.8) – thermomechanical benchmark for a steady Newtonian channel flow with variable thermal conductivity; solution of 2D Stokes, continuity, temperature and advection equations with finite-differences and marker-in-cell technique using external functions *Stokes_Continuity_solver_Couette.m, Temperature_solver.m*.

Program 56: **Variable_viscosity_convection_irregular_grid.m** (Figs. 16.9, 16.10) – thermomechanical benchmark for thermal convection with constant and temperature- and depth-dependent viscosity; solution of 2D Stokes, continuity, temperature and advection equations with finite-differences and marker-in-cell technique on regular/irregular grid using external functions *Stokes_Continuity_solver_grid.m, Temperature_solver_grid.m*; nearly steady-state temperature distribution for 1a, 1c and 2a cases can be loaded from data files *data_1a_regular.txt, data_1c_regular.txt, data_2a_regular.txt, data_1a_irregular.txt, data_1c_irregular.txt, data_2a_irregular.txt*.

Program 57: **Stress_buildup.m** (Fig. 16.11) – mechanical benchmark for stress build-up in a visco-elastic incompressible Maxwell body; solution of 2D Stokes, continuity and advection equations with finite-differences and marker-in-cell technique using external function *Stokes_Continuity_solver_grid.m*.

Program 58: **Slab_deformation.m** (Fig. 16.12, Exercise 16.2) – mechanical benchmark for recovery of the original shape of an elastic slab; solution of 2D Stokes, continuity and advection equations with finite-differences and marker-in-cell technique using external function *Stokes_Continuity_solver_grid.m*.

Program 59: **Sandbox_shortening_ratio.m** (Fig. 16.14) – mechanical visco-elasto-plastic benchmark for numerical sandbox shortening experiment; solution of 2D Stokes, continuity and advection equations with finite-differences and

marker-in-cell technique using external function ***Stokes_Continuity_solver_
sandbox.m***.

Chapter 17

Program 60: **Subducting_slab_bending.m** (Fig. 17.1) – thermomechanical
visco-elasto-plastic numerical model for spontaneous bending of subducting oceanic
slab; the model uses external functions ***Stokes_Continuity_solver_sandbox.m,
Temperature_solver_grid.m***.

Program 61: **Subduction.m** (Figs. 17.2, 17.3) – thermomechanical visco-elasto-plastic
numerical model for spontaneous retreating oceanic subduction; the model uses
external functions ***Stokes_Continuity_solver_sandbox.m, Temperature_
solver_grid.m***.

Program 62: **Extension.m** (Fig. 17.4) – thermomechanical visco-elasto-plastic numerical
model for oceanic lithosphere extension; the model uses external functions
Stokes_Continuity_solver_sandbox.m, Temperature_solver_grid.m.

Program 63: **Collision.m** (Figs. 17.5, 17.6) – thermomechanical visco-elasto-plastic
numerical model for post-subduction continental collision; the model accounts for
erosion/sedimentation processes and uses external functions
Stokes_Continuity_solver_sandbox.m, Temperature_solver_grid.m.

Program 64: **Collision_and_breakoff.m** (Fig. 17.7) – thermomechanical
visco-elasto-plastic numerical model for slab breakoff during continental collision;
the model accounts for erosion/sedimentation processes and uses external functions
Stokes_Continuity_solver_sandbox.m, Temperature_solver_grid.m.

Program 65: **Intrusion_emplacement.m** (Fig. 17.9) – thermomechanical
visco-elasto-plastic numerical model for trans-lithospheric mafic-ultramafic intrusion
emplacement into the crust; equilibrium melt fraction for different rocks is computed
with external function ***Melt_fraction.m***; the model also accounts for
erosion/sedimentation processes and uses external functions
Stokes_Continuity_solver_sandbox.m, Temperature_solver_grid.m.

Program 66: **Mantle_convection.m** (Fig. 17.12, 17.13) – thermomechanical
visco-elasto-plastic numerical model for mantle convection with phase changes; the
phase changes are treated based on Gibbs free energy minimisation approach with
pre-computed density and enthalpy maps in *P–T* space; these maps are loaded with
external function ***loading_database.m*** from data files ***m895_ro, m895_hh, morn_ro,
morn_hh***; pre-computed non-steady temperature distribution can be loaded from
data file ***convection.txt;*** the model also uses external functions
Stokes_Continuity_solver_sandbox.m, Temperature_solver_grid.m.

Program 67: **Core_formation.m** (Fig. 17.14) – thermomechanical visco-elasto-plastic
numerical model for the deformation of a self-gravitating iron–silicate planetary
body; gravity field is computed with external function ***Poisson_solver_
planet_grid.m***; phase changes in the silicate component are treated based on Gibbs
free energy minimization approach with pre-computed density and enthalpy maps in
P–T space; these maps are loaded with external function ***loading_database.m*** from
data files ***m895_ro, m895_hh, morn_ro, morn_hh***; the model also uses external
functions ***Stokes_Continuity_solver_sandbox.m, Temperature_solver_grid.m***.

References

Albers, M. (2000) A local mesh refinement multigrid method for 3D convection problems with strongly variable viscosity. *Journal of Computational Physics*, **160**, 126–50.

Amestoy, P., Duff, I., Koster, J. and L'Excellent, A. (2001) A fully asynchronous multifrontal solver using distributed dynamic scheduling. *SIAM Journal on Matrix Analysis and Applications*, **23** (1), 15–41.

Anderson, O. L. (1995) *Equations of State for Solids in Geophysics and Ceramic Science*. Oxford University Press.

Andrews, E. R. and Billen, M. I. (2009) Rheologic controls on the dynamics of slab detachment, *Tectonophysics*, **464**, 60–9.

Asphaug, E., Agnor, C. B. and Williams, Q. (2006) Hit-and-run planetary collisions, *Nature*, **439**, 155–60, doi: 10.1038/nature04311.

Bagdassarov, N. S. and Fradkov, A. S. (1993) Evolution of double diffusion convection in a felsic magma chamber. *Journal of Volcanology and Geothermal Research*, **54**, (3–4), 291–308.

Barazangi, M., Isacks, B. L., Oliver, J., Dubois, J. and Pascal, G. (1973) Descent of lithosphere beneath New Hebrides, Tonga–Fiji and New Zealand: evidence for detached slabs. *Nature*, **242** (5393), 98–101.

Baumann, C., Gerya, T. V. and Connolly, J. A. D. (2009) Numerical modelling of spontaneous slab breakoff dynamics during continental collision. In *Advances in Interpretation of Geological Processes: Refinement of Multi-scale Data and Integration in Numerical Modelling*. Geological Society of London Special Publication, (in press).

Baumgardner, J. R. (1985) Three-dimensional treatment of convective flow in the Earth's mantle. *Journal of Statistical Physics*, **39**, 501–11.

Beaumont, C., Jamieson, R. A., Nguyen, M. H. and Lee, B. (2001) Himalayan tectonics explained by extrusion of a low-viscosity crustal channel coupled to focused surface denudation. *Nature*, **414**, 738–42.

Belytschko, T., Liu, W. K. and Moran, B. (2000) *Nonlinear Finite Elements for Continua and Structures*. John Wiley & Sons.

Benz, W., Slattery, W. and Cameron, A. G. W. (1986) The origin of the moon and the single-impact hypothesis.1. *Icarus*, **66**, 515.

Ben-Zion, Y. (2008) Collective behavior of earthquakes and faults: continuum-discrete transitions, evolutionary changes and corresponding dynamic regimes. *Review of Geophysics*, **46**, RG4006, doi:10.1029/2008RG000260.

Bercovici, D. (ed.) (2007) *Mantle Dynamics. Treatise on Geophysics, Volume 7,* (editor-in-chief: Gerald Schubert), Elsevier.

Berganrz, G. W. (2000) On the dynamics of magma mixing by reintrusion: implications for pluton assembly processes. *Journal of Structural Geology,* **22,** 1297–309.

Berman, R. G. (1988) Internally-consistent thermodynamic data for minerals in the system Na2O-K2O-CaO-MgO-FeO-Fe2O3-Al2O3-SiO2-TiO2-H2O-CO2. *Journal of Petrology,* **29,** 445–522.

Berner, H., Ramberg, H. and Stephanson, O. (1972) Diapirism in theory and experiment. *Tectonophysics,* **15,** 197–218.

Best, M. G. and Christiansen, E. H. (2001) *Igneous Petrology.* Blackwell Science.

Birch, F. 1947. Finite elastic strain of cubic crystals. *Physical Review,* **71,** 809–24.

Bird, P. (1978) Finite elements modeling of lithosphere deformation: The Zagros collision orogeny, *Tectonophysics,* **50,** 307–36.

Bittner, D. and Schmeling, H. (1995) Numerical modeling of melting processes and induced diapirism in the lower crust. *Geophysical Journal International,* **123,** 59–70.

Blankenbach, B., Busse, F., Christensen, U., *et al.* (1989) A benchmark comparison for mantle convection codes, *Geophysical Journal International,* **98** (1), 23–38.

Boris, J. P. and Book, D. L. (1973) Flux-Corrected Transport. I. SHASTA, A Fluid transport algorithm that works, *Journal of Computational Physics,* **11,** 38–69.

Brace, W. F., Kohlstedt, D. T. (1980) Limits on lithospheric stress imposed by laboratory experiments. *Journal of Geophysical Research,* **85,** 6248–52.

Braun, J. and Sambridge, M. (1997) Modelling landscape evolution on geological time scales: a new method based on irregular spatial discretization: *Basin Research,* **9,** 27–52.

Braun, J., Thieulot, C., Fullsack, P., *et al.* (2008) DOUAR: A new three-dimensional creeping flow numerical model for the solution of geological problems. *Physics of the Earth and Planetary Interiors,* **171,** 76–91.

Buiter, S. J. H., Govers, R. and Wortel, M. J. R. (2002) Two-dimensional simulations of surface deformation caused by slab detachment. *Tectonophysics,* **354,** 195–210.

Buiter, S. J. H., Babeyko, A., Yu., Ellis, S., *et al.* (2006) The numerical sandbox: Comparison of model results for a shortening and an extension experiment. In Buiter, S. J. H. and Schreurs, G. (eds.) 2006. *Analogue and Numerical Modelling of Crustal-Scale Processes.* Geological Society, London, Special Publications, **253,** 29–64.

Burg, J.-P. and Gerya, T. V. (2005) The role of viscous heating in Barrovian metamorphism of collisional orogens: thermomechanical models and application to the Lepontine Dome in the Central Alps. *Journal of Metaphorphic Geology,* **23,** 75–95.

Burg, J.-P., Bodinier, J.-L., Gerya, T., *et al.* (2009) Translithospheric mantle diapirism: geological evidence and numerical modelling of the Kondyor zoned ultramafic complex (Russian Far-East). *Journal of Petrology,* **50,** 289–321.

Burov, E. B. and Cloetingh, S. (1997) Erosion and rift dynamics: new thermomechanical aspects of post-rift evolution of extensional basins. *Earth and Planetary Science Letters,* **150,** 7–26.

Burov, E. and Poliakov, A. (2001) Erosion and rheology controls on synrift and postrift evolution: Verifying old and new ideas using a fully coupled numerical model. *Journal of Geophysical Research–Solid Earth,* **106** (B8), 16461–81.

Burov, E., Jolivet, L., Le Pourhiet, L. and Poliakov, A. (2001) A thermomechanical model of exhumation of high pressure (HP) and ultra-high pressure (UHP) metamorphic rocks in Alpine-type collision belts, *Tectonophysics,* **342,** 113–36.

Burov, E., Jaupart, C. and Guillou-Frottier, L. (2003) Ascent and emplacement of buoyant magma bodies in brittle-ductile upper crust. *Journal of Geophysical Research–Solid Earth*, **108**, Article Number: 2177.

Busse, F. H., Christensen, U., Clever, R., *et al.* (1994) 3-D convection at infinite Prandtl number in Cartesian geometry – a benchmark comparison. *Geophysical and Astrophysical Fluid Dynamics*, **75**, 39–59.

Byerlee, J. D. (1978) Friction of rocks. *Pure Applied Geophysics*, **116**, 615–26.

Canup, R. M. (2004) Simulations of a late lunar-forming impact. *Icarus*, **168**, 433–56.

Canup, R. M. and Asphaug, E. (2001) Origin of the Moon in a giant impact near the end of the Earth's formation. *Nature*, **412**, 708.

Caricchi, L., Burlini, L., Ulmer, P., *et al.* (2007) Non-Newtonian rheology of crystal-bearing magmas and implications for magma ascent dynamics. *Earth and Planetary Science Letters*, **264**, 402–19.

Carslaw, H. S. and Jaeger, J. C. (1986) *Conduction of Heat in Solids*. Oxford University Press.

Castro, A. and Gerya, T. V. (2007) Magmatic implications of mantle wedge plumes: Experimental study. *Lithos*, **103**, 138–48.

Chambers, J. E. (2001) Making More Terrestrial Planets. *Icarus*, **152**, 205–24.

Chambers, J. E. and Wetherill, G. W. (1998) Making the terrestrial planets: N-body integrations of planetary embryos in three dimensions. *Icarus*, **136**, 304–27.

Chapman, B., Jost, G. and Van Der Pas, R. (2007) *Using OpenMP: Portable Shared Memory Parallel Programming*. The MIT Press.

Chemenda, A. I., Burg, J.-P. and Mattauer, M. (2000) Evolutionary model of the Himalaya-Tibet system: geopoem based on new modelling, geological and geophysical data. *Earth and Planetary Science Letters*, **174**, 397–409.

Chen, S., Zhang, H., Yuen, D., Zhang, S., Zhang, J., Shi, Y. (2008) Volume rendering visualization of 3D spherical mantle convection with an unstructured mesh. *Visual Geosciences*, **13**, 97–104.

Chopin, C. (1984) Coesite and pure pyrope in high-grade blueschists of the Western Alps: A first record and some consequences. *Contributions to Mineralogy and Petrology*, **86**, 107–18.

Chopin, C. (2003) Ultrahigh-pressure metamorphism: tracing continental crust into mantle, *Earth and Planetary Science Letters*, **212**, 1–14.

Christensen, U. (1982) Phase boundaries in finite amplitude mantle convection. *Geophysical Journal of the Royal Astronomical Society*, **68**, 487–97.

Christensen, U. R. and Yuen, D. A. (1985) Layered convection induced by phase changes. *Journal of Geophysical Research*, **90**, 10291–300.

Chung, W.-Y. and Kanamori, H. (1976) Source process and tectonic implications of the Spanish deep-focus earthquake of March 29, 1954. *Physics of the Earth and Planetary Interiors*, **13** (2), 85–96.

Clauser, C. and Huenges, E. (1995) Thermal conductivity of rocks and minerals. In *Rock Physics and Phase Relations*. AGU Reference Shelf 3. (ed. Ahrens, T. J.), American Geophysical Union, pp. 105–26.

Clemens, J. D. and Mawer, C. K. (1992) Granitic magma transport by fracture propagation. *Tectonophysics*, **204**, 339–60.

Cloetingh, S. A. P. L., Ziegler, P. A., Bogaard, P. J. F., *et al.* (2007) TOPO-EUROPE: The geoscience of coupled deep Earth-surface processes. *Global and Planetary Change*, **58**, 1–118.

Cloos, M. (1982) Flow melanges – numerical modeling and geologic constraints on their origin in the Franciscan subduction complex, California. *Geological Society of America Bulletin*, **93**, 330–45.

Cohen, R. E. (ed.) (2005) *High-Performance Computing Requirements for the Computational Solid Earth Sciences.* www.geo-prose.com/computational_SES.html

Connolly, J. A. D. (2005) Computation of phase equilibria by linear programming: a tool for geodynamic modeling and an application to subduction zone decarbonation. *Earth and Planetary Science Letters*, **236**, 524–41.

Connolly, J. A. D. and Kerrick, D. M. (1987) An algorithm and computer program for calculating composition phase diagrams, CALPHAD, **11**, 1–55.

Connolly, J. A. D. and Podladchikov, Y. Y. (1998) Compaction-driven fluid flow in viscoelastic rock. *Geodinamica Acta*, **11**, 55–84.

Cserepes, L., Rabinowicz, M. and Rosemberg-Borot, C. (1988) Three-dimensional infinite Prandtl number convection in one and two layers and implications for the Earth's gravity field. *Journal of Geophysical Research*, **93**, 12009–25.

Dabrowski, M., Krotkiewski, M. and Schmid, D. W. (2008) MILAMIN: MATLAB-based finite element method solver for large problems. *Geochemistry, Geophysics, and Geosystems*, **9**, Q04030, doi:10.1029/2007GC001719.

Daignières, M., Fremond, M. and Friaa, A. (1978) Modèle de type Norton-Hoff généralisé pour l'étude des déformations lithosphériques (exemple: la collision Himalayenne), *Comptes Reudus Hebdomadaires des Séances de l'Academie de Sciences*, **268B**, 371–74.

Damon, M., Kameyama, M. C., Knox, M., *et al.* (2008) Interactive visualization of 3D mantle convection. *Visual Geosciences*, **13**, 49–57.

Davies, G. F. (1985) Heat Deposition and Retention in a Solid Planet Growing by Impacts, *Icarus*, **63**, 45–68.

Davies, J. H. and Von Blanckenburg, F. (1995) Slab breakoff: a model of lithospheric detachment and its test in the magmatism and deformation of collisional orogens. *Earth and Planetary Science Letters*, **129**, 85–102.

de Capitani, C. and Brown, H. (1987) The computation of chemical equilibrium in complex systems containing non-ideal solid solutions. *Geochimica et Cosmochimica Acta*, **51**, 2639–52.

DePaolo, D. J., Cerling, T. E., Hemming, S. R., *et al.* (2008) *Origin and Evolution of Earth: Research Questions for a Changing Planet.* Committee on Grand Research Questions in the Solid-Earth Sciences, Board on Earth Sciences and Resources, Division on Earth and Life Studies, National Research Council of the National Academies, The National Academies Press, Washington, DC.

Deubelbeiss, Y. and Kaus, B. J. P. (2008) Comparison of Eulerian and Lagrangian numerical techniques for the Stokes equations in the presence of strongly varying viscosity. *Physics of the Earth and Planetary Interiors*, **171**, 92–111.

Dobrzhinetskaya, L. F., Eide, E. A., Larsen, R. B., *et al.* (1995) Microdiamond in high-grade metamorphic rocks of the Western Gneiss Region, Norway. *Geology*, **2**, 597–600.

Dorogokupets, P. I. and Karpov, I. K. (1984) *Thermodynamics of Minerals and Mineral Equilibria.* Nauka (in Russian).

Driesner, T. and Geiger S. (2007) Numerical simulation of multiphase fluid flow in hydrothermal systems. *Fluid-Fluid Interactions*, **65**, 187–212.

Driesner, T., Geiger S. and Heinrich C. A. (2006) Modeling multiphase flow of H2O-NaCl fluids by combining CSP5.0 with SoWat2.0. *Geochimica et Cosmochimica Acta*, **70** (18), A147–A147.

Ellenberger, F. (1994) *Histoire de la Géologie. La grande éclosion et ses prémices.* Petite collection d'histoire des sciences, 2. Technique et Documentation (Lavoisier), Paris.

Elsasser, W. M. (1963) Early history of the Earth. In Geiss, J. and Goldberg, E. (eds.) *Earth Science and Meteoritics.* North-Holland, pp. 1–30.

Evans, B. and Goetze, C. (1979) The temperature variation of hardness of olivine and its implication for polycrystalline yield stress. *Journal of Geophysical Research*, **84**, 5505–24.

Faccenda, M., Burlini, L., Gerya, T. V. and Mainprice, D. (2008a) Fault-induced seismic anisotropy by hydration in subducting oceanic plates. *Nature*, **455**, 1097–101.

Faccenda, M., Gerya, T. V. and Chakraborty, S. (2008b) Styles of post-subduction collisional orogeny: Influence of convergence velocity, crustal rheology and radiogenic heat production. *Lithos*, **103**, 257–87.

Faccenna, C., Bellier, O., Martinod, J., Piromallo, C. and Regard, V. (2006) Slab detachment beneath eastern Anatolia: a possible cause for the formation of the North Anatolian fault. *Earth and Planetary Science Letters*, **242**, 85–97.

Fedorenko, R. P. (1964) The speed of convergence of one iterative process. USSR *Journal of Computational Mathematics and Mathematical Physics*, **4** (3), 227–35.

Fornberg, B. (1995) *A Practical Guide to Pseudospectral Methods*. Cambridge University Press.

Frehner, M., Schmalholz, S. M., Saenger, E. H. and Steeb, H. (2008) Comparison of finite difference and finite element methods for simulating two-dimensional scattering of elastic waves. *Physics of the Earth and Planetary Interiors*, **171**, 112–21.

Fuchs, K., Bonjer, K.-P., Bock, G., *et al.* (1979) The Romanian earthquake of March 4, 1977: II, aftershocks and migration of seismic activity. *Tectonophysics*, **53** (3–4), 225–47.

Gerya, T. V. and Burg, J.-P. (2007) Intrusion of ultramafic magmatic bodies into the continental crust: Numerical simulation. *Physics of the Earth and Planetary Interiors*, **160**, 124–42.

Gerya, T. V. and Maresch, W. V. (2004) Metapelites of the Kanskiy granulite complex, (Eastern Siberia): kinked P-T paths and geodynamic model. *Journal of Petrology*, **45**, 1393–412.

Gerya, T. V. and Stoeckhert, B. (2006) 2-D numerical modeling of tectonic and metamorphic histories at active continental margins. *International Journal of Earth Sciences*, **95**, 250–74.

Gerya, T. V. and Yuen, D. A. (2003a) Characteristics-based marker-in-cell method with conservative finite-differences schemes for modeling geological flows with strongly variable transport properties. *Physics of the Earth and Planetary Interiors*, **140**, 293–318.

Gerya, T. V. and Yuen, D. A. (2003b) Rayleigh–Taylor instabilities from hydration and melting propel cold plumes at subduction zones. *Earth and Planetary Science Letters*, **212**, 47–62.

Gerya, T. V. and Yuen, D. A. (2007) Robust characteristics method for modelling multiphase visco-elasto-plastic thermo-mechanical problems. *Physics of the Earth and Planetary Interiors*, **163**, 83–105.

Gerya, T. V., Perchuk, L. L., Van Reenen, D. D. and Smit, C. A. (2000) Two-dimensional numerical modeling of pressure-temperature-time paths for the exhumation of some granulite facies terrains in the Precambrian. *Journal of Geodynamics*, **29**, 17–35.

Gerya, T. V., Maresch, W. V., Willner, A. P., Van Reenen, D. D. and Smit, C. A. (2001) Inherent gravitational instability of thickened continental crust with regionally developed low- to medium-pressure granulite facies metamorphism. *Earth and Planetary Science Letters*, **190**, 221–35.

Gerya, T. V., Stoeckhert, B. and Perchuk, A. L. (2002) Exhumation of high-pressure metamorphic rocks in a subduction channel – a numerical simulation. *Tectonics*, **21**, Article Number: 1056.

Gerya, T. V., Yuen, D. A. and Maresch, W. V. (2004a) Thermomechanical modeling of slab detachment. *Earth and Planetary Science Letters*, **226**, 101–116.

Gerya, T. V., Yuen, D. A. and Sevre, E. O. D. (2004b) Dynamical causes for incipient magma chambers above slabs. *Geology*, **32**, 89–92.

Gerya, T. V., Perchuk, L. L., Maresch, W. V. and Willner, A. P. (2004c) Inherent gravitational instability of hot continental crust: implication for doming and diapirism in granulite facies terrains. In *Gneiss Domes in Orogeny*, edited by D. Whitney and C. Teyssier and C.S. Siddoway, GSA Special Paper **380**, 97–115.

Gerya, T. V., Podlesskii, K. K., Perchuk, L. L. and Maresch, W. V. (2004d) Semi-empirical Gibbs free energy formulations for minerals and fluids. *Physics and Chemistry of Minerals*, **31** (7), 429–55.

Gerya, T. V., Connolly, J. A. D., Yuen, D. A., Gorczyk, W. and Capel, A. M. (2006) Sesmic implications of mantle wedge plumes. *Physics of the Earth and Planetary Interiors*, **156**, 59–74.

Gerya, T. V., Connolly, J. A. D. and Yuen, D. A. (2008a) Why is terrestrial subduction one-sided? *Geology*, **36**, 43–6.

Gerya, T. V., Perchuk, L. L. and Burg, J.-P. (2008b) Transient hot channels: Perpetrating and regurgitating ultrahigh-pressure, high-temperature crust-mantle associations in collision belts. *Lithos*, **103**, 236–56.

Golabek, G. J., Gerya, T. V. and Tackley, P. J. (2008a) Rheological controls on the terrestrial core formation mechanism. *European Planetary Science Congress Abstracts*, **3**, EPSC2008-A-00 087.

Golabek, G. J., Schmeling, H. and Tackley, P. J. (2008b) Earth's core formation aided by flow channelling instabilities induced by iron diapirs. *Earth and Planetary Science Letters*, **271**, 24–33.

Gorczyk, W., Gerya, T. V., Connolly, J. A. D., Yuen, D. A. and Rudolph, M. (2006) Large-scale rigid-body rotation in the mantle wedge and its implications for seismic tomography. *Geochemistry, Geophysics, and Geosystems*, **7**, doi:10.1029/2005GC001075.

Gorczyk, W., Guillot, S., Gerya, T. V. and Hattori, K. (2007a) Asthenospheric upwelling, oceanic slab retreat and exhumation of UHP mantle rocks: insights from Greater Antilles. *Geophysical Research Letters*, **34**, Article Number: L21309.

Gorczyk, W., Gerya, T. V., Connolly, J. A. D. and Yuen, D. A. (2007b) Growth and mixing dynamics of mantle wedge plumes. *Geology*, **35**, 587–90.

Gupta, A. (2000) WSMP: Watson Sparse Matrix Package (Part-II: direct solution of general sparse systems). Technical Report RC 21888 (98472), IBM T.J. Watson Research Center, Yorktown Heights, NY.

Gustafsson, B. (2008) *High Order Finite-Difference Methods for Time-dependent PDE*, Springer-Verlag.

Hager, B. H., O'Connell, R. J. (1981) A simple global model of plate dynamics and mantle convection. *Journal of Geophysical Research*, **86**, 4843–67.

Hall, C. E., Gurnis, M., Sdrolias, M., Lavier, L. L. and Muller, R. D. (2003) Catastrophic initiation of subduction following forced convergence across fractures zones. *Earth and Planetary Science Letters*, **212**, 15–30.

Hansen, U., Yuen, D. A. (1988) Numerical simulations of thermal-chemical instabilities at the core–mantle boundary. *Nature*, **334**, 237–40.

Helgeson, H. C., Delany, J. M., Nesbitt, H. W. and Bird, D. K. (1978) Summary and critique of the thermodynamic properties of rock-forming minerals. *American Journal of Science*, **278A**.

Hess, P. C. (1989) *Origin of Igneous Rocks*. Harvard University Press.

Hirschmann, M. M. (2000) Mantle solidus: Experimental constraints and the effects of peridotite composition. *Geochemistry, Geophysics, and Geosystems*, **1** (10), 1042, doi:10.1029/2000GC000070.

Hofmeister, A. M. (1999) Mantle values of thermal conductivity and the geotherm from phonon lifetimes. *Science*, **283**, 1699–706.

Holland, T. J. B. and Powell, R. (1990) An enlarged and updated internally consistent thermodynamic data set with uncertainties and correlations: the system K2O–Na2O–CaO–MgO–FeO–Fe2O3–Al2O3–TiO2–SiO2–C–H2–O2. *Journal of Metamorphic Geology*, **8**, 309–43.

Holland T. J. B. and Powell R. (1998) Internally consistent thermodynamic data set for phases of petrlogical interest. *Journal of Metamorphic Geology*, **16**, 309–44.

Honda, R., Mizutani, H. and Yamamoto, T. (1993) Numerical simulation of Earth's core formation. *Journal of Geophysical Research*, **98**, 2075–89.

Houseman, G. (1988) The dependence of convection planform on mode of heating. *Nature*, **332**, 346–9.

Ida, S., Nakagawa, Y. and Nakazawa, K. (1987) The Earth's core formation due to the Rayleigh-Taylor instability. *Icarus*, **69**, 239–48.

Isacks, B. and Molnar, P. (1969) Mantle earthquake mechanisms and the sinking of the lithosphere. *Nature*, **223**, 1121–4.

Jamieson, R. A., Beaumont, C., Nguyen, M. H. and Lee, B. (2002) Interaction of metamorphism, deformation, and exhumation in large convergent orogens. *Journal of Metamorphic Geology*, **20**, 9–24.

Johannes, W. (1985) The significance of experimental studies for the formation of migmatites. In Ashworth, V. A. (ed.), *Migmatites*, Blackie, pp. 36–85.

Kadlec, B., Dorn, G., Tufo, H. and Yuen, D. (2008) Interactive 3-D computation of fault surfaces using level sets. *Visual Geosciences*, **13**, 133–8.

Kameyama, M., Yuen, D. A., Karato, S. (1999) Thermal-mechanical effects of low-temperature plasticity (the Peierls mechanism) on the deformation of a viscoelastic shear zone. *Earth and Planetary Science Letters*, **168**, 159–72.

Karato, S. (2008) *Deformation of Earth Materials*. Cambridge University Press.

Karato, S. and Wu, P. (1993) Rheology of the upper mantle: a synthesis. *Science*, **260**, 771–8.

Karato, S., Riedel, M. R. Yuen, D. A. (2001) Rheological structure and deformation of subducted slabs in the mantle transition zone: implications for mantle circulation and deep earthquakes. *Physics of the Earth and Planetary Interiors*, **127**, 83–108.

Karniadakis, G. E., Kirby, R. M. (2003) *A Seamless Approach to Parallel Algorithms and their Implementation*. Cambridge University Press.

Karpov, I. K., Kiselev, A. I. and Letnikov, F. A. (1976) *Computer Modeling of Natural Mineral Formation*. 'Nedra' Press (in Russian).

Katayama, I. and Karato, S. (2008) Low-temperature, high-stress deformation of olivine under water-saturated conditions. *Physics of the Earth and Planetary Interiors*, **168**, 125–33.

Katz, R. F. (2008) Magma dynamics with the enthalpy method: benchmark solutions and magmatic focusing at mid-ocean ridges. *Journal of Petrology*, **49**, 2099–121.

Kaus, B. J. P. and Becker, T. W. (2007) Effects of elasticity on the Rayleigh-Taylor instability: implications for large-scale geodynamics. *Geophysical Journal International*, **168**, 843–62.

Kaus, B. J. P. and Podladchikov, Y. Y. (2006) Initiation of localized shear zones in viscoelastoplastic rocks. *Journal of Geophysical Research-Solid Earth*, **111**, Article Number: B04412.

Kaus, B. J. P. and Schmalholz, S. M. (2006) 3D finite amplitude folding: Implications for stress evolution during crustal and lithospheric deformation. *Geophysical Research Letters*, **33**, Article Number: L14309.

Kaus, B. J. P., Connolly, J. A. D., Podladchikov, Y. Y. and Schmalholz, S. M. (2005) Effect of mineral phase transitions on sedimentary basin subsidence and uplift. *Earth and Planetary Science Letters*, **233**, 213–28.

Kaus, B. J. P., Gerya, T. V., Schmid, D. W. (eds.) (2008a) Recent advances in computational geodynamics: Theory, numerics and applications. *Physics of the Earth and Planetary Interiors*, **171**, Issue: 1–4, Special Issue.

Kaus, B. J. P., Steedman, C. and Becker, T. W. (2008b) From passive continental margin to mountain belt: Insights from analytical and numerical models and application to Taiwan. *Physics of the Earth and Planetary Interiors*, **171**, 235–251.

Keondzhyan, V. P. and Monin, A. S. (1977) Continental drift and large-scale wandering of the Earth's pole. *Izvestiya Physics of the Solid Earth*, **13**, 760–72.

Keondzhyan, V. P. and Monin, A. S. (1980) Compositional convection in the Earth's Mantle. *Dokladi Akademii Nauk SSSR*, **253**, 78–81.

Khan, A., Connolly, J. A. D. and Olsen, N. (2006) Constraining the composition and thermal state of the mantle beneath Europe from inversion of long-period electromagnetic sounding data. *Journal of Geophysical Research*, **111**, B10102.

Kocks, U. F., Argon, A. S. and Ashby, M. F. (1975) Thermodynamics and kinetics of slip. *Progress in Materials Science*, **19**, 1–291.

Kooi, H. and Beaumont, C. (1994) Escarpment evolution on high-elevation rifted margins – insights derived from a surface processes model that combines diffusion, advection, and reaction. *Journal of Geophysical Research-Solid Earth*, **99** (B6), 12191–209.

Krotkiewski, M., Dabrowski, M. and Podladchikov, Y. Y. (2008) Fractional Steps methods for transient problems on commodity computer architectures. *Physics of the Earth and Planetary Interiors*, **171**, 122–36.

Kundu, P. K. and Cohen, I. M. (2002) *Fluid Mechanics*. Academic.

Landau, L. D. and Lifshitz, E. M. (1987) *Fluid Mechanics*. 2nd English edition. Pergamon Press.

Larsen, T. B., Yuen, D. A. and Malevsky, A. V. (1995) Dynamical consequences on fast subducting slabs from a self-regulating mechanism due to viscous heating in variable viscosity convection. *Geophysical Research Letters*, **22**, 1277–80.

LeVeque, R. (2002) *Finite Volume Methods for Hyperbolic Problems*. Cambridge University Press.

Levin, V., Shapiro, N., Park, J. and Ritzwoller, M. (2002) Seismic evidence for catastrophic slab loss beneath Kamchatka. *Nature*, **418**, 763–7.

Lin, J.-R., Gerya, T. V., Tackley, P. J., Yuen, D. A., Golabek, G. J. (2009) Protocore destabilization during planetary accretion: Influence of a deforming planetary surface. *Icarus*, (in press).

Liou, J. G., Tsujimori, T., Zhang, R. Y., Katayama, I. and Maruyama, S. (2004) Global UHP Metamorphism and Continental Subduction/Collision: The Himalayan Model. *International Geology Review*, **46**, 1–27.

Longo A., Vassalli M., Papale P. and Barsanti M. (2006) Numerical simulation of convection and mixing in magma chambers replenished with CO2-rich magma. *Geophysical Research Letters*, **33** (21).

Lynch, D. R. (2005) *Numerical Partial Differential Equations for Environmental Scientists and Engineers: A Practical First Course*. Springer-Verlag.

Machetel, P., Rabinowicz, M. and Bernardet, P. (1986) Three-dimensional convection in spherical shells. *Geophysical and Astrophysical Fluid Dynamics*, **37**, 57–84.

MacPherson, G. J., Davis, A. M. and Zinner, E. K. (1995) The distribution of aluminium-26 in the early Solar system – A reappraisal. *Meteoritics*, **30**, 365–86.

Marsh, B. D. (1982) On the mechanics of igneous diapirism, stoping, and zone melting. *American Journal of Science*, **282**, 808–55.

Massonne, H.-J. (1999) A new occurrence of microdiamonds in quartzofeldspathic rocks of the Saxonian Erzgebirge, Germany, and their metamorphic evolution, Proc. 7th Int. Kimberlite Conf., 533–9.

Melnik, O. (2000) Dynamics of two-phase conduit flow of high viscosity gas-saturated magma: large variations of sustained explosive eruption intensity. *Bulletin of Volcanology*, **62**, 153–70.

Melnik, O. and Sparks, R. S. J. (1999) Nonlinear dynamics of lava dome extrusion. *Nature*, **402**, 37–41.

Melosh, H. J. (1990) Giant impacts and the thermal state of the early Earth. In Newsome, H. E. and Jones, J. H. (eds.) *Origin of the Earth*, Oxford University Press, pp. 69–83.

Melosh, H. J. (2008) Did an impact blast away half of the martian crust? *Nature Geoscience*, **1**, 412–14.

Miller, S. A., Collettini, C., Chiaraluce, L., *et al.* (2004) Aftershocks driven by a high-pressure CO2 source at depth. *Nature*, **427**, 724–7.

Minear, J. W., Toksöz, M. N. (1970) Thermal regime of a downgoing slab and new global tectonics. *Journal of Geophysical Research*, **75**, 1397–419.

Mishin, Y. A., Gerya, T. V., Burg, J.-P. and Connolly, J. A. D. (2008) Dynamics of double subduction: Numerical modeling. *Physics of the Earth and Planetary Interiors*, **171**, 280–95.

Moresi, L., Zhong, S., Gurnis, M. (1996) The accuracy of finite element solutions of Stokes' flow with strongly varying viscosity. *Physics of the Earth and Planetary Interiors*, **97**, 83–94.

Moresi, L., Dufour, F., Mühlhaus, H.-B. (2003) A Lagrangian integration point finite element method for large deformation modeling of viscoelastic geomaterials. *Journal of Computational Physics*, **184**, 476–97.

Moresi, L., Quenette, S., Lemiale, V., *et al.* (2007) Computational approaches to studying non-linear dynamics of the crust and mantle. *Physics of the Earth and Planetary Interiors*, **163**, 69–82.

Murakami, M., Hirose, K., Kawamura, K., Sata, N. Ohishi, Y. (2004) Post-perovskite phase transition in MgSiO3. *Science*, **304**, 855–8.

Murnaghan, F. D. (1944) The compressibility of media under extreme pressures. *Proceedings of the National Academy of Sciences*, **30**, 244–7.

Nikolaeva, K., Gerya, T. V. and Connolly, J. A. D. (2008) Numerical modelling of crustal growth in intraoceanic volcanic arcs. *Physics of the Earth and Planetary Interiors*, **171**, 336–56.

Oganov, A. R. and Ono, S. (2004) Theoretical and experimental evidence for a post-perovskite phase of MgSiO3 in Earth's D″ layer. *Nature*, **430**, 445–8.

Oldenburg, C. M., Spera, F. J. and Yuen, D. A. (1990) Self-organization in convective magma mixing. *Earth-Science Reviews*, **29**, 331–48.

Papale, P. (1999) Strain-induced magma fragmentation in explosive eruptions. *Nature* **397**, 425–8.

Papale, P. (2001) Dynamics of magma flow in volcanic conduits with variable fragmentation efficiency and nonequilibrium pumice degassing. *Journal of Geophysical Research*, **106**, 11043–65.

Pascal, G., Dubois, J., Barazangi, M., Isacks, B. L. and Oliver, J. (1973) Seismic velocity anomalies beneath the New Hebrides island arc: evidence for a detached slab in the upper mantle. *Journal of Geophysical Research*, **78** (29), 6998–7004.

Patankar, S. V. (1980) Numerical Heat Transfer and Fluid Flow. McGraw-Hill.

Pergler, T. and Matyska, C. (2008) A hybrid spectral and finite element method for coseismic and postseismic deformation. *Physics of the Earth and Planetary Interiors*, **163**, 122–48.

Petford, N., Cruden, A. R., McCaffrey, K. J. and Vigneresse, J.-L. (2000) Granite magma formation, transport and emplacement in the Earth's crust. *Nature*, **408**, 669–73.

Petrini, K., Connolly, J. A. D. and Podladchikov, Y. Y (2001) A coupled petrological-tectonic model for sedimentary basin evolution: the influence of metamorphic reactions on basin subsidence. *Terra Nova*, **13**, 354–59.

Pinkerton, H. and Stevenson, R. J. (1992) Methods of determining the rheological properties of magmas at subliquidus temperatures. *Journal of Volcanology and Geothermal Research*, **53**, 47–66.

Pitcher, W. S. (1979) The nature, ascent and emplacement of granitic magma. *Journal of the Geological Society (London)*, **136**, 627–62.

Poli, S. and Schmidt, M. W. (2002) Petrology of subducted slabs. *Annual Review of Earth and Planetary Sciences*, **30**, 207–35.

Popov, A. A. and Sobolev, S. V. (2008) SLIM3D: A tool for three-dimensional thermomechanical modeling of lithospheric deformation with elasto-visco-plastic rheology. *Physics of the Earth and Planetary Interiors*, **171**, 55–75.

Pysklywec, R. N. (2006) Surface erosion control on the evolution of the deep lithosphere. *Geology*, **34**, 225–8.

Ramberg, H. (1968) Instability of layered system in the field of gravity. *Physics of the Earth and Planetary Interiors*, **1**, 427–74.

Ramberg, H. (1981) The role of gravity in orogenic belts. In McClay, K. R., Price, N. J. (eds.), *Thrust and Nappe Tectonics*. Geol. Soc. Special Publication, London, pp. 125–40.

Ranalli, G. (1995) *Rheology of the Earth*. Chapman & Hall.

Ranero, C. R., Phipps Morgan, J. and Reichert, C. (2003) Bending-realted faulting and mantle serpentinization at the Middle America trench. *Nature*, **425**, 367–73.

Ranero, C. R., Villaseñor, A., Phipps Morgan, J. and Weinribe, W. (2005) Relationship between bend-faulting at trenches and intermediate-depth seismicity. *Geochemistry, Geophysics and Geosystems*, **6**, doi:10.1029/2005GC000997.

Regenauer-Lieb, K. and Yuen, D. A. (2008) Multiscale Brittle-Ductile Coupling and Genesis of Slow Earthquakes. *Pure and Applied Geophysics*, **165**, 523–43.

Revenaugh, J. and Parsons, B. (1987) Dynamic topography and gravity anomalies for fluid layers whose viscosity varies exponentially with depth. *Geophysical Journal of the Royal Astronomical Society*, **90**, 349–68.

Ricard, Y., Bercovici, D. and Schubert, G. (2001) A two-phase model for compaction and damage 2. Applications to compaction, deformation, and the role of interfacial surface tension. *Journal of Geophysical Research*, **106**, 8907–24.

Ricard, Y., Šrámek, O. and Dubuffet, F. (2009) Runaway core-mantle segregation of terrestrial planets, *Earth and Planetary Science Letters*, in revision.

Richter, F. M. (1973) Finite amplitude convection through a phase boundary. *Geophysical Journal of the Royal Astronomical Society*, **35**, 265–76.

Richter, F. M. (1978) Mantle convection models. *Annual Review of Earth and Planetary Sciences*, **6**, 9–19.

Rosen, O. M., Zorin, Y. M. and Zayachkovsky, A. A. (1972) A find of a diamond linked with eclogites of the Precambrian Kokchetav massif. *Dokladi Akademii Nauk SSSR*, **203**, 674–76 (in Russian).

Rudolph, M. L., Gerya, T. V. Yuen, D. A. and DeRosier, S. (2004) Visualization of multiscale dynamics of hydrous cold plumes at subduction zones. *Visual Geosciences*, doi.org/10.1007/s10069–004-0017–2.

Ruprecht, P., Bergantz, G. W. and Dufek, J. (2008) Modeling of gas-driven magmatic overturn: Tracking of phenocryst dispersal and gathering during magma mixing. *Geochemistry, Geophysics, and Geosystems*, **9**, Article Number: Q07017.

Sacks, P. E. and Secor, D. T. (1990) Delamination in collisional orogens. *Geology*, **18**, 999–1002.

Samuel, H., Tackley, P. J. (2008) Dynamics of core formation and equilibration by negative diapirism. *Geochemistry, Geophysics, and Geosystems*, **9**, Q06011, doi:10.1029/2007GC001896.

Scambelluri, M. and Philippot, P. (2001) Deep fluids in subduction zones. *Lithos*, **55**, 213–27.

Schenk, O. and Gärtner, K. (2004) Solving unsymmetric sparse systems of linear equations with PARDISO. *Journal of Future Generation Computer Systems*, **20**, 475–87.

Schenk, O. and Gärtner, K. (2006) On fast factorization pivoting methods for symmetric indefinite systems. *Electronic Transactions on Numerical Analysis*, **23**, 158–79.

Schmeling, H. (1987) On the relation between initial conditions and late stages of Rayleigh-Taylor instabilities. *Tectonophysics*, **133**, 65–80.

Schmeling, H. (2000) Partial melting and melt segregation in a convecting mantle. In Bagdassarov, N., Laporte, D. and Thompson, A. B. (eds.) *Physics and Chemistry of Partially Molten Rocks*. Kluwer Academic Publisher, pp. 141–78.

Schmeling, H., Babeyko, A. Y. Enns, A., *et al.* (2008) A benchmark comparison of spontaneous subduction models – Towards a free surface. *Physics of the Earth and Planetary Interiors*, **171**, 198–223.

Schmid, D. W. and Podladchikov, Y. Y. (2003) Analytical solutions for deformable elliptical inclusions in general shear. *Geophysical Journal International*, **155**, 269–88.

Schmidt, M. W. and Poli, S. (1998) Experimentally based water budgets for dehydrating slabs and consequences for arc magma generation. *Earth and Planetary Science Letters*, **163**, 361–79.

Schubert, G. (1992) Numerical models of mantle convection. *Annual Review of Fluid Mechanics*, 1992. **24**, 359–94.

Schubert, G., Yuen, D. A. and Turcotte, D. L. (1975) Role of phase transitions in a dynamic mantle. *Geophysical Journal of the Royal Astronomical Society*, **42**, 705–35.

Scott, D. R. and Stevenson, D. J. (1986) Magma ascent by porous flow. *Journal of Geophysical Research*, **91**, 9283–96.

Senshu, H., Kuramoto, K. and Matsui, T. (2002) Thermal evolution of a growing Mars. *Journal of Geophysical Research*, **107**, E12, 5118, doi:10.1029/2001JE001819.

Shabana, A. A. (2008) *Computational Continuum Mechanics*. Cambridge University Press.

Shukla, K. N. (2005) *Mathematical Principles of Heat Transfer*. Begell House Inc.

Simakin, A. and Botcharnikov R. (2001) Degassing of stratified magma by compositional convection. *Journal of Volcanology and Geothermal Research*, **105**, 207–24.

Smith, D. C. (1984) Coesite in clinopyroxene in the Caledonides and its implications for geodynamics. *Nature*, **310**, 641–4.

Sobolev, N. V. and Shatsky, V. S. (1990) Diamond inclusions in garnets from metamorphic rocks: an environment for diamond formation. *Nature*, **343**, 742–5.

Sobolev, S. V. and Babeyko, A. Y. (1994) Modeling of mineralogical composition, density and elastic-wave velocities in anhydrous magmatic rocks. *Surveys in Geophysics*, **15**, 515–44.

Sobolev, S. V. and Babeyko, A. Y. (2005) What drives orogeny in the Andes? *Geology*, **33**, 617–20.

Sobolev, A. V., Hofmann, A. W., Sobolev, S. V. and Nikogosian, I. K. (2005) An olivine-free mantle source of Hawaiian shield basalts. *Nature*, **434**, 590–7.

Souza de Neto, E. A., Periæ, D. and Owen, D. R. J. (2009) *Computational Methods for Plasticity: Theory and Applications*, Wiley.

Spakman, W., Wortel, M. J. R. and Vlaar, N. J. (1988) The Hellenic subduction zone: a tomographic image and its geodynamic implications. *Geophysical Research Letters*, **15**, 60–3.

Spera, F. J., Oldenburg, C. M., Christensen, C., Todesco, M. (1995) Simulations of convection with crystallization in the system $KAlSi2O6$-$CaMgSi2O6$: Implications for compositionally zoned magma bodies. *American Mineralogist*, **80** (11–12), 1188–207.

Spiegelman, M. and Kelemen, P. B. (2003) Extreme chemical variability as a consequence of channelized melt transport. *Geochemistry, Geophysics, and Geosystems*, **4**, Art. No. 1055.

Stevenson, D. J. (1981) Models of the Earth's core. *Science*, **214**, 611–19.

Stevenson, D. J. (2008) A planetary perspective on the deep Earth. *Nature*, **451**, 261–5.

Stixrude, L. and Bukowinski, M. S. T. (1990) Fundamental thermodynamic relations and silicate melting with implications for the constitution of D″. *Journal of Geophysical Research*, **95**, 19311–25.

Stixrude, L. and Lithgow-Bertelloni, C. (2005) Mineralogy and elasticity of the oceanic upper mantle: Origin of the low-velocity zone. *Journal of Geophysical Research*, **110**, B03204.

Stoeckhert, B. and Gerya, T. V. (2005) Pre-collisional high pressure metamorphism and nappe tectonics at active continental margins: a numerical simulation. *Terra Nova*, **17**, 102–10.

Tackley, P. J. (1993) Effects of strongly temperature-dependent viscosity on time-dependent, 3-dimensional models of mantle convection. *Geophysical Research Letters*, **20**, 2187–90.

Tackley, P. J. (2000) Self-consistent generation of tectonic plates in time-dependent, three-dimensional mantle convection simulations Part 1: Pseudo-plastic yielding. *Geochemistry, Geophysics, and Geosystems*, **1**, Paper No 2000GC000036.

Tackley, P. J. (2008) Modelling compressible mantle convection with large viscosity contrasts in a three-dimensional spherical shell using the yin-yang grid. *Physics of the Earth and Planetary Interiors*, **171**, 7–18.

Tikhonov, A. N. and Samarsky, A. A. (1972) *Equations of Mathematical Physics*. Nauka (in Russian).

Tonks, W. B. and Melosh H. J. (1992) Core formation by Giant Impacts, *Icarus*, **100**, 326–46.

Toro, E. F. (1999) *Riemann Solvers and Numerical Methods for Fluid Dynamics*. Springer-Verlag.

Torrance, K. E. and Turcotte, D. L. (1971) Thermal convection with large viscosity variations. *Journal of Fluid Mechanics*, **47**, 113–25.

Turcotte, D. L., Schubert, G. (2002) *Geodynamics*. Cambridge University Press.

Ueda, K., Gerya, T. and Sobolev, S. V. (2008) Subduction initiation by thermal-chemical plumes: Numerical studies. *Physics of the Earth and Planetary Interiors*, **171**, 296–312.

van Keken, P. E., King, S., Schmeling, H., *et al.* (1997) A comparison of methods for the modeling of thermochemical convection. *Journal of Geophysical Research*, **102**, 22477–95.

van Keken, P. E., Currie, C., King, S. D., *et al.* (2008) A community benchmark for subduction zone modeling. *Physics of the Earth and Planetary Interiors*, **171**, 187–97.

Vance, D., Bickle, M., Ivy-Ochs, S. and Kubik, P. W. (2003) Erosion and exhumation in the Himalaya from cosmogenic isotope inventories of river sediments. *Earth and Planetary Science Letters*, **206**, 273–88.

Vasilyev, O. V., Podladchikov, Y. Y. and Yuen, D. A. (1998) Modeling of compaction driven flow in poro-viscoelastic medium using adaptive wavelet collocation method. *Geophysical Research Letters*, **25**, 3239–42.

Vasilyev, O. V., Gerya, T. V. and Yuen, D. A. (2004) The application of multidimensional wavelets to unveiling multi-phase diagrams and in situ physical properties of rocks. *Earth and Planetary Science Letters*, **223**, 49–64.

von Blanckenburg, F. and Davies, J. H. (1995) Slab breakoff: a model for syncollisional magmatism and tectonics in the Alps, *Tectonics*, **14**, 120–31.

Wada, K., Kokubo, E. and Makino, J. (2006) High-resolution simulations of a Moon-forming impact and postimpact evolution. *Astrophysical Journal*, **638**, 1180–86.

Warren, C. J., Beaumont, C. and Jamieson, R. A. (2008) Modelling tectonic styles and ultra-high pressure (UHP) rock exhumation during the transition from oceanic subduction to continental collision, *Earth and Planetary Science Letters*, **267**, 129–45.

Weinberg, R. B. and Shmelling, H. (1992) Polydiapirs: multiwavelength gravity structures. *Journal of Structural Geology*, **14**, 425–36.

Wesseling, P. (1992) *An Introduction to Multigrid Methods*. John Wiley & Sons.

Willett, S. D. (1999) Orogeny and orography: The effects of erosion on the structure of mountain belts. *Journal of Geophysical Research-Solid Earth*, **104** (B12), 28957–81.

Woidt, W. D. (1978) Finite-element calculations applied to salt dome analysis. *Tectonophysics*, **50** (2–3), 369–86.

Wong, A Ton, S. Y. M. and Wortel, M. J. R. (1997) Slab detachment in continental collision zones: an analysis of controlling parameters. *Geophysical Research Letters*, **24** (16), 2095–98.

Wortel, M. J. R. and Spakman, W. (1992) Structure and Dynamics of Subducted Lithosphere in the Mediterranean Region. *Proceedings of the Koninklijke Nederlandse Akademie van Wetenschappen*, **95**, pp. 325–47.

Wortel, M. J. R. and Spakman,W. (2000) Geophysics – subduction and slab detachment in the Mediterranean-Carpathian region. *Science*, **290**, 1910–17.

Xie, S. and Tackley, P. J. (2004a) Evolution of helium and argon isotopes in a convecting mantle. *Physics of the Earth and Planetary Interiors*, **146**, 417–39.

Xie, S. and Tackley, P. J. (2004b) Evolution of U-Pb and Sm-Nd systems in numerical models of mantle convection, *Journal of Geophysical Research*, **109**, B11204, doi:10.1029/2004JB003176.

Xu, P. F., Sun, R. M., Liu, F. T., Wang, Q. and Cong, B. (2000) Seismic tomography showing, subduction and slab breakoff of the Yangtze block beneath the Dabie–Sulu orogenic belt. *Chinese Science Bulletin*, **45**, 70–4.

Yamato, P., Burov, E., Agard, P., Le Pourhiet, L. and Jolivet, L. (2008) HP-UHP exhumation during slow continental subduction: Self-consistent thermodynamically and thermomechanically coupled model with application to the Western Alps. *Earth and Planetary Science Letters*, **271**, 63–74.

Yoshioka, S., Wortel, M. J. R. (1995) Three-dimensional numerical modeling of detachment of subducted lithosphere. *Journal of Geophysical Research*, **100**, 20233–44.

Yoshioka, S., Yuen, D. A. and Larsen, T. B. (1995) Slab weakening: thermal and mechanical consequences for slab detachment. *Island Arc*, **40**, 89–103.

Yuen, D. A., Balachandar, S. and Hansen, U. (2000) Modelling mantle convection: A significant challenge in geophysical fluid dynamics. In Fox, P. A. and Kerr, R. M. (eds.) *Geophysical and Astrophysical Convection*, Gordon and Breach Science Publishers, pp. 257–94.

Yuen, D. A. and Zhang, H. (eds.) (2007) Computational challenges in the earth sciences. *Physics of the Earth and Planetary Interiors*, **161**, 1–4, special issue.

Zhong, S. (1996) Analytic solutions for Stokes' flow with lateral variations in viscosity. *Geophysical Journal International*, **124**, 18–28.

Zhong, S. and Gurnis, M. (1994) The role of plates and temperature-dependent viscosity in phase change dynamics. *Journal of Geophysical Research*, **99**, 15903–17.

Zhong, S. J., Yuen, D. A. and Moresi, L. N. (2007) Numerical methods in mantle convection. In Bercovici, D. (ed.) *Treatise in Geophysics*, Volume 7, (editor-in-chief: Gerard Schubert), *Elsevier*, pp. 227–52.

Zhong, S., McNamara, A., Tan, E., Moresi, L. and Gurnis, M. (2008) A benchmark study on mantle convection in a 3-D spherical shell using CitcomS. *Geochemistry, Geophysics and Geosystems*, **9**, Q10017, doi:10.1029/2008GC002048.

Zhu, G., Gerya, T., Yuen, D., *et al.* (2009) 3D dynamics of hydrous thermal-chemical plumes in intra-oceanic subduction zones. *Geochemistry, Geophysics and Geosystems* (in press).

Zienkiewicz, O. C., Taylor, R. L. and Zhu, J. Z. (2005) *The Finite Element Method: Its Basis and Fundamentals*, sixth edition. Butterworth and Heinemann Inc.

Zlotnik, S., Fernandez, M., Diez, P. and Verges, J. (2008) Modelling gravitational instabilities: slab break–off and Rayleigh–Taylor diapirism. *Pure and Applied Geophysics*, **165**, 1491–510.

Index

Printed in the United States
By Bookmasters